WJTA-IMCA Conference and Expo 2015

New Orleans, Louisiana, USA
2 - 4 November 2015

ISBN: 978-1-5108-2126-2

Printed from e-media with permission by:

Curran Associates, Inc.
57 Morehouse Lane
Red Hook, NY 12571

Some format issues inherent in the e-media version may also appear in this print version.

Printed by Curran Associates, Inc. (2016)

For permission requests, please contact WaterJet Technology Association-Industrial & Municipal Cleaning Association at the address below.

WaterJet Technology Association-Industrial & Municipal Cleaning Association
906 Olive Street – Suite 1200
St. Louis, MO 63101-1448
USA

Phone: (314) 241-1445
Fax: (314) 241-1449

www.wjta.org

Additional copies of this publication are available from:

Curran Associates, Inc.
57 Morehouse Lane
Red Hook, NY 12571 USA
Phone: 845-758-0400
Fax: 845-758-2633
Email: curran@proceedings.com
Web: www.proceedings.com

Presentation Index

Tuesday, November 3

Session A – Waterjet Studies

Session B – Waterjet Cleaning

Session C – Design

Wednesday, November 4

Session D – Advanced Applications and Studies

Session F – Environmental and Safety

Session E – Basic Studies

Session G – Water and Mechanical

2015 WJTA-IMCA Conference and Expo
November 2-4 • New Orleans, Louisiana

Paper

3D PRINTING APPLICATIONS IN WATERJET NOZZLE FABRICATION

E. Charrier & H. B. Miller
Colorado School of Mines
Golden, Colorado, USA

ABSTRACT

Conventional manufacturing methods have historically limited the ability to economically fabricate a wide variety of potential nozzle designs for use in commercial and research applications. Nozzles manufactured using subtractive methods (conventional machine tools) are often severely constrained in terms of possible internal geometries, orifice placement, and the configuration of flow paths, and routinely experience poor performance and inefficiencies associated with tool marks and abrupt changes in the nozzles internal annulus. The recent development of reliable and affordable additive manufacturing methods is providing an opportunity to overcome these challenges. Additive manufacturing, commonly referred to as 3D printing, builds solid parts through sequential layering. As such, it is now possible to produce nozzles and other intricate components with almost any internal geometry using materials ranging from many types of plastics to steel and titanium. This paper presents the current state of this technology and its novel applicability in waterjet research and component manufacturing.

Organized and Sponsored by WJTA®-IMCA®

1.0 INTRODUCTION

Traditionally, the production of one piece metal parts has been dominated by subtractive processes, forging, and casting. Subtractive processes, such as conventional milling or turning a part on a lathe, remove material to produce the finished part. Forging uses dynamic impacts to force hot metal into specific geometries. Castings are made by pouring molten metal into a mold where it hardens. Due to geometry limitations and surface roughness, subtractive manufacturing is the dominant process used in producing waterjet nozzles. However, subtractive manufacturing still has significant disadvantages due to the constraints that it places on nozzle design.

There are three major drawbacks with subtractive processes. First, curves are difficult at best. A simple two dimensional curve (such as a filleted corner in a pocket) may require a tool change. Compound curves are far worse and may require a specialized tool and a state of the art multi-axis CNC mill if they are even possible. Second, tool geometry severely restricts the internal geometry of machined parts. It is not possible to position a tool inside the part to make many of the features that would be easy to cut on the outside of a part. Third, machining processes often leave tool marks and may have abrupt transitions where tool changes or sequential drilling has occurred.

Additive manufacturing drastically mitigates these drawbacks. By building the part in layers, a 3D printer has unrestricted access to the inside of the part. If a 3D printer can print a feature, it can print it almost anywhere. While there are still some design rules for additive manufacturing, 3D printers are capable of producing almost any surface geometry.

Despite these advantages, additive manufacturing still has significant drawbacks. 3D printers tend to produce rougher surfaces than subtractive processes. Furthermore, the deposition of material often leads to undesirable material properties. Until recently, the severity of these drawbacks was sufficient to preclude the use of additive manufacturing in many precision applications.

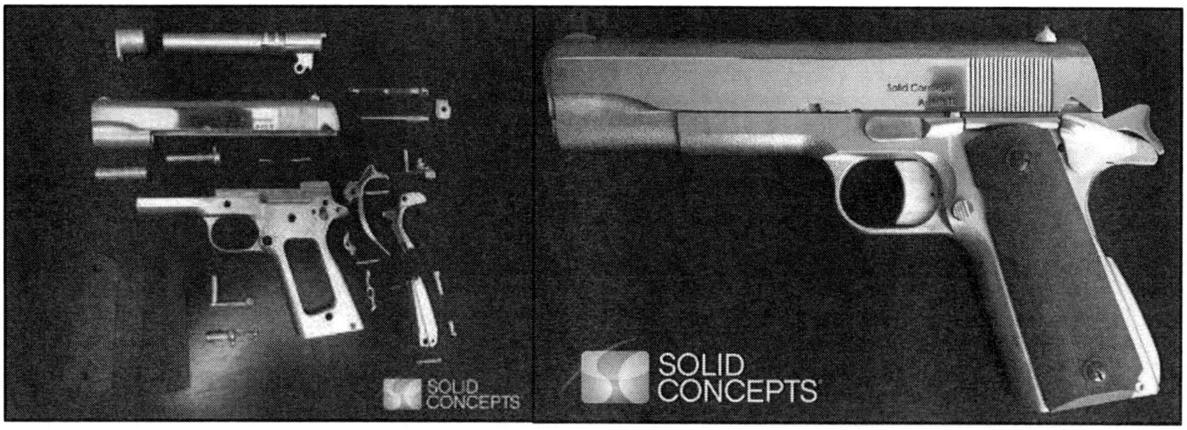

Figure 1: Fully functional 3D printed 1911 by Solid Concepts (Stratasys).

However, additive manufacturing technology is advancing rapidly. As of 2015, the capabilities of additive manufacturing have reached the point where 3D printed metal parts can often be used as an alternative of machined parts. In 2013, NASA tested a 3D printed injector in a rocket test that produced 20.000 pounds of force (NASA). During the same year, Solid Concepts 3D printed a fully functional 1911 pistol in .45ACP (Stratasys). The Solid Concepts 1911 has since successfully fired over a thousand rounds (Stratasys). As these milestones indicate, additive manufacturing is now a viable option for high stress applications.

2.0 LIMITS OF SUBTRACTIVE MANUFACTURING

Subtractive manufacturing is constrained by tool geometry, tool access, and the CNC mill's degrees of freedom. When a part's geometry is a good match for the tool and the mill has the appropriate degrees of freedom, a mill or lathe can be very effective. However, complicated parts may run up against one or more of these constraints. These constraints then force the designer to adopt a less ideal geometry. In waterjet applications, these geometry compromises are often found in the interior geometry of nozzles. Nozzle manufacturing is of particular interest because it runs up against all three constraints.

Tool geometry determines the shape of a cut for a given tool path. For example, an end mill is going to make a cut that has a flat bottom and straight sides. This tool geometry is useful for cutting slots with straight sides in a part. However, if a fillet is desired on the bottom of the pocket, the tool will need to be swapped for another tool with the appropriate end geometry. Due to the widespread availability of tools designed for common and simple geometries, tool geometry is not an issue for many jobs. Tool geometry becomes a limiting factor for designs with more exotic geometries.

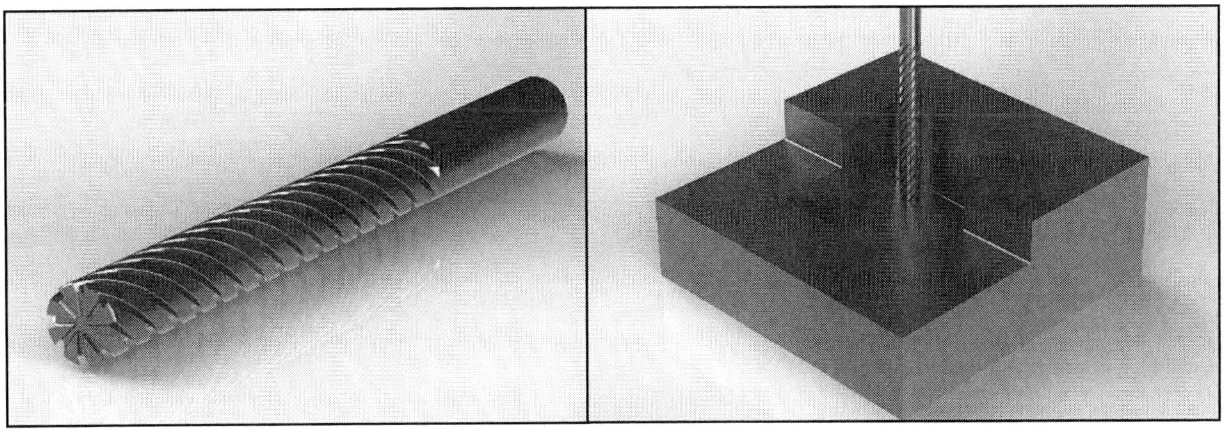

Figure 2: End Mill and its use to cut a pocket in an aluminum block. Note that the end mill produces a flat bottomed pocket with straight sides and filleted corners.

A machine's degrees of freedom determine its possible toolpaths. For CNC milling machines, three degrees of freedom is the minimum. A milling machine with three degrees of freedom can move its tool to any point in its workspace. However, the alignment of the tool's axis cannot be changed. With three degrees of freedom, a mill can easily cut straight sided features like the pocket shown in Figure 2. The addition of two more degrees of freedom would allow a toolpath that could chamfer the top edge of the pocket. It is important to note that additional degrees of freedom do not fully translate into freedom from geometric constraints. While it is easy to add a toolpath for a chamfer or a fillet, a fillet with a variable radius is much more difficult. At 5 degrees of freedom, subtractive manufacturing can push the technology further at the cost of increased complexity and more complicated toolpaths.

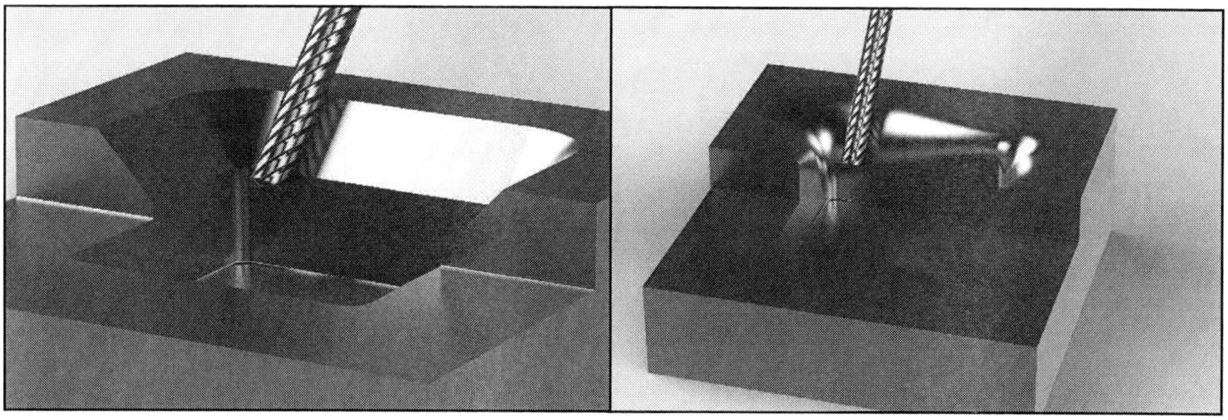

Figure 3: A chamfer, which is possible with a 5 axis CNC mill and an arbitrary variable radius fillet which cannot be exactly machined (though it could be approximated with enough tool passes).

Access to the surface to be machined is critical to subtractive machining. This is what makes internal geometries more difficult. As access is reduced, the possible tool paths and tool geometries are reduced. Operations that would be easy with stock tooling on the surface become difficult or impossible in internal cavities. Complicated surfaces and curved passages are impossible for all practical purposes.

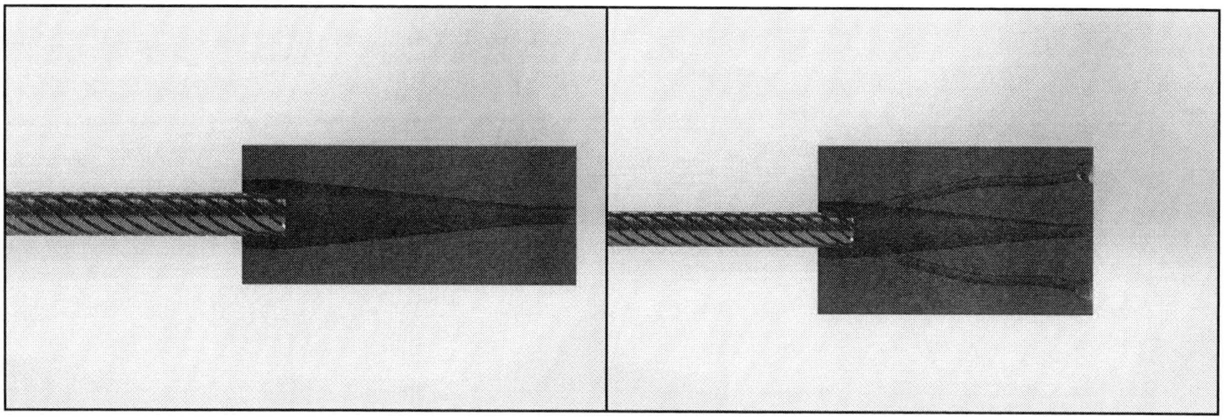

Figure 4: The pictured tool cannot produce the internal chamfer, although specialized tools could stull produce the geometry on the left. The curved passageways in the right model cannot realistically be produced through subtractive manufacturing.

The access problem compounds subtractive manufacturing's issues with surface finish, tool marks, and abrupt transitions. It is much harder to mitigate tool marks or polish a surface in a difficult to access passageway. The confined space also means that the tool geometry is the main parameter that can be varied. The only way to add some features with stock tooling is through sequential drilling. This means tool changes and increases the likelihood of abrupt transitions between operations. High quality custom tooling can mitigate the abrupt transitions, but is often cost prohibitive.

All three of these constraints come into play in waterjet nozzle design. The geometry of readily available tooling constrains the shape of cuts that can be made. The confined geometry and nature of the available tooling restrict the degrees of freedom and possible tool paths. Effectively, these constraints limit the tooling in nozzle construction to pre 5 axis CNC machining technology. Tool movement is limited to a single axis. Tool geometry is usually limited to cylindrical and conical bits. The end result is that nozzles are often lacking basic internal features like fillets that could improve efficiency.

Unfortunately, the situation in subtractive manufacturing is not likely to improve significantly for waterjet nozzle construction. Subtractive manufacturing is a mature technology, where most of the recent improvements have been in automation and control, not the fundamental process. The constraints involved in machining waterjet nozzles have effectively negated most of the recent technological advances. The ability to quickly design a part in a CAD program like SolidWorks, program the toolpaths in a program like Mastercam, and then export the G-code (machine movement commands) to a 5 axis CNC mill, does not overcome the shortcomings of subtractive manufacturing.

3.0 ADDITIVE MANUFACTURING TECHNOLOGY

In 3D printing, the part is built by adding material to form the desired geometry. The vast majority of additive processes build parts in two dimensional layers. These layers are generated from the CAD file and are analogous to contour lines on a map. The print will start at the bottom layer and add material within the contour lines. There are two general paradigms of material deposition. In the first, material is only added where it is needed and all of the material is bonded to the part. In the second, the material covers the entire layer but is only bonded within the contour of the part.

The 3D printing process starts with a CAD model in a program like SolidWorks. The file is then sent to a program called a slicer, which is often integrated with the 3D printer's control software. The slicer cuts the CAD model into layers and generates the tool paths required to build the layers. These paths are typically exported as G-code, which contains the low level commands that drive the printer's components. Once the slicer has generated the tool paths for the print, the printer control software will home the system and then begin running the print.

It is important to note that the capability of going straight from CAD to a plastic part has a significant impact on prototyping. One of the authors (Charrier) uses a consumer grade 3D printer as part of a mechanical engineering consulting business. The fast turnaround time afforded by the 3D printer enables the business to implement client feedback and test several iterations of the design within a day of receiving the feedback. This has significantly reduced lead times during the design phase and lead to a more refined product.

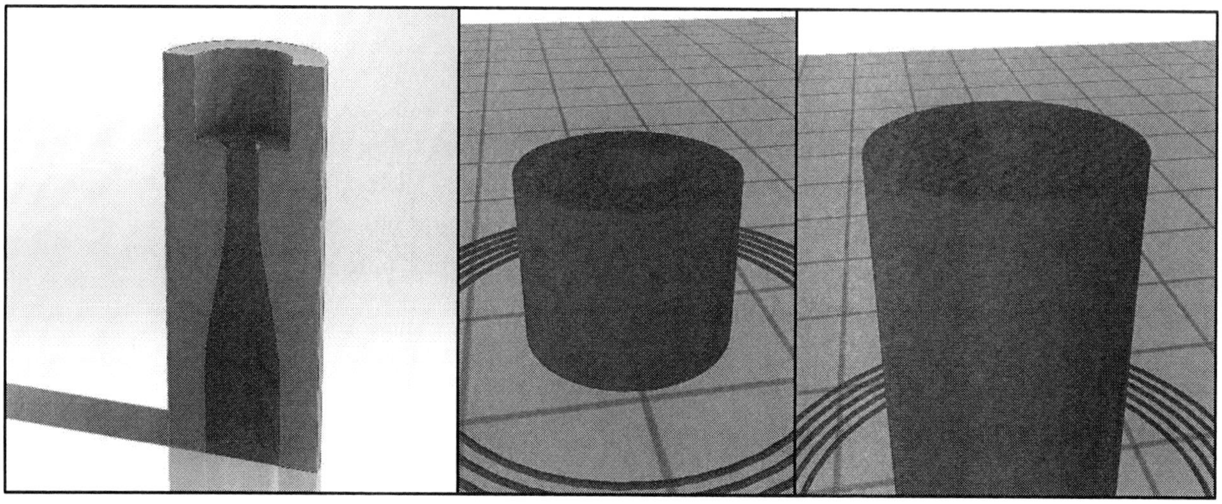

Figure 5: 3D printing workflow. A section view of the model in SolidWorks is shown on the left and middle and right images show different layers of the part. Note that the path the extruder will follow is indicated by the lines in the surface of the part. This part should take approximately 40 minutes to print with an extruder speed of 60mm/s.

Breaking 3D printed parts into discrete layers has two significant impacts on part geometry. First, this layered structure is what allows 3D printed parts to approximate almost any geometry. While the height of a layer is fixed, the path is variable. This allows the printer to follow the edge of the part and produce the approximated geometry. Second, the layers guarantee that there will be surface roughness. The magnitude of the roughness is dependent on the thickness of the layers, the accuracy of printer's movements, the material, the temperature, and other factors. Many printers are capable of varying the layer height depending on the requirements of the job.

Layer thicknesses vary greatly depending on the technology employed. Consumer grade 3D printers, such as Robo 3D's R1, are typically capable of printing 0.1mm to 0.3mm layers. Higher end professional models are capable of far higher resolution. For example, Stratasys Direct is advertising layer thicknesses as low as 40-60 microns for their direct metal laser sintering service (Stratasys). Stratasys Polyjet systems are capable of producing layers as fine as 16 microns (Stratasys). Protolabs is currently offering direct metal laser sintering layer thicknesses of 20 microns (Protolabs). All of these are currently commercially available products or services.

It is important to note that 3D printing can have a significant impact on the material properties of a part. First, building a part in layers means that there will be numerous small joints throughout the part. The material properties of these joints are often different from the properties of the rest of the part or a conventionally manufactured part of the same material. Typically, the joint is weaker than the surrounding material. This tends to be an issue with consumer grade plastic 3D printers, where part failure as a consequence of delamination is common.

Second, many 3D printing processes involve heating a region of the part. This means uneven cooling of the part is likely. Parts approaching the size of the build area on consumer grade 3D printers often experience significant warping of the extremities. Even if visible warping does not occur, uneven cooling can result in residual stresses in the part.

Third, 3D printing – especially with consumer grade printers – can result in uneven material deposition rates. If the filament (feedstock) is too small or extruded too slowly, an insufficient volume of plastic is extruded and can lead to void spaces inside a part. These voids cause stress concentrations and can cause the part to be susceptible to fluid penetration. If the filament is too thick or is extruded to fast it can lead to too much filament being deposited. Too much filament will cause the part to expand in that region. This problem primarily exists in the lower end plastic printers and is a function of incorrect settings or varying feedstock.

Finally, 3D printing may result in varying and potentially undesirable grain structures in a material. In processes where the material is melted locally, most material will be melted once and go through several heating and cooling cycles as the laser or extruder moves beside and above a point. The cooling rate is not fully controlled and these cycles can change the material properties. With some materials, a metal part can be annealed to mitigate this, but this is not true of all materials.

The problems with 3D printing processes can typically be mitigated by selecting the right material and process. Quick proof of concept prototypes can be run on almost any 3D printer. For example, the Colorado School of Mines uses consumer grade printers to produce ABS parts. Far higher quality printers and printing services are available from companies like Stratasys Direct. These printers can mitigate most process related problems. The key is to match the material and process to the application.

4.1 Additive Manufacturing Methods

The recent explosion in the availability of consumer grade 3D printers was caused by the expiration of U.S. patent US5121329A in 2009 (Crump). This was the core patent for fused deposition modeling (FDM) and lead to the founding of Stratasys (Crump, Brewster). Once the patent expired, low cost consumer oriented companies, like MakerBot, flooded the FDM market and drove prices down. Thanks to these companies, FDM printers make up the vast majority of consumer grade 3D printers on the market. It is now possible to buy a complete FDM 3D printer for approximately $800. The printer used by Charrier is a slightly modified Robo 3D R1, which retails for the same amount.

FDM printers work by extruding a bead of molten plastic filament. The extruder, which has 3 degrees of freedom, fills in the part's contour in each layer. When a layer is complete, the extruder moves up to the next layer and builds that contour on top of the preceding layer. Where there is an overhang, the printer will build a support structure to support the part.

A second wave of consumer grade 3D printers has emerged using stereolithography. The core patents on stereolithography were developed by 3D Systems during the 1980s and early 1990s (Brewster, Hull). Stereolithography works by using light to selectively cure resin in a tank. This allows intricate 3D shapes to be formed within the tank. Formlabs is playing a role analogous to Makerbot in bringing stereolithography to consumers.

Polyjet printing was developed by an Israeli company, Objet Geometries, around 2000 and is still covered by U.S. patents (Gothait). The Polyjet process works by spraying resin droplets on the model surface and immediately curing the resin with a laser or other light source. As necessary, a polyjet printer will use support material for overhangs. Objet was acquired by Stratasys in 2012 (Bloomberg). Stratasys Polyjet systems are producing layers as fine as 16 microns (Stratasys).

Selective laser sintering is used to 3D print metal and plastic parts. The technology was developed by DTM, which was then acquired by 3D Systems (Krassenstein, Deckard). The key patent on laser sintering expired in 2014 (Deckard). Selective laser sintering works by spreading a thin layer of metal powder over the work area and then sintering the areas within the part geometry with a laser. When that is complete, another layer of powder is spread across the build area and the cycle is repeated.

Laser metal deposition is a versatile technology that can be used to build, repair, or coat parts. The process works by melting the surface of a material and then blowing metal powder into the melt pool (RPM Innovations Inc.). This allows laser metal deposition machines work without any support material and makes it far more useful for maintenance and surfacing operations. According to LPW, which provides powder for laser metal deposition, the process may be called Laser Metal Deposition, Direct Metal Deposition. Direct Laser Deposition, Laser Engineered Net Shaping, Laser Cladding, Laser Deposition Welding, and Powder Fusion Welding (LPW Technology).

Electron beam additive manufacturing (EBAM) was patented in 2001 by Arcam Ab (Andersson). EBAM is similar laser metal deposition. Metal powder is dispensed onto the surface and then the powder is fused to by the electron beam. EBAM machines are capable of working with a wide variety of metals and have large work areas. Sciaky's EBAM machine can build parts up to 5.8m x 1.2m x 1.2m and can work with titanium, steel, Inconel, tantalum, tungsten, niobium and other metals (Sciaky).

Despite all of its advantages in geometry, 3D printing is still behind subtractive manufacturing in terms of surface finish and quality. However, this shortcoming can be overcome by using a hybrid process. It is now possible to combine 3D printing with a milling machine. DMG Mori has developed a combination laser deposition welding and 5 axis CNC machine, the LASERTEC 65 3D. This system can automatically switch between laser and milling operation and achieve machining quality surface over most geometries (DMG MORI).

4.2 Surface Capabilities

In general, the surface roughness of a 3D printed surface should be less than the layer thickness. The roughness will vary depending on the material, the process, and the geometry of the surface. Major 3D printing services like Stratasys direct include surface roughness in their general tolerances. For example, Stratasys' direct metal laser sintering surface roughness is 350 Ra -

µinch, which is approximately 9 microns (Stratasys). This is approximately one fifth of the 40 micron layer thickness in the Stratasys process and between the ISO N11 and N10 standards.

It is possible to improve the surface finish of metal 3D printed parts. Most major printing companies, such as Stratasys Direct offer a variety of surface finishes on completed parts. Stratasys Direct offers media blasting, tumbling, and hand polishing for its parts (Stratasys). Stratasys claims that surfaces can be improved to a super mirror finish – 0.025 Ra µmeter (Stratasys). It is likely that small internal volumes will be significantly more difficult to improve. In addition to the surface treatments, secondary processes including CNC machining, chemical etching, and wire EDM can be performed on parts (Stratasys).

5.0 CONCLUSION

3D printing is a rapidly developing field. Many of the core patents have expired over the last five years, increasing the competitiveness of the landscape. New venders and service providers are pressuring the established companies into improving their offerings and decreasing prices. 3D printing services, such as Stratasys Direct, are now taking orders for prototypes and small production runs. Any company with CAD capabilities can order a 3D printed part and expect delivery within a week or two.

For waterjet applications, most 3D printers offer a tradeoff in capabilities. 3D printing can produce far more complicated geometries at the cost of increased surface roughness. For certain applications, this trade off may already be worth the increased roughness. However, the field is advancing quickly so that these disadvantages will be further mitigated. One of the latest mitigations was the introduction of DMG Mori's hybrid laser deposition welding and 5 axis CNC machine. This demonstrates that it is possible combine the quality of machined surfaces with the geometry capabilities of 3D printing.

Not all designs or applications will benefit from 3D printing, and it is not a solution to every problem. However, it is likely that some designs will benefit from smoother internal geometry and other previously unfeasible designs will see production. More importantly, the ability to explore exotic geometries will likely lead to innovation.

The widespread availability of 3D printing as a service has lowered the barriers to entry. Anyone with a CAD file and a credit card can order metal parts for around $5 per cubic centimeter of material. While presently available services may not offer all of the recent advances, these commercial vendors offer services with enough features to be useful in many applications.

ACKNOWLEDGEMENTS

The authors wish to thank Dr. Fun-Den Wang for his contributions in support of this research.

REFERENCES

Andersson, Lars-Erik. "Device and arrangement for producing a three-dimensional object." Patent US7537722. April 27, 2001. <http://www.google.com/patents/US7537722>

Bloomberg. "Company Overview of Objet Ltd." Bloomberg. Bloomberg, n.d. Web. August 8, 2015. <http://www.bloomberg.com/research/stocks/private/snapshot.asp?privcapId=232132>

Brewster, Signe. "Why 2014 Could Be the Year of the light based 3D printer." Gigamo. Gigamo, January 14, 2014. <https://gigaom.com/2014/01/14/why-2014-could-be-the-year-of-the-light-based-3d-printer/>

Brewster, Signe. "Why you won't see a laser sintering 3D printer on your desk anytime soon." Gigamo. Gigamo, April 25, 2014. Web. August 8, 2015. <https://gigaom.com/2014/04/25/why-you-wont-see-a-laser-sintering-3d-printer-on-your-desk-anytime-soon/>

Crump, Scott. "Apparatus and method for creating three-dimensional objects." Patent US5121329. October 30, 1989. <http://www.google.com/patents/US5121329>

Deckard, Carl. "Apparatus for producing parts by selective sintering." Patent US5597589. May 31, 1994. <https://www.google.com/patents/US5597589>

DMG MORI. "ALL IN 1: Laser Deposition Welding & Milling – Additive Manufacturing in Milling Quality." DMG MORI. DMG MORI, n.d. Web. August 8, 2015. <http://us.dmgmori.com/products/lasertec/lasertec-additivemanufacturing/lasertec-65-3d>

Gothait, Hanan. "Apparatus and method for three dimensional model printing." Patent US6259962. March 1, 2001. <https://www.google.com/patents/US6259962?dq>

Krassenstein, Eddie. "Laser Sintering 3D Printing May Now Take Off with a Very Important Patent Expiring Today." 3DPrint. 3D Print, January 28, 2014. Web. August 8, 2015. <http://3dprint.com/387/laser-sintering-3d-printing-may-now-take-off-with-a-very-important-patent-expiring-today/>

LPW Technology. "Laser Metal Deposition" LPW Technology. LPW Technology, n.d. Web. August 8, 2015. <http://www.lpwtechnology.com/technical-information/laser-metal-deposition/>

NASA. "NASA Tests Limits of 3-D Printing with Powerful Rocket Engine Check." NASA. NASA, August 27, 2013. Web. August 8, 2015. <http://www.nasa.gov/exploration/systems/sls/3d-printed-rocket-injector.html>

Protolabs. "Fineline: Direct Metal Laser Sintering." Proto Labs. Proto Labs, n.d. Web. August 8, 2015. <http://www.protolabs.com/additive-manufacturing/direct-metal-laser-sintering/design-guidelines>

RPM Innovations Inc. “Laser Deposition Technology.” RPM Innovations Inc. RPM Innovations Inc., n.d. Web. August 8, 2015. <http://www.rpm-innovations.com/laser_deposition_technology>

Sciaky Inc. “Make Metal Parts Faster & Cheaper Than Ever with Electron Beam Additive Manufacturing (EBAM) Technology.” Sciaky. Sciaky, n.d. Web. August 8, 2015. <http://www.sciaky.com/additive-manufacturing/electron-beam-additive-manufacturing-technology>Stratasys. “Between the Layer Lines: Understanding material nuances from metal 3D printing.” Stratasys Direct Manufacturing. Stratasys, November 7, 2013. Web. August 8, 2015. <https://www.stratasysdirect.com/blog/material-nuances-metal-3d-printing/>

Stratasys. “Direct Metal Laser Sintering (DMLS): Design Guideline.” Stratasys Direct Manufacturing. Stratasys, n.d. Web. August 8, 2015. <https://www.stratasysdirect.com/resources/direct-metal-laser-sintering-dmls/>

Stratasys. “Objet1000 Plus.” Stratasys. Stratasys, n.d. Web. August 8, 2015. <http://www.stratasys.com/3d-printers/production-series/objet1000-plus>

Stratasys. *World’s First 3D Printed Metal Gun.* Stratasys Direct Manufacturing. Stratasys, November 7, 2013. Web. August 8, 2015. <https://www.stratasysdirect.com/blog/worlds-first-3d-printed-metal-gun/>

Stratasys. “World’s First 3D Printed Metal Gun.” Stratasys Direct Manufacturing. Stratasys, November 7, 2013. Web. August 8, 2015. <https://www.stratasysdirect.com/blog/worlds-first-3d-printed-metal-gun/>

Stratasys. “World’s First 3D Printed Metal Gun Manufactured by Solid Concepts.” Stratasys Direct Manufacturing. Stratasys, November 7, 2013. Web. August 8, 2015. <https://www.stratasysdirect.com/newsroom/worlds-first-3d-printed-metal-gun-manufactured-solid-concepts/>

2015 WJTA-IMCA Conference and Expo
November 2-4 • New Orleans, Louisiana

Paper

ABRASIVE WATERJET MACHINING EFFECTS ON KERF QUALITY IN THIN FIBER METAL LAMINATE

M. Ramulu, R. Pahuja, M. Hashish*, V. Isvilonanda,
University of Washington,
Seattle, Washington, U.S.A.

*Flow International Corporation,
Kent, Washington, U.S.A.

ABSTRACT

High temperature Fiber Metal Laminate – Titanium/Graphite (Ti/Gr) is an advanced material system, developed to meet the high temperature requirements in aerospace applications. High specific strength and stiffness of composite core along with its protection from aggressive environment by tough titanium alloy sheets, qualify FMLs for a promising alternative material where metallic and composites overcome each other's limitations. However, industrial employability of this three phase system is often limited by the machining challenges posed by the difference in material removal mechanisms of Titanium alloy, PIXA thermoplastic polyimide resin and graphite fibers. An experimental investigation was conducted to evaluate the machinability of 1 mm thick Ti/Gr laminate sheets through Abrasive Waterjet (AWJ) machining process in terms of kerf characteristics and material removal rate. The parametric influence of AWJ operating variables on machining performance was studied by systematically measuring operating variables (traverse speed and Abrasive flow rate) using fully crossed Design of experiment (DOE) scheme, and statistically analyzing using ANOVA (Analysis of variance) technique. Empirical models were developed to quantify these effects and predict the influence of process parameters on material removal rate, kerf taper, entry damage width and overcut in straight cutting of Ti/Gr sheets.

Organized and Sponsored by WJTA®-IMCA®

1. INTRODUCTION

Hybrid composite laminates, also known as Fiber Metal Laminates (FML) is a group of engineered materials consisting of polymer matrix composites (PMC) plies interspersed with metal foils. The capability to withstand aggressive environment, high temperature; to resist oxidation and moisture ingress while maintaining structural integrity are few reasons why FMLs are known as new age engineered materials. This hybrid material system was first developed by Delft Research Group as ARALL® (Aramid Fiber Aluminum Laminate) followed by GLARE® (Glass Fiber Aluminum Reinforced Epoxy). However development of Titanium Graphite laminate (TiGr) by The Boeing Research Group along with NASA Langley Research Center has outperformed both the previously developed material systems in terms of fatigue strength, tensile strength and Inter-Laminar Shear Strength (ILSS). Bourlegat et al. [1] reported 95% improvement of tensile strength and 37% improvement in ILSS of TiGr when compared to GLARE®. TiGr is composed of alternate layers of graphite reinforced thermoplastic composite (PIXA-M) and titanium sheets. It is widely characterized as a futuristic material and finding employability in aerospace applications to address the future needs where aircrafts could high skin temperatures (177 °C) at supersonic speeds >3 mach. The directional properties of carbon composite (better fatigue characteristics, high stiffness-to-weight, and strength-to-weight ratio) with added benefits of Titanium alloy sheets (high tensile and flexural strength, high stiffness) coupled with properties rendered by hybrid nature of the system such as protection of composite core from heat and moisture, better distribution of point load and enhanced compression after impact (CAI) strength [2-5] by titanium sheets.

Although, TiGr laminates are generally molded/autoclaved to a near net shape, secondary machining is often required. Machining of composites alone poses a great challenge, which becomes even worse in a three phase material system – hybrid composites, where the non-homogeneity and anisotropy of composites together with the difference in the removal mechanism for all the three different phases aggravate machining problem. Machining of composite laminates using conventional processes is often limited by excessive tool wear and thermal distortion, fiber pull out, matrix cratering, delamination, fraying, machining dust and burr formation from metal foils [6-14]. Several investigators have reported that up to 60% of the rejected composite parts come from secondary machining defects [10-14], foregrounding the need and importance of more precise and non-traditional methods in composite machining.

Several non-traditional methods have been studied for the machining of composites. Ultrasonic machining is widely employed for brittle matrix composites (ceramic composites). Electrical discharge machining (EDM) has been employed to machine electrically conductive composite materials. However, objectionable surface integrity, high delamination due to arching and low MRR account for its inaptness in machining fiber metal laminates. Laser cutting is well suited for contour trimming of carbon fiber reinforced plastics because of point-sized geometry and multidirectional cutting ability. However, uneven kerf width and decrease in static strength of components due to thermal cracking and large HAZ restricts its use. Abrasive Waterjet machining is a widely used unconventional alternative because of properties such as no thermal damage to the workpiece, high cutting speed, wide working range and environmental friendly. Several FEM, analytical and experimental models have been developed to predict the effect of process parameters on the performance of machining ceramics and composites. However, the AWJ machining of laminates is seldom studied due to the complexity of the cutting process. Paul et al.

[15] studied the material removal process and quantified the effect of different parameters on GLARE®.

In this study, the machinability of TiGr is investigated and determined in terms of Material Removal Rate (MRR), kerf characteristics – overcut, damage width and taper as performance parameters of straight profile thin TiGr sheets. AWJ process variables are evaluated and optimized for high machinability. ANOVA is applied to statistically quantify the response of these variables and develop regression models to establish a nexus between these variables and machining response.

2. EXPERIMENT

2.1 Experimental Setup and Conditions

TiGr specimens were obtained from The Boeing Company in the form of $228 \times 28 \times 1$ mm, each composed of 8 plies with layers of graphite reinforced polymer matrix composite (PIXA-M) and 117 μm thick titanium alloy foils (Ti-15V-3Cr-3Sn-3Al) stacked together using autoclave consolidation process. The ply stacking sequence is Ti,90°,Ti,0°,0°,Ti,90°,Ti as depicted in Figure 1(c). Reinforced fibers in the composite were IM-7 intermediate modulus carbon fibers and the matrix used was LARC-IAX polymer. The mechanical properties of Titanium, fiber, and matrix material are depicted in Table 1.

A Flow Water Jet Pro model WJP1313 (CNC controlled) AWJ machine equipped with vacuum assist was used. Table 2. provides the description of experimental conditions and constant parameters used in this study.

2.2 Design and Methodology

The experiment was designed according to Full Factorial Scheme of Design of Experiments (DOE). The design matrix is depicted in Table 4. Three levels of traverse speed and abrasive flow rate were selected, as mentioned in Table 3, on the basis of previous investigations [4, 10, 16]. Nine straight cuts of approximately 22 mm long were machined at constant jet pressure setting (275.8 MPa) with 8 mm spacing between each cut, as depicted in Figure 1(b). Analysis of Variance (ANOVA) was performed to study the effect of individual process parameters and their interaction on machining responses.

The cutting process is depicted in Figure 1 (a) and the responses studied in the process are defined as follows.

Overcut is defined as the variation of predefined and actual dimension of cut for both entry and exit side and was calculated according to Equation (1). Overcut dimension is important in order to assign dimensional compensation to the NC program and achieve a tighter geometric tolerances.

$$Overcut = (W_{cut} - d_{tool})/2 \tag{1}$$

Kerf Taper is defined as the difference between entry and exit kerf width averaged over thickness over the workpiece. Kerf width was measured by optical microscopy.

$$Kerf\ Taper = (d_{entry} - d_{exit})/2t \quad (2)$$

Damage width is the width of discolored region on Titanium alloy face sheet around the cut, observed under optical microscope.

$$Damage\ width = (D - d)/2 = (Entry\ Overcut - Exit\ Overcut)/t \quad (3)$$

D is the dimension of discolored region across the cut feature and d is the dimension of feature machined. The reason for reporting only entry damage width is the poor visibility of exit damage under optical microscopy due to its negligible size.

Analysis of Variance (ANOVA) was performed using Design-Expert v7.1.5 software with 95% confidence level in order to analyze and determine the effect of traverse speed and AFR on the machining response.

3. RESULTS AND DISCUSSION

3.1 Kerf Surface Observations

As shown in Figure 2, rounded edge along the entry side of the jet was identified as the Initial damage region, followed by Smooth cutting and Rough cutting region as previously observed by Ramulu et al. [10, 14, 17]. Higher visible damage and distortion was observed in the middle composite plies. No delamination was observed in the chosen domain of parameters.

SEM inspection was carried on to observe the mechanism of material removal for titanium, graphite fiber and thermoplastic matrix as different phases. Specimen machined with lowest traverse speed (60 mm/min) and highest abrasive flow rate (9.78 g/s) showed wear tracks and signs of shear deformation on titanium ply (Figure 3a) concluding that Titanium was cut by ductile shearing, abrasive plowing, and scratching action, as observed by Seo et al. [18]. Further within PIXA-M composite, LARC-IAX matrix material and IM-7carbon fiber also subjected to different material removal mechanisms. The matrix material was removed by shearing and plastic deformation while fibers were cut by micro chipping, brittle fracture, and bending failure. The bending and chipping can be easily observed in Figure 3(b), depicting PIXA-M ply machined at 720 mm/min traverse speed and 3.35 g/s AFR. Thickness of the specimen was small enough that a smooth cutting region was extended throughout the depth. In some location, an abrasive particle plowed through the material, fractured, and embedded to the bottom of the track was detected and as shown in Figure 3(c).

The surface characteristics for cut surfaces were investigated. It was observed that using high traverse speed and low AFR resulted in more topologically irregular surface. However, entry damage on the top titanium ply was smaller. Cleaner and smoother cut surface was obtained at low traverse speed and high AFR. At this condition, shallower grooves and smaller titanium burrs were observed. Within the range of experimental conditions, all plies remained intact and no delamination was observed. Average surface roughness for the best and worst surface condition is

0.636 and 0.899 μm respectively. Surface profiles for both conditions are shown in Figure 4. The effect of cutting geometry on surface roughness was also investigated by comparing the straight cut surface roughness with trepanned surface roughness. The surface roughness for trepanned surfaces was observed between 0.662 and 0.671 μm.

3.2 Statistical Analysis of Machining Responses

The observed machining responses are depicted in Table 4. The effect of process parameters on aforementioned machining responses is statistically analyzed by employing ANOVA (Analysis of variance) and the responses are modelled with regression analysis, governed by equations depicted in Table 5. In order to account for the credibility of analysis, the model was verified by plotting diagnostic curves i.e. normal probability and residual plots. A straight line in normal probability plot verified the distribution of residuals as Gaussian. A Residual vs Run plot was used to check for any consistent trends and thus periodically sustained error(s) in the experimentation. A random scatter of data points verified the absence of any time-related lurking variable(s) which could be traced back and blocked for a sound analysis. Residual versus Predicted value plot verified the model assumption of constant variance across the range of predictions. A constant variance is characterized by a randomly scattered data within the control limits and absence of any pattern. Besides, a leverage versus run was graphed to indicate the influence of each point in the model. If a point depicts a leverage of 1.0, then the point controls the model and model must go through that point. The plots for MRR model analysis are included in this discussion. However, similar observations were made for other model plots. Figure 5(a, b) depicts near normal distribution of residuals and absence of time dependent experimentation errors. Figure 6(a) depicts a constant variance model. A leverage versus run plot, shown in Figure 6(b) indicates the absence of one point dominance in the model.

The ANOVA response models are shown in Figure 7 and listed in Table 5. Figure 7(a) indicates that an increasing jet traverse speed and decreasing abrasive mass flow rate contributed to higher MRR which can be explained with the reduced cutting time at the respective parameter settings. MRR was found to change significantly with traverse speed which is supported by its low P-value (7.80E-07) in the model, as depicted in Table 6. The parametric influence on overcut is shown in Figure 7(b, c). Overcut is observed to be more at jet entry side in comparison to exit side. This can be accounted to initial high energy abrasive jet at the entry side of the cut in the cutting wear zone which reduced with the depth of penetration. Model also delineated higher overcut at high AFR and low traverse speed. The kerf width was observed within the range of 1.15-1.3 times nozzle diameter (447-647 μm) at entry side and 0.92-1.16 times nozzle diameter (458-580 μm) at exit side.

Kerf taper was observed to be positive and within the absolute range of 0.065-0.115 mm/mm in the domain of chosen parameter values. Jet traverse speed exhibited a negative effect on the kerf taper, whereas Abrasive flow rate exhibited a positive effect, as depicted in Figure 7(d). ANOVA analysis showed traverse speed (P-value = 2.53E-03) and AFR (P-value = 2.58E-02) to be significantly contributive to the kerf taper response. Similar to overcut response, entry damage was observed within 59-286 μm, and showed positive trend when AFR was increased and traverse speed was reduced.

A low traverse speed increases jet exposure time and resulted in lesser amount of high energy abrasive particles impacting the work piece. This induced smoother kerf wall, smaller kerf taper, larger kerf width, and larger initial damage width. Effect of increasing AFR was similar to slowing down the traverse speed as increasing AFR would mean larger amount of abrasive particles exposed to the work piece per unit time performing the cutting operation. However, a balance between these two process variables was found critical.

4. SUMMARY AND CONCLUSION

The straight cutting of Titanium Graphite fiber metal laminate sheet with Abrasive Water jet cutting was studied in this work. The effect of AWJ process parameters- jet traverse speed and abrasive flow rate on the cutting characteristics and process performance was investigated based on a fully crossed design of experiment scheme. ANOVA was employed for parametric optimization and predictive response models were established and verified based on the regression based statistical analysis for a chosen range of process variables. Machinability of TiGr was evaluated in terms of cutting time, material removal rate (MRR), overcut, taper and damage as machining responses. Traverse speed was found as a dominant factor in controlling these responses. The effect of slowing traverse speed was found similar to increasing AFR. Both result in larger amount of high energy abrasive particles impact the work piece per unit time. High MRR, small overcut, and damage could be obtained by using high traverse speed. However, high AFR should be used in order to get smooth surface finish and reduce kerf taper.

The scope of the study is limited to thin sheets and the inspection of the effect of only two, but significant parameters – AFR and traverse speed. However, the variability in cutting performance may be achieved with the alteration of other feasible process variables viz. standoff distance, pressure, and nozzle geometry. The models cannot be projected to machining thicker laminate material where the effects of fiber layup and cutting profile are expected to be more pronouncedly influencing the jet energy distribution and its dissipation throughout the penetration depth, accounting for more sensitive control of the process parameters.

5. REFERENCES

1. Le Bourlegat, L. R., Damato, C. A., da Silva, D. F., Botelho, E. C., & Pardini, L. C. (2010). Processing and mechanical characterization of titanium-graphite hybrid laminates. Journal of Reinforced Plastics and Composites, 29 (22), 3392–3400. Doi: 10.1177/0731684410377541.
2. Burianek DA, Giannakopoulos AE and Spearing SM. Modeling of facesheet crack growth in titanium–graphite hybrid laminates, Part I. Eng Fract Mech 2003; 70:775–798.
3. Burianek, D.A., Fatigue Damage in Titanium Graphite Hybrid Laminates, Master of Science Thesis, Department of Aeronautics and Astronautics, Massachusetts Institute of Technology, 1998
4. Ramulu, M., Stickler, P.B., and McDevitt, N.S., Influence of Processing Methods on the Tensile and Flexure Properties of High Temperature Composites, Composite Science and Technology, Vol. 64, 2004, pp. 1763-1772

5. Edward, L., Johnson, W.S., and Lowthere, S.E., An Evaluation of Two Fabrication Methods for Hybrid Titanium Composite Laminates, Composite Materials: Testing and Design, Vol. 13, 1997, pp. 202-214
6. Kim, D., and Ramulu, M., Cutting and Drilling Characteristics of Hybrid Titanium Composite Laminate, Transactions of NAMRI/SME, Vol. 33, 2005, pp. 445-452
7. Hashish, M."AWJ Trimming of 787 Composite Stingers," 2010 SAMPE Proceedings, Seattle, WA. www.sampe.orb 2010
8. Hashish., M., "AWJ Trimming and Drilling of Composite for Aircraft Structures, "Proceedings of the 9th Pacific Rim International Conference on Water Jetting Technology, November 20-23, 2009, Koriyama, Japan
9. Hashish, M., "Machining of Advanced Composites with Abrasive-Waterjets", Manufacturing Reviews, Vol. 2, No. 2, 1989, pp. 142-150
10. Arola, D., and Ramulu, M., A study of Kerf Characteristics in Abrasive Waterjet Machining of Graphite/Epoxy Composite, Journal of Engineering Materials and Technology, Vol. 118, April 1996, pp. 256-265
11. Connor, I., M. Hashish and M. Ramulu, "Abrasive Waterjet Machining of Aerospace structural Sheet and Thin Plate Materials", Proceedings of 2003 WJTA Conference
12. Ramulu, M., Jenkins, M.G., and Guo, Z., Abrasive Water Jet Machining Mechanisms in Continuous-Fiber Ceramic Composites, Journal of Composites Technology & Research, JCTRER, Vol. 23, No. 2, April 2001, pp. 82-91
13. Hashish, M., (1995) "Waterjet Machining of Advanced Composites," Materials and Manufacturing Processes, Marcel Decker, NY, Vol. 10, No 6, pp 1129-1152.
14. Hamatani, G. and Ramulu, M., Machinability of High Temperature Composites by Abrasive Waterjet, Journal of Engineering Materials and Technology, Vol. 112, Oct. 1990, pp. 381-386
15. Paul, S., Hoogstrate, A.M., Van Praag, R., 2002. Abrasive water jet machining of glass fibre metal laminates. J. Eng. Manuf. Part B 216, 1459–1469.
16. Edward, L., Johnson, W.S., and Lowthere, S.E., An Evaluation of Two Fabrication Methods for Hybrid Titanium Composite Laminates, Composite Materials: Testing and Design, Vol. 13, 1997, pp. 202-214.
17. Seo, Y.W., Ramulu, M., and Hashish M., Cost Analysis of Abrasive Waterjet Cutting: Thin Sheet Materials, Proceedings of SAMPE, Fall Conference 2005, (37th ISTC).

6. NOMENCLATURE

V	Traverse speed in (mm/min)
$\dot{m}$	Abrasive flow rate (AFR) (g/s)
t	Time (s)
V	Volume of material removed (mm^3)
SOD	Standoff distance (mm)
MRR	Material removal rate (mm^3/s)
$\mathrm{W_{cut}}$	Width of cut (mm)
$\mathrm{d_{tool}}$	Tool diameter (mm)
d_{entry}	Jet entry diameter (mm)
d_{exit}	Jet exit diameter (mm)
C_0-C_5	Empirical constants

7. TABLES

Table 1. Mechanical properties of the constituent materials of TiGr [16]

Properties	Material		
	Ti-15-3	**LARC-IAX**	**IM7 Carbon fiber**
Longitudinal modulus (GPa)	100.61	3.33	272.2
Transverse modulus (GPa)	-	-	13.58
Shear modulus (GPa)	33.75	1.25	200
Poisson's ratio	0.36	0.33	0.25
Yield strength (MPa)	1307	67.92	-
Ultimate strength (MPa)	1459	105	36612
Coefficient of thermal expansion (10^{-6}/k)	8.21	10	-
Percent elongation (%)	4.1	6	1.8

Table 2. Constant parameters and Experimental conditions for AWJ cutting experiments

Parameter (constant)	**Description**
Orifice	Sapphire, Ø 0.381 mm
Nozzle	Length 76.2 mm, dia. Ø 1.016 mm
Grit type	Garnet
Grit size	Mesh 80
Impact angle	90°

Table 3. Range of variables cutting experiments

Variable	**Range**
Traverse speed (mm/min)	60, 390, 720
Abrasive flow rate (g/s)	3.53, 6.23, 9.78

Table 4. Experimental design matrix with observed responses

Experiment No.	Run Order	Parameters		Measurement data		
		Traverse speed (mm/min)	Abrasive flow rate (g/s)	MRR (mm^3/min)	Entry damage	Kerf Taper
1	2	3.53	60	0.8000	0.136	0.0744
2	5	3.53	390	2.8015	0.077	0.0930
3	6	3.53	720	3.5904	0.059	0.1155
4	4	6.23	60	0.8156	0.193	0.0695
5	1	6.23	390	2.8234	0.106	0.0890
6	3	6.23	720	3.6182	0.097	0.0959
7	9	9.78	60	0.8210	0.286	0.0656
8	7	9.78	390	2.8897	0.165	0.0832
9	8	9.78	720	3.7216	0.155	0.0930

Table 5. Mathematical models for AWJ straight cut machining operation performed on TiGr laminate.

Performance Parameter	Model expression	Coefficient of determination R^2
$\mathrm{MRR}(V, \dot{m})$	$0.34956 + 8.48267 \times 10^{-3}V - 0.015744\dot{m} - 5.60514 \times 10^{-6}V^2 - 1.34073 \times 10^{-3}\dot{m}^2 + 2.74886 \times 10^{-5}V\dot{m}$	1.00
$\mathrm{Entry\ overcut}(V, \dot{m})$	$0.65138 - 2.40986 \times 10^{-4}V - 6.93315 \times 10^{-4}\dot{m} + 1.91307 \times 10^{-7}V^2 + 1.81951 \times 10^{-4}\dot{m}^2 - 8.76375 \times 10^{-7}V\dot{m}$	0.9987
$\mathrm{Exit\ overcut}(V, \dot{m})$	$0.56341 - 3.31102 \times 10^{-4}V + 5.3243 \times 10^{-3}\dot{m} + 2.17325 \times 10^{-7}V^2 - 1.95583 \times 10^{-4}\dot{m}^2 + 2.32201 \times 10^{-6}V\dot{m}$	0.9989
$\mathrm{Kerf\ Taper}(V, \dot{m})$	$0.086289 + 8.80089 \times 10^{-5}V - 5.94505 \times 10^{-3}\dot{m} - 2.52525 \times 10^{-8}V^2 + 3.73667 \times 10^{-4}\dot{m}^2 - 3.12938 \times 10^{-6}V\dot{m}$	0.9740
$\mathrm{Entry\ Damage}(V, \dot{m})$	$0.10651 - 3.42218 \times 10^{-4}V + 0.013565\dot{m} + 3.52 \times 10^{-7}V^2 + 7.05547 \times 10^{-4}\dot{m}^2 - 1.31855 \times 10^{-5}V\dot{m}$	0.9892

Table 6. P value test as depiction of factor significance in each of the five response models

Factor	P-value				
	MRR	Entry overcut	Exit overcut	Entry damage	Kerf Taper
Model	7.80E-07	1.53E-04	1.29E-04	3.76E-03	1.39E-02
$\dot{m}$	4.44E-03	8.95E-03	1.48E-03	1.38E-03	2.58E-02
V	1.07E-07	2.40E-05	1.98E-05	1.76E-03	2.53E-03
$\dot{m} \times V$	2.05E-02	3.76E-01	1.45E-01	1.03E-01	2.09E-01
$\dot{m}^2$	2.47E-01	2.54E-01	3.60E-01	4.77E-01	3.02E-01
V^2	6.92E-06	4.54E-04	8.50E-04	1.92E-02	4.09E-01

8. GRAPHICS

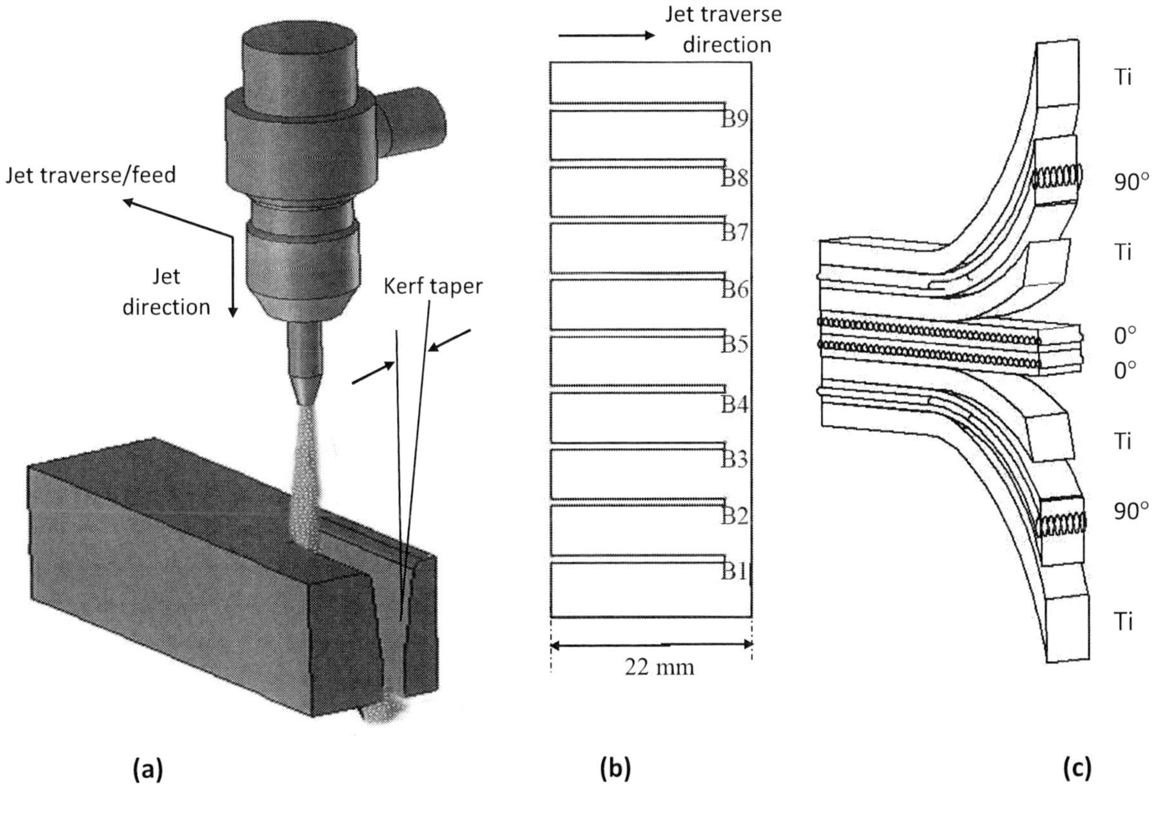

Figure 1. (a) Ply stacking sequence of fiber metal laminate workpiece; (b) Cutting pattern undertaken in the present study; (c) AWJ cutting process depicting kerf taper.

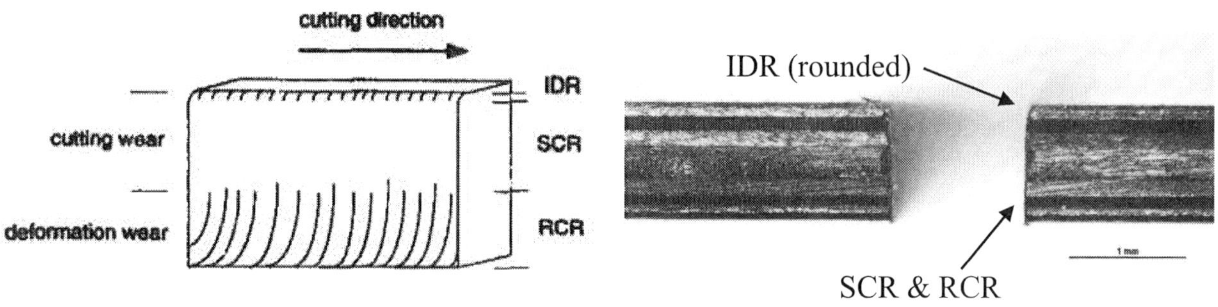

(a) (b)

Figure 2. Identified wear regions (a) schematic representation, (b) cross-section of AWJ cut specimen at 60 mm/min with 3.73 g/s AFR.

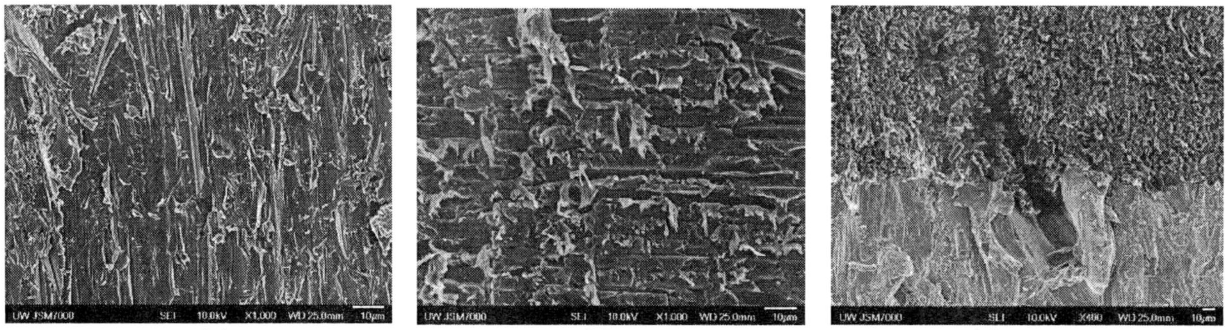

(a) (b) (c)

Figure 3. SEM images (a) Titanium ply (b) 90 degree PIXA-M composite ply (c) embedded grit in titanium ply at 720 mm/min traverse speed and 3.35 g/s AFR.

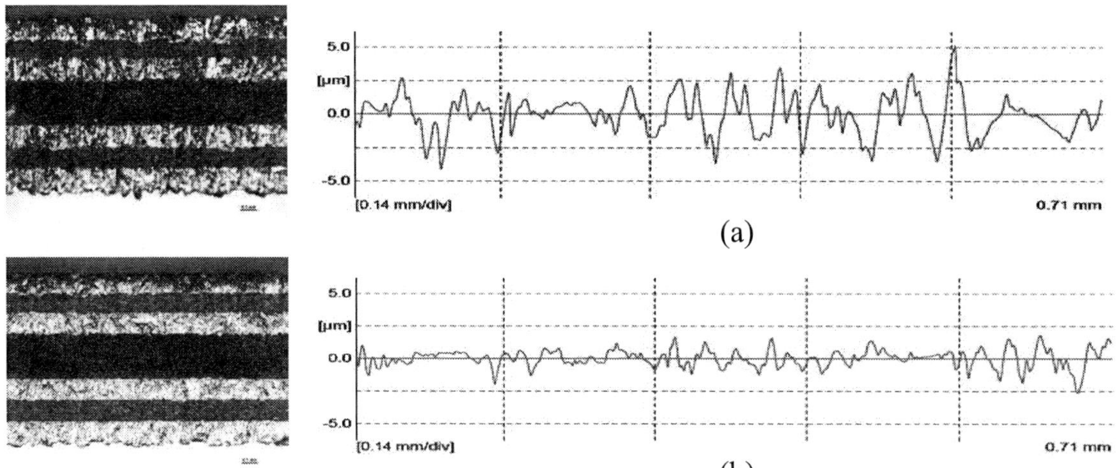

(a)

(b)

Figure 4. AWJ machined surface (left) and surface roughness profiles along the thickness direction (right) for (a) Worst AWJ cut with traverse speed = 720 mm/min and AFR = 3.53 g/s; (b) Best AWJ cut with traverse speed = 60 mm/min and AFR = 9.78 g/s;

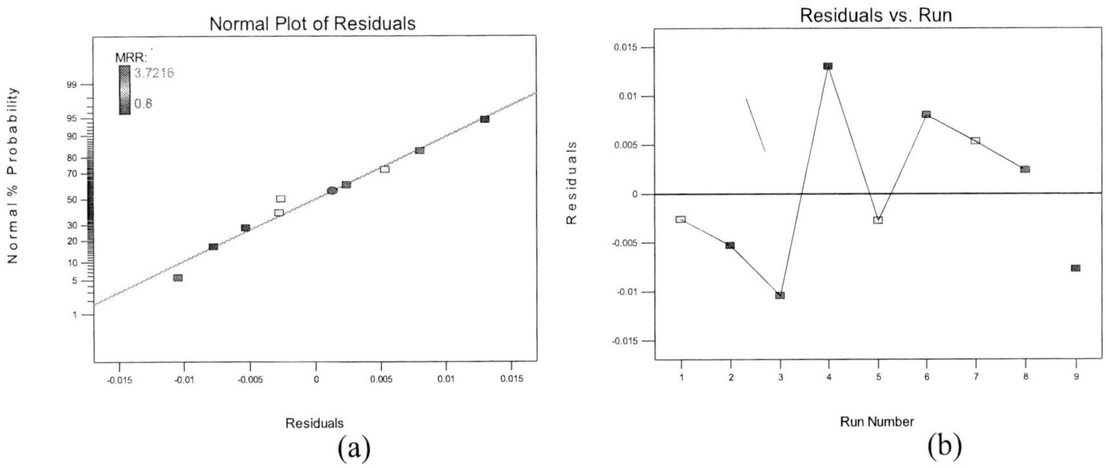

Figure 5. (a) Normal plot of residuals; (b) Residuals versus Run plot.

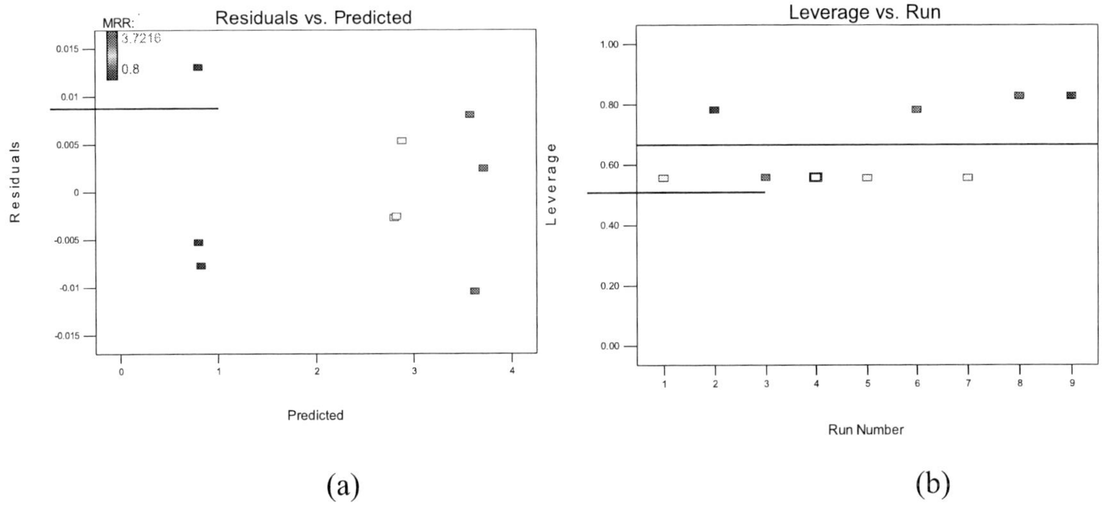

Figure 6. (a) Residual versus predicted; (b) Leverage versus Run plot.

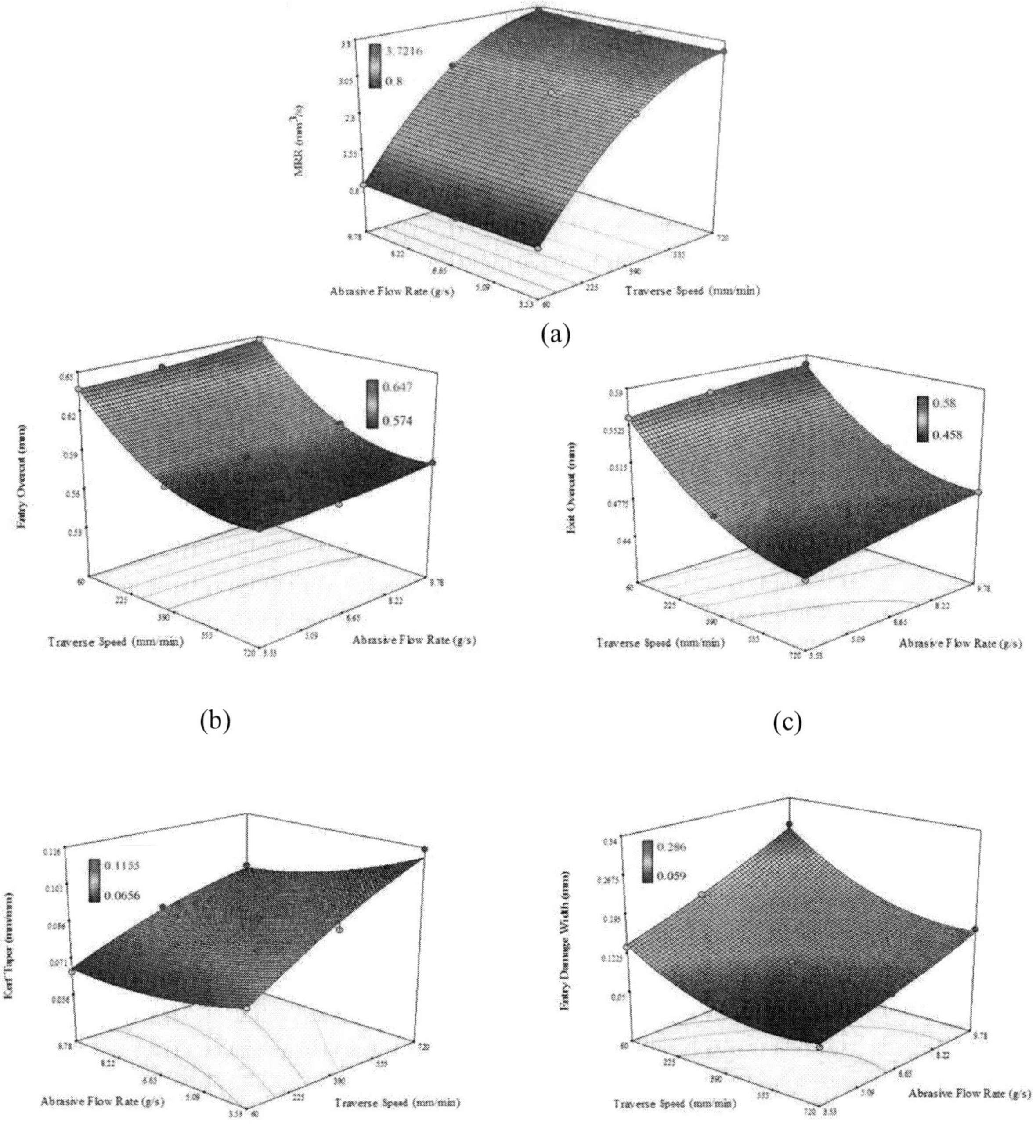

Figure 7. Effect of traverse speed and AFR on (a) Material Removal Rate (MRR); (b) Entry overcut; (c) Exit overcut; (d) Kerf taper; (e) Entry damage width during slit cutting operation.

2015 WJTA-IMCA Conference and Expo
November 2-4 • New Orleans, Louisiana

Paper

3D NEAR NET SHAPING OF HARD TO MACHINE MATERIALS VIA ABRASIVE WATERJET CONTROLLED-DEPTH MILLING

F. Faltin, K. Flögel, E. Uhlmann
Technische Universität Berlin
Institute for Machine Tools and Factory Management
Berlin, Germany

ABSTRACT

Nickel alloys e.g. Inconel 718 have outstanding mechanical, thermal and corrosive properties. Despite its wide range of use Inconel 718 can still be considered as a hard to cut material using conventional cutting processes e.g. milling or turning. The abrasive waterjet (AWJ) technology has the capability to machine these hard to cut materials reliably with reduced mechanical and thermal damages at the same time. Being able to create notches with predictable depths, high aspect ratios and low predictable notch base irregularities via AWJ can lead to a substantial increase of machining performance of hard to cut materials. Thus, 3D near net shaping as a roughening process prior to conventional cutting processes is possible. The goal of this investigation was to analyze the correlation between AWJ settings parameters and the notch depth as well as the notch base irregularity when machining Inconel 718. A prediction model including the main effects, interactions and quadratic effects was developed using design of experiments.

Organized and Sponsored by WJTA®-IMCA®

1. INTRODUCTION

The constant development of new materials with enhanced mechanical or chemical properties results in great challenges for the manufacturing technology. These improved material properties often lead to difficulties for cutting processes e.g. long machining times or high tool wear. An economic industrial application of these materials is only possible with an adequate manufacturing method. Despite its wide range of use Inconel 718 can still be considered as a hard to cut material. Inconel 718 shows a good long period creep resistance at temperatures up to 700 °C, resistance to oxidation up to 1,000 °C and excellent corrosive properties at high and low temperatures. Due to this combination Inconel 718 is used in a variety of applications e.g. aircraft engines or cryogenic tanks.

One of the advantages of the abrasive waterjet technology is its wide range of machinable materials. Even hard to cut materials can be economically and reliably processed with AWJ with reduced mechanical or thermal damages. At the moment the industrial use of AWJ is limited to machining operations where the workpiece is completely penetrated e.g. cutting or drilling operations.

In the late 1980s Hashish introduced the abrasive waterjet controlled depth milling technology (AWJ-CDM) to create notches in different materials using multiple passes with a high feed velocity. He identified the feed velocity as a critical setting parameter for the notch depth irregularity and suggested a minimum value of 960 mm/min [1]. He also concluded that already induced irregularities in the notch base cannot be corrected by further passes but rather aggravated [2]. Therefore, the initial state of the surface being machined is of importance and must be considered at the beginning of the machining operation. Laurinat et al. [3], Öjmertz et al. [4] and Kong et al. [5] presented mathematical models to describe the notch depth for small notch depths (< 3 mm). In order to use AWT-CDM as a pre-shaping process of bulk material as shown in Figure 1 [6], a greater notch depth with predictable depth and irregularity in the base is necessary. Also roundings at the sides of the workpiece must be prevented. Fowler [7] suggested the use of so called “butt masks” at the sides of the workpiece as shown in Figure 2.

The goal of this investigation was to provide an empirical model to describe the notch depth and notch base irregularity as a function of the setting parameters for high aspect ratios (50:1) in Inconel 718. By using design of experiments (DoE) main effects as well as parameter interactions and quadratic effects could be included in the model. This model was then applied to create a notch with a depth of 50 mm with a notch base irregularity of less than 0.5 mm. The roundings at the sides of the test specimen were not part of the investigation.

2. EXPERIMENTAL INVESTIGATION

2.1 Technical objective and applied statistical method

The scope of this investigation was to develop an empirical model of the average notch depth t_K and its irregularity in the base Δt_K for notches with high aspect ratios (50:1) in nickel alloy Inconel 718 using AWJ-CDM, Figure 3. In screening experiments significant parameters and adequate parameter ranges were determined. The investigated parameters were the jet pressure p, the feed velocity v_f, the abrasive flow rate $\dot{m}_A$ and the number of passes z. The stand-off distance was kept constant at $s_0 = 2$ mm. A central composite design (CCD) followed by a linear regression was used to analyze main effects but also parameter interactions and second order effects on the target values. The applied CCD consists of three successive experimental steps. A factorial design with two parameter setting levels is used to determine main effects and parameter interactions. Including a center point indicates a nonlinear behavior. A set of axial points are used to analyze the curvature of the function. By using linear regression empirical models can be developed. The general mathematical equations for the target values are:

$$y = \beta_0 + \sum_{i=1}^{k} \beta_i x_i + \sum_{i<j=2}^{k} \beta_i x_i x_j + \sum_{i=1}^{k} \beta_{ii} x_i^2 + \epsilon \tag{1}$$

The CCD is described in detail by Myers et al. [8]. The structure of the design is shown in Figure 4. The investigated parameters and parameter ranges are summarized in Table 1.

2.2 Test setup

For the experimental investigation a six-axis robot was used for jet manipulation with a maximum traverse speed of up to 2.5 m/s. The water pressure was realized with a 45 kW hydraulic high pressure intensifier delivering up to 600 MPa and a maximum flow rate of 2.5 l/min. The abrasive flow rate could continuously be varied from 50 to 500 g/min. A standard combination of orifice with a diameter of 0.25 mm and focus with a diameter of 0.76 mm and a length of 76 mm were selected. Garnet sand Mesh 80 was used as abrasive. The tested material was nickel alloy Inconel 718 (2.4668). The used machining system is shown in Figure 5.

2.3 Measurement method

To measure the notch depth and its irregularity in the base, the test specimens were separated along the notch base after the experiments using wire EDM. These target values were determined by the optical measuring device FRT microProf 100 using white light interferometry. A range of 30 mm x 3 mm with a resolution of 300 pixels x 300 pixels was measured. An example of a measured notch depth irregularity is shown in Figure 6.

2.4 Results for the notch depth

The analysis of the data for AWJ-CDM on Inconel 718 specimen showed that all four investigated parameters have a significant main effect on the notch depth t_K. Furthermore, two factor parameter interactions and quadratic effects could be observed by the use of CCD, Figure 7. Using linear regression an empirical model could be developed to predict the notch depth t_K. The coefficient of determination showed to be excellent ($R^2 = 0.99$). The empirical model to describe the notch depth t_K is:

$$t_K = 28.94 + 6.95\,p - 6.51\,v_f + 2.08\,\dot{m}_A + 2.24\,z - 1.41\,p\,v_f + 1.47\,p\,\dot{m}_A + 0.50\,p\,z - 0.50\,v_f\,\dot{m}_A - 0.51\,v_f\,z + 1.55\,v_f^2 - 1.35\,z^2 \tag{2}$$

Figure 8 shows examples the modeled notch depths for the pressure, feed velocity, abrasive flow rate and number of passes. For this representation the remaining parameters are kept constant on "center-level".

As expected, with rising pressure and number of passes the notch depth increased in a linear behavior. By raising the feed velocity the notch depth was reduced in a regressive behavior. For the abrasive flow rate a maximum notch depth at about 150 g/min was found. This location of the maximum was dependent on the used pressure. Furthermore, two factor parameter interactions and quadratic effects could be observed by the use of CCD. An example for the parameter interaction is shown in Figure 9, left. Increasing the pressure and lowering the feed velocity at the same time resulted in a higher notch depth than it would be expected when examining the parameters independently.
In detail, the significant effects are:

Main effects	**Parameter interactions:**		
Pressure p	Pressure p	x	feed velocity v_f
Feed velocity v_f	Pressure p	x	abrasive flow rate $\dot{m}_A$
Abrasive flow rate $\dot{m}_A$	Pressure p	x	number of passes z
Number of passes z	Feed velocity v_f	x	abrasive flow rate $\dot{m}_A$
	Feed velocity v_f	x	number of passes z

Quadratic effects:
(Feed velocity v_f)^2
(Abrasive flow rate $\dot{m}_A$)^2

For using the model to calculate the target value, the parameter settings must be entered in standardized form. An example to create a notch depth of t_K = 50 mm is shown in Figure 10. The deviation of the calculated and achieved depth was in a range of 0.1 mm.

2.3 Results for the notch depth irregularity

As in the analysis of the notch depth, the behavior of the notch depth irregularity showed to be nonlinear. The parameters pressure p, feed velocity v_f and the abrasive flow rate $\dot{m}_A$ have a significant main effect on the target value. The number of passes did not show to have a significant effect within the tested parameter range. Also one two-factor interaction and quadratic effects could be observed, Figure 11. Using the same mathematical analysis as for the notch depth an empirical model to describe the notch depth irregularity Δt_K as a function of the setting parameters was obtained. The coefficient of determination showed to be good ($R^2 = 0.82$). The equation to predict the notch depth irregularity Δt_K is:

$$\Delta t_K = 0.54 + 0.25\,p - 0.20\,v_f - 0.19\,m_A - 0.13\,p\,v_f + 0.11\,p^2 + 0.13\,v_f^2 + 0.14\,m_A^2 \quad (3)$$

Figure 8 shows examples the modeled notch depth irregularity for the pressure, feed velocity, abrasive flow rate and number of passes. For this representation the remaining parameters are again kept constant on "center-level".

With rising pressure the notch depth irregularity increased in a linear behavior. Raising the feed velocity resulted in a much lower notch depth irregularity. Remarkably, the irregularity increased with lower abrasive flow rate. This behavior can be explained by the use of a relatively long abrasive feeding tube, due to the use of a robot for jet guidance. Due to air vibrations inside of the abrasive feeding tube the abrasive did not enter the cutting head continuously, but in waves. This effect was stronger for low abrasive flow rates and resulted in an increased notch depth irregularity which was aggravated with every pass as explained in [2]. One two factor interaction and quadratic effects could be observed. The parameter interaction of pressure and feed velocity is shown in Figure 9, right. Lowering the pressure and increasing the feed velocity at the same time resulted in a much lower notch depth irregularity than it would be expected when analyzing the main effects only. In detail, the significant effects are:

Main effects
Pressure p
Feed velocity v_f
Abrasive flow rate $\dot{m}_A$

Quadratic effects:
(Pressure p)^2
(Feed velocity v_f)^2
(Abrasive flow rate $\dot{m}_A$)^2

Parameter interactions:
Pressure p x feed velocity v_f

To calculate the target value, the parameter settings must again be entered in the equation in standardized form. An example to create a notch depth of $t_K = 50$ mm is shown in Figure 10. The deviation of the calculated and achieved notch depth irregularity was in a range of 1.3 mm.

3. CONCLUSION

The investigation showed that for estimating the notch depth and notch depth irregularity not only main effects but also parameter interactions as well as quadratic effects have a significant influence on the target values and must be considered. By using a central composite design followed by a regression analysis an empirical model to estimate the notch depth and notch base irregularity for high aspect ratios in Inconel 718 using AWJ-CDM was generated. This model was then applied for a machining example creating a notch with $t_K = 50.1$ mm with an irregularity in the base of $\Delta t_K = 0.3$ mm. By overlapping these notches 3D near net shaping prior to conventional cutting processes would be an adequate machining solution for hard to cut materials.

4. ACKNOWLEDGMENTS

This paper is based on results acquired in the project DFG UH 100/165-1, which is kindly supported by the German Research Foundation (DFG).

5. REFERENCES

[1] Hashish, M.: Milling with abrasive-waterjets: A preliminary investigation. In: Hood, M.; Dornfield, D. (Hrsg.): Proceedings of the Fourth U.S. Water Jet Conference (1987), S. 1 - 20.

[2] Hashish, M.: An investigation of Milling With Abrasive-Waterjets. Journal of Engineering for Industry 111 (1989) 2, S. 158 - 166.

[3] Laurinat, A.; Louis, H.; Meier-Wiechert, G.: A model for milling with abrasive water jets. In: Hashish, M. (Hrsg.): Proceedings of the 7th American Water Jet Conference (1993) 1, S. 119 - 139.

[4] Öjmertz, K. M. C.: Abrasive waterjet milling: An experimental investigation. In: Hashish, M. (Hrsg.): Proceedings of the 7th American Water Jet Conference (1993) 2, S. 777 - 791.

[5] Kong, M.C.; Anwar, S.; Billingham, J.; Axinte, D. A.: Mathematical modelling of abrasive waterjet footprints for arbitrarily moving jets - Part I - single straightpaths. International Journal of Machine Tools and Manufacture 53 (2012) 1, S. 58 - 68.

[6] Faltin, F.: Bearbeitungsstrategie zur materialeffizienten Vorkonturierung von Nickelbasislegierungen mittels Wasserabrasivstrahltechnologie. In: Uhlmann, E. (Hrsg.): Tagungsband zur 7. Berliner Runde - Neue Konzepte für Werkzeugmaschinen, Berlin, 22.-23. März 2012, Fraunhofer IPK, Berlin, S. 183 - 192. ISBN: 978-3-9814405-5-3.

[7] Fowler, G.: Abrasive Water-jet-controlled Depth Milling of Titanium Alloys. PhD Thesis, University of Nottingham, UK, 2003.

[8] Myers, R. H.; Montgomery, D. C.; Anderson-Cook, C. M.: "Response Surface Methodology: Process and Product Optimization Using Designed Experiments", 3rd Edition, January 2009.

6. NOMENCLATURE

Parameter	Symbol	Unit
Orifice diameter	d_D	mm
Focus diameter	d_F	mm
Modeled notch depth	$t_{K,m}$	mm
Modeled notch depth irregularity	$\Delta t_{K,m}$	mm
Abrasive flow rate	$\dot{m}_A$	g/min
Coefficient of determination	R^2	-
Correlation coefficient	β_0	-
Correlation coefficient	β_i	-
Correlation coefficient	ε	-
Feed velocity	v_f	mm/min
Notch depth	t_K	mm
Notch depth irregularity	Δt_K	mm
Number of passes	z	-
Parameter setting	x_i	-
Parameter setting	x_j	-
Pressure	p	MPa
Pixels	px	-

Stand-off distance	s_0	mm
Target value	y	-

Abbreviation

AWJ	Abrasive waterjet
CCD	Central composite design
CDM	Controlled depth milling
DoE	Design of experiments
EDM	Electro-discharge machining

7. TABLES

Table 1. Investigated parameters and parameter ranges using DoE-CCD

Parameter	**Symbol**	**Unit**	**- α**	**- 1**	**0**	**+ 1**	**+ α**
Pressure	p	MPa	100	125	150	175	200
Feed velocity	v_f	mm/min	2000	2750	3500	4250	5000
Abrasive flow rate	$\dot{m}_A$	g/min	50	88	125	163	200
Number of passes	z	-	250	275	300	325	350

8. GRAPHICS

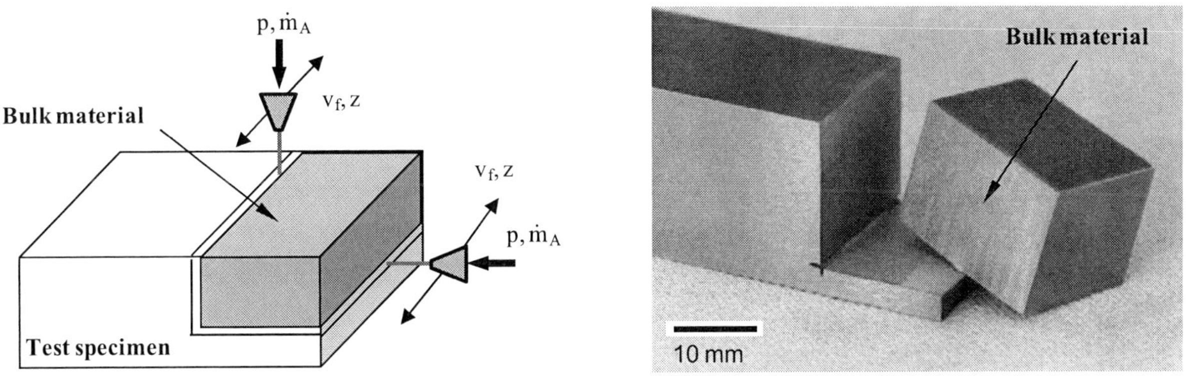

Figure 1. Removal of bulk material with AWJ-CDM [6]

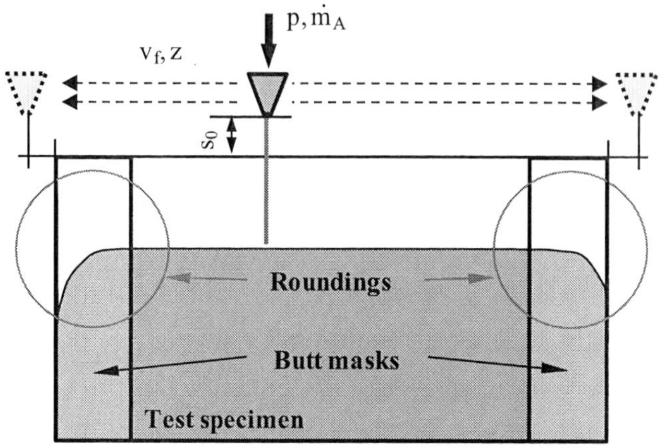

Figure 2. Removal of bulk material with AWJ-CDM using butt masks

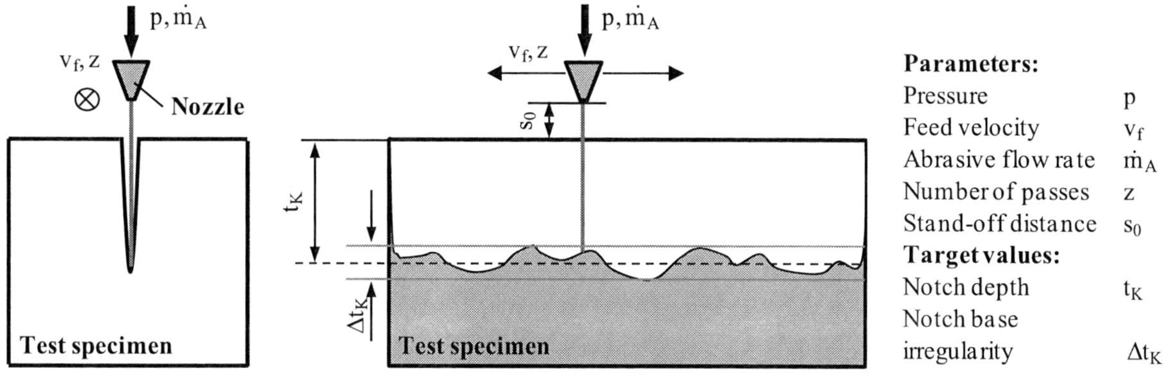

Figure 3. Schematic representation of the target values

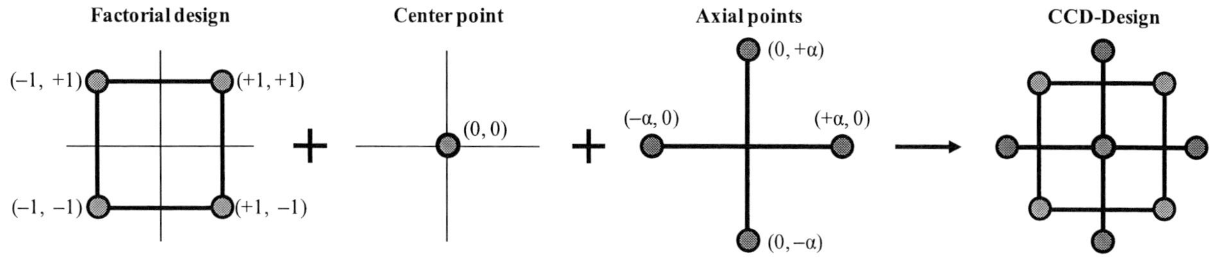

Figure 4. Structure of a central composite design

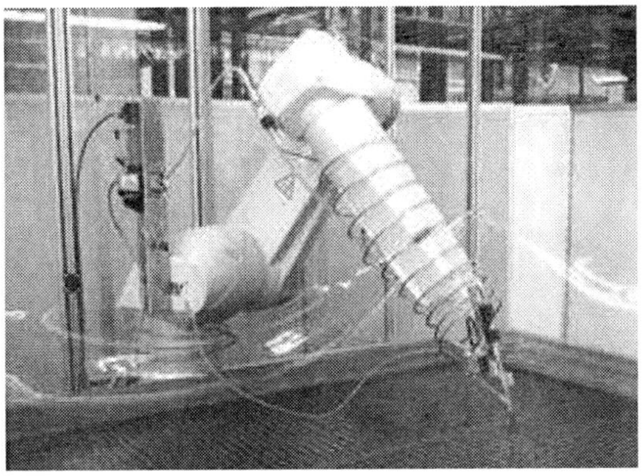

MaximatorJET HRX 160L
Working space:
2.200 x 1.200 x 900 mm³
Jet guiding system:
Stäubli RX160L
Water pump:
UHDE HPS 6045
Abrasive dosing system
STM Mini Hopper

Figure 5. Robot guided AWJ-machine: MaximatorJET HRX 160L

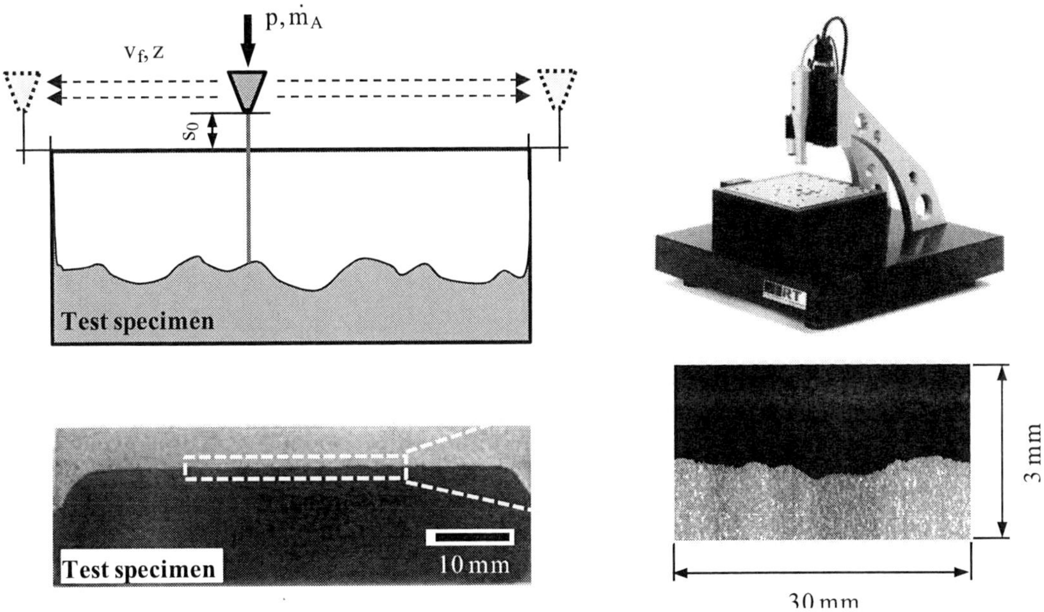

Figure 6. Measuring method

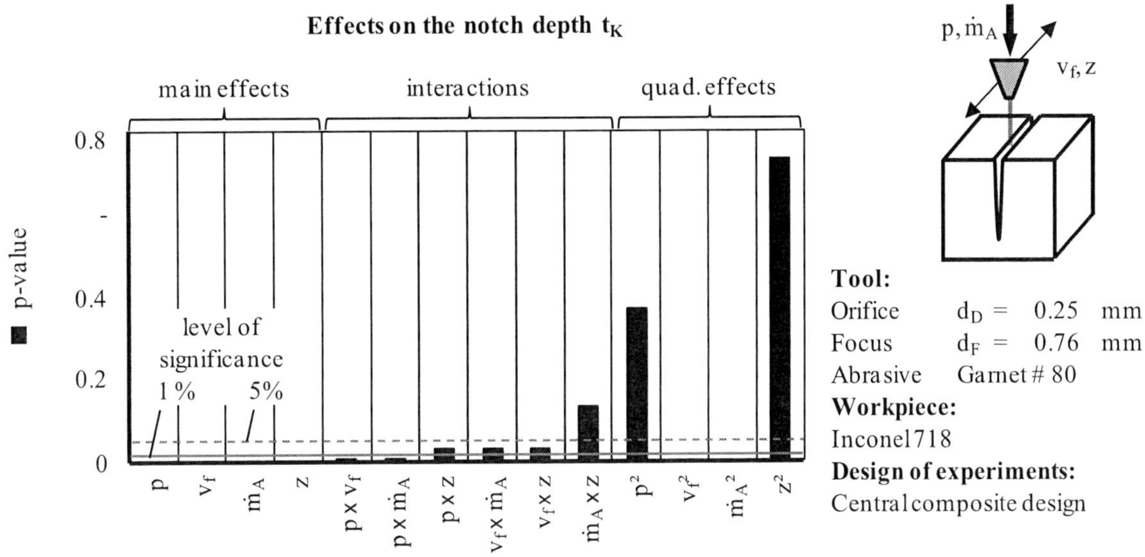

Figure 7. Main effects, interactions and quadratic effects on the notch depth t_K

Tool:				
Orifice	d_D	=	0.25	mm
Focus	d_F	=	0.76	mm
Abrasive	Garnet #80			
Workpiece:				
Inconel 718				
Parameters on center-level:				
Pressure	p	=	150	MPa
Feed velocity	v_f	=	3500	mm/min
Abrasive flow rate	$\dot{m}_A$	=	125	g/min
Number of passes	z	=	300	-

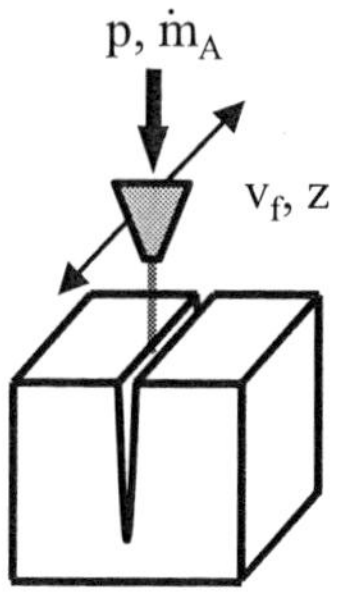

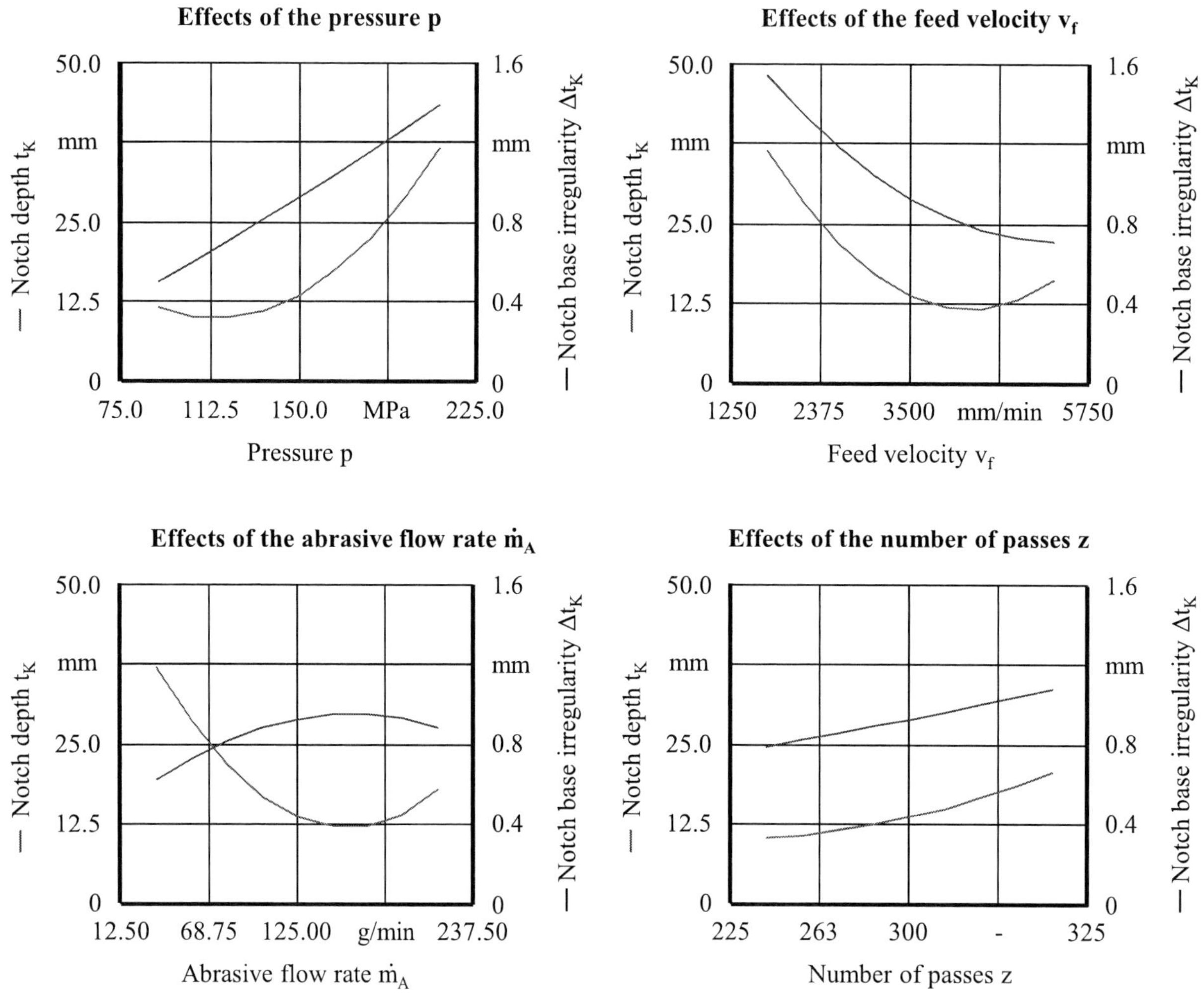

Figure 8. Effects on the notch depth t_K and notch depth irregularity Δt_K

Tool:			
Orifice	d_D	=	0.25 mm
Focus	d_F	=	0.76 mm
Abrasive	Garnet # 80		
Workpiece:			
Inconel 718			
Parameters:			
Abrasive flow rate	$\dot{m}_A$	=	125 g/min
Number of passes	z	=	300 -

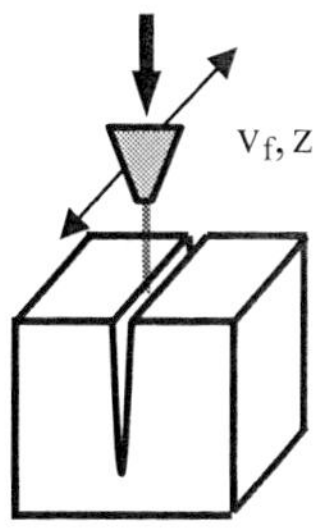

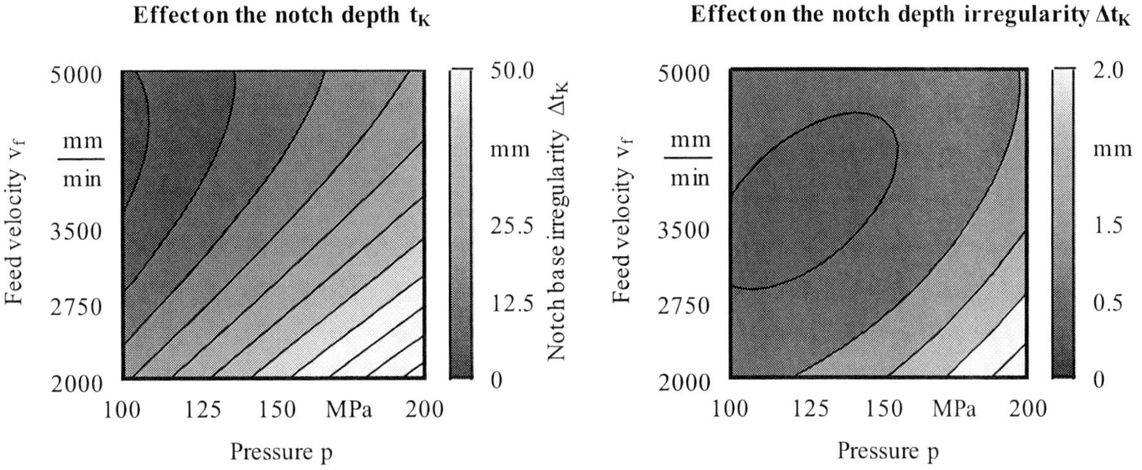

Figure 9. Contourplots of the interactions on the notch depth t_K and notch depth irregularity Δt_K

Tool:				
Orifice	d_D	=	0.25	mm
Focus	d_F	=	0.76	mm
Abrasive	Garnet #80			
Workpiece:				
Inconel 718				
Parameters:				
Pressure	p	=	188	MPa
Feed velocity	v_f	=	3330	mm/min
Abrasive flow rate	m_A	=	163	g/min
Number of passes	z	=	340	-
Results:				
Modeled notch depth	$t_{K,m}$	=	50.0	mm
Notch depth	t_K	=	50.1	mm
Modeled notch base irregularity	$\Delta t_{K,m}$	=	1.6	mm
Notch base irregularity	Δt_K	=	0.3	mm

Figure 10. Machining example of AWJ-CDM of Inconel 718

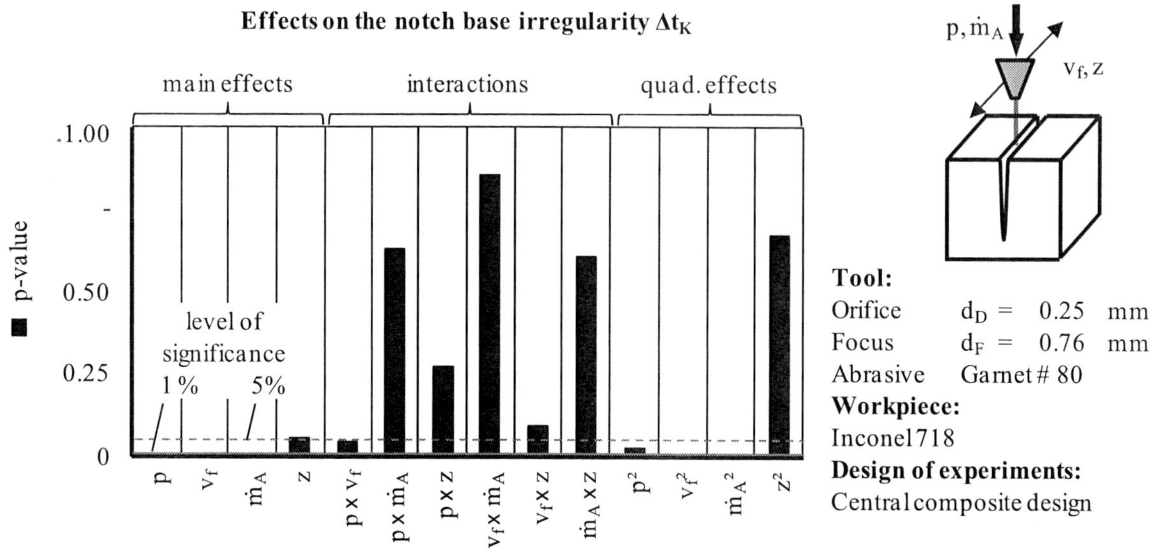

Figure 11. Main effects, interactions and quadratic effects on the notch base irregularities Δt_K

2015 WJTA-IMCA Conference and Expo
November 2-4 • New Orleans, Louisiana
Paper

ABRASIVE KINETIC CUTTING POWER:

A FUNCTION OF POWER CONVERSION

Pete Miles, Axel Henning

OMAX Corporation

21409 - 72nd Avenue South

Kent, Washington 98032 USA

ABSTRACT

The overall cutting performance of any abrasive waterjet comes down to maximizing the abrasive kinetic power of the cutting jet. In its simplest form, the abrasive kinetic power is a function of the abrasive mass flow rate and the velocity of the abrasive particles. The parameters that are typically adjusted in most a cutting processes are the abrasive mass flow rates, orifice size, and operating pressures, because they affect the mass flow rate and velocities of the abrasive particles. Though when individually increasing the orifice diameter or operating pressure has been shown to result in increased cutting speeds, what is often neglected is the fact that these changes result in an increase in the hydraulic power in the system. This paper will show that cutting speeds directly related to the overall power consumption of the pump and not a sole function of individual process parameters. This paper will also show how increasing the efficiency of the pump increases the available hydraulic power to the cutting nozzle. Thus, maximizing the abrasive kinetic cutting power is achieved by maximizing the overall hydraulic power being supplied to the cutting head.

Organized and Sponsored by WJTA®-IMCA®

1 INTRODUCTION

The cutting of virtually any known material with abrasive waterjets has led to become a widely accepted manufacturing technology since its introduction the 1980s. Today, applications of abrasive waterjet cutting can be found in many different industries and range from producing very small high precision parts to making rough separation cuts of 6+" steel plates, from singulating tiny electronic components to medical surgery research. Advancements in understanding the physics of the abrasive waterjet cutting process continues to further advance the state of the art in predictive modeling and motion control software of the abrasive waterjet cutting process [1, 2, 3].

Currently, the most common parameter that is used to evaluate abrasive waterjet cutting performance is the operating pressure of the pump because it is the easiest parameter to adjust by simply varying either the pump's RPM or adjusting pressure regulators. But in reality the jet pressure is an indirect measure of the overall hydraulic power being delivered to the workpiece for removing material. Hydraulic power is the product of pressure and flow rate. For a constant power rating, any increase in pressure requires a proportional decrease in flow rates. Higher pressures are desirable because it drives the velocity of the abrasive particles higher which increases the kinetic energy of each particle. Higher flow rates are desirable since it can carry and accelerate more abrasive particles which increases the abrasive kinetic power.

Single parameter comparisons can be misleading when it doesn't hold the energy conversion constant. For example, in a pressure effects study the orifice diameter is often held constant, and the jet pressure is increased to study its effects. But this also results in an increase in flow rates, and in order to maintain the flow rate at the desired pressure, the pump is required consume more power. Hence two parameters are actually being increased and when one parameter isn't being observed, the results are often attributed to the parameter that was observed.

The purpose of this paper is to study the cutting performance by evaluating the abrasive kinetic power of the cutting jet while maintaining a constant power consumption from a standard 37kW (50 HP) pump. Because the cutting results have non-linear relationships to each when varying the process parameters, and the choice of pumping hardware have their own impact on final cutting results, the overall power input consumption of the pump was held constant for these tests so that comparisons can be made when converting electrical input energy to abrasive kinetic power and its ultimate impact on cutting performance.

2 PUMPING TECHNOLOGIES

In all manufacturing processes, the volumetric removal rates are a direct function of how much energy is delivered to the workpiece to remove material. The greater the energy input, the greater the volumetric material removal rates. Abrasive waterjets are no different, the more power that can be put into the cutting beam, the greater the cutting speed. The heart of the cutting process comes from the pumps.

Currently there are three main pumping technologies available to the waterjetting industry. Hydraulic intensifiers and direct drive crank shaft style pumps have been the main workhorses for the ultra-high pressure waterjet cutting and cleaning industries since the 1970s. Direct drive crank shaft style pumps generating pressures upwards to 420 MPa (60 kpsi), and hydraulic intensifiers upwards to 620 MPa (90 kpsi). Since 2008 electric servo pumps entered the market that are capable of generating pressures upwards to 450 MPa (66 kpsi). Though they are all capable of generating high pressures at a wide range of flow rates, they are all not the same in delivering required power to the cutting nozzle.

The overall pump power ratings are driven by international electrical codes (CE) that are based on the pump's maximum electrical power draw from the electrical grid supply, and all pump manufacturers comply to these international codes. Industrial electric motors are designed to be at their peak efficiency near their peak output power rating. Modern AC electric motors are achieving electrical efficiencies greater than 97%, whereas high powered DC motors have efficiencies upwards around 90%.

Table 1. Pump technology efficiency

Pumping Technology	Efficiency
Direct Drive [5]	80-92%
Electric Servo [8]	71-77%
Hydraulic Intensifier [10]	60-70%

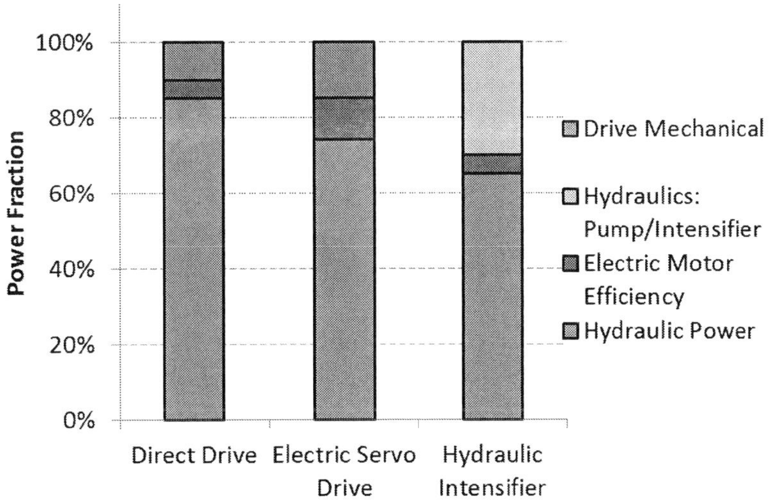

Figure 1. Pump sub-system power fractions within a pump

Table 1 shows various ranges of overall efficiencies of different pumping technologies. Mechanical efficiencies for crank shaft pumps are in the 87 to 93%. Mechanically electric servo pumps are also very efficient since they are using precision ball screws, linear anti-rotation bearings, and roller/thrust bearings with individual efficiencies in the 95 to 98% efficiency for overall mechanical efficiencies around 85%. Hydraulic intensifiers have 2 sources of efficiency losses, hydraulic pump and the double acting intensifier. Herbig [8] showed that theoretically axial hydraulic pumps combined with double acting intensifiers can have efficiencies up to 70%. But because the hydraulic pump is constantly dumping excess flow of oil to the reservoir, the overall efficiency can range from 60 to 70% for operating in the 350 MPa to 400 MPa range. But that is not practical. Peak efficiency occurs at maximum flow and maximum pressure for hydraulic systems. Figure 1 graphically shows power fractions of the various major components within a pump. The remaining hydraulic power is the power being delivered to the nozzle.

The compressibility of the water is often not considered when calculating theoretical waterjet velocities, but as the pressure of the water increases the density of the water also increases which then reduces the theoretical velocity of the waterjet, see Figure 1, Eqns. (1, 2), Figure 1. One of the factors that help keep the efficiencies of the direct drive and electric servo pumps at a higher efficiency levels is the ability to recover much of the energy required to compress the water in the high pressure cylinders. At 620 MPa (90 kpsi), the water is almost 18% compressible. Since Direct drive and servo drive pumps are closed systems, the energy from the decompressing of the water is recovered by helping to maintain the angular momentum of the crank shaft by pushing on the plunger at the beginning of the return trip. Whereas in a hydraulic system, the stored energy in the compressed fluid is sent to the hydraulic drain of the pump, and is lost. Equation (1) shows how the density of the of the water increases as the fluid pressure increases. The compressibility coefficients are B=365 MPa and K=2215 MPa.

$$\rho_w = \rho_o \left(1 + \frac{p}{B}\right)^{\frac{B}{K}} \tag{1}$$

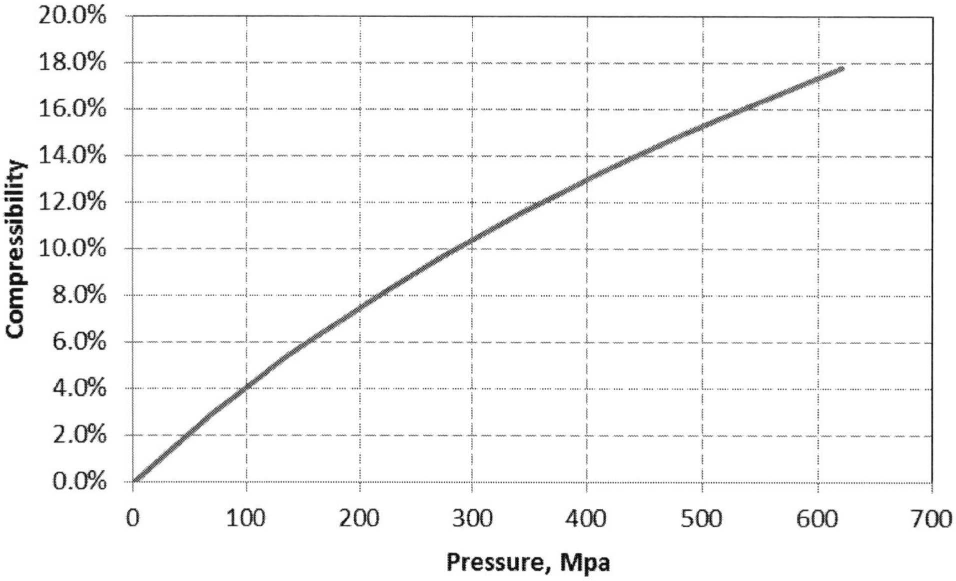

Figure 2. Fluid Compressibility

3 CONVERTING INPUT POWER TO ABRASIVE KINETIC POWER

Ultimately the kinetic power of the abrasive jet exiting the abrasive waterjet nozzle that does all the work in removing material from a workpiece. When looking at the equation for the abrasive kinetic power, it would naturally appear that increasing the abrasive mass flow rate, ma, and/or the abrasive velocity, v_a, will result faster cutting speeds. But in actual practice, increasing either of these properties alone doesn't lead to expected improvements in cutting speeds. One of the main reasons for this is due to the difficulty of efficiently converting the input hydraulic power into high velocity abrasive particles. Bernoulli's equation is used to determine the conservation of nozzle's input pressure into a jet velocity.

$$v_w = \sqrt{\frac{2p}{\rho_w}} \tag{2}$$

$$\dot{m}_w = \frac{\pi}{4} d_o^2 \rho_w C_D v_w \tag{3}$$

Here it can be seen that the velocity of the waterjet is a function of the nozzle's input pressure, and that the mass flow rate of the water is a function of the jet's velocity, and thus it is also a function of the nozzle's input pressure. But the water's mass flow rate is also a function of the nozzle's orifice diameter. The maximum orifice diameter in a waterjet system is pump limited.

$$d_{o,\ Max} = \sqrt{\frac{2^{1.5}\sqrt{\rho_w}}{\pi C_D} \frac{\eta_{pump} P_{pump}}{p^{1.5}}} \tag{4}$$

$$d_{o,\ Max} \propto \frac{\sqrt{\eta_{pump} P_{pump}}}{p^{0.75}} \tag{5}$$

$$v_w = \left(\frac{2^{1.5} \eta_{pump} P_{pump}}{\pi d_o^2 \rho_w C_D} \right)^{\frac{1}{3}} \tag{6}$$

The orifice is the most critical component in the waterjet nozzle for converting the pump's electrical power into kinetic power of the waterjet, and yet it is often over looked. It is the one component that controls both the pressure and flow rate through the nozzle. The orifice is what limits the flow rate so that pressure can be developed in the system. It is the orifice that controls the flow rate at any desired operating pressure. Because the orifice is the central component in the energy conversion, the waterjet velocity can be represented as a function of the pump's electrical power, efficiency, and orifice diameter.

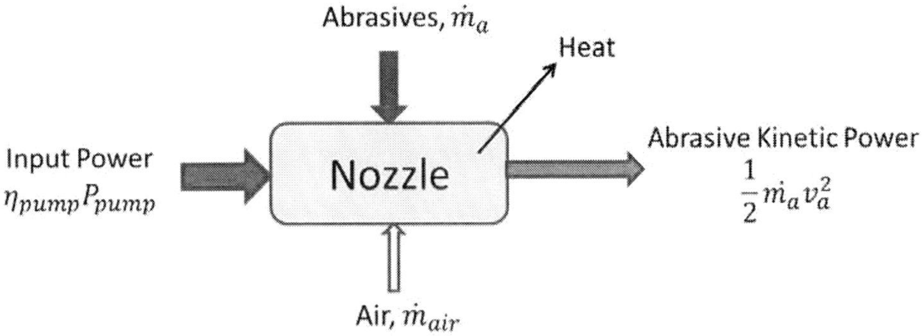

Figure 3. Power conversion within an abrasive waterjet nozzle

The standard operating practice for most waterjet job shops is to choose an office size that will allow the pump to operate near its maximum capacity. Though pressure is the one operating parameter that is the easiest to adjust, ultimately, the only thing that the operator is actually adjusting is the amount of power the pump is drawing.

Once the waterjet is created, it will then accelerate the air and abrasive particles to create the abrasive waterjet for cutting applications. The abrasive kinetic power is a function of the abrasive mass flow rate and the velocity of the abrasive particles.

$$P_{abrasive} = \frac{1}{2}\dot{m}_a v_a^2 \tag{7}$$

The biggest unknown in the abrasive waterjet cutting process is determining the exact velocity of the abrasive particles leaving the mixing tube because it takes a certain amount of time to accelerate the abrasive particles through the mixing tube. One method that has been successfully used to measure average abrasive particle velocities is the use of the Dual Disk Anemometer (DDA) [4, 9]. The DDA measures the abrasive particles by measuring the time of flight between two high speed spinning disks. By knowing the speed of the disks and the spacing between the disks, the particle velocities can be measured.

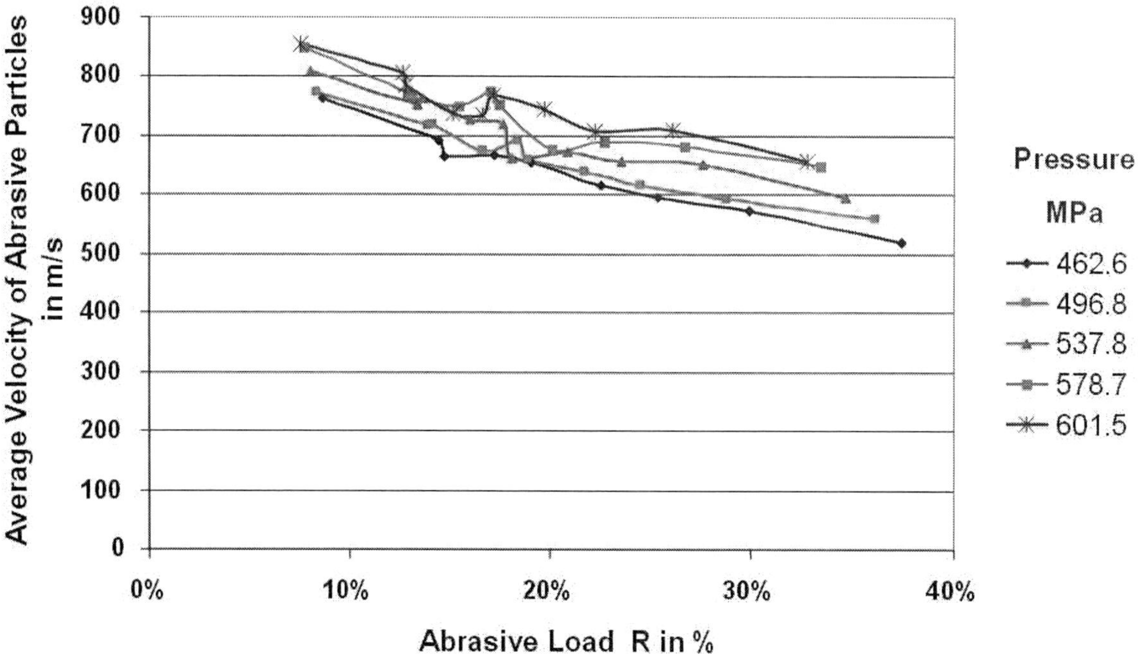

Figure 4. Abrasive particle velocities at various pressures and abrasive mass loadings, R. (d_o = 250μm, d_m = 764μm, and l_m = 100 mm)

Figure 4 shows measured average particle velocities using the DDA for various operating pressures and abrasive mass loadings. The abrasive mass loading, R, is defines as the ratio of the abrasive mass flow rate and the waterjet mass flow rate, $R = \dot{m}_a/\dot{m}_w$. It can be seen that the abrasive particle velocities appear to increase proportionally with jet pressure and it can also be seen that as the abrasive mass loading increases, the abrasive particle velocities appear to be inversely proportional to the abrasive mass loading. Because the waterjet velocity is proportional to the jet pressure, the abrasive particle velocity can be estimated as a function of the waterjet velocity and abrasive mass loading.

$$v_a = \Psi_a v_W \tag{8}$$

$$\Psi_a = \frac{a}{R+b} = \frac{v_a}{v_w} \tag{9}$$

The abrasive speed ratio, Ψ_a, can be thought of as the momentum transfer efficiency. The b term cannot be zero since if only one abrasive particle is being accelerated through the nozzle it will never approach the speed of the initial waterjet. And b will always be greater than a because the abrasive particle velocities can never be greater than the waterjet velocity.

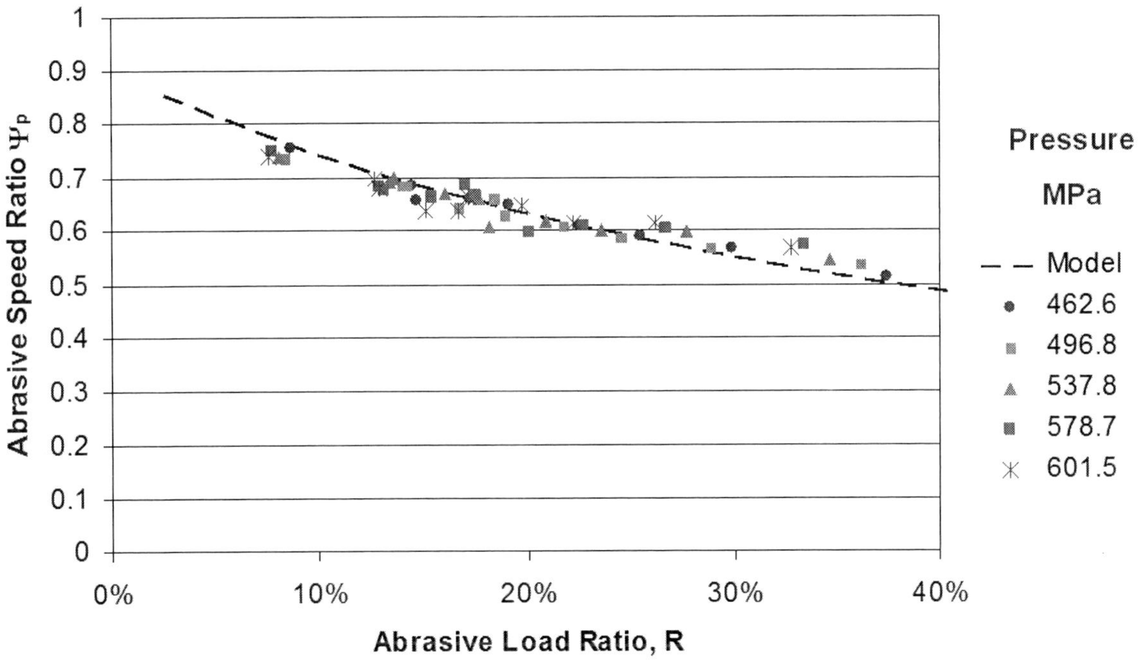

Figure 5. Abrasive speed ratio for various operating pressures and abrasive mass loading, R.

Using a regression analysis the coefficients a=0.423 and b=0.470 was found to be a good measure of the abrasive particle velocities. Figure 5 shows the abrasive speed ratio for the data shows in Figure 4. The dashed line shows the model fit of abrasive speed ratio function. In traditional waterjet cutting applications the abrasive mass loading, R, is around 12.5%. This equates to an abrasive speed ratio of 71% or an abrasive accelerating efficiency of 71%.

The abrasive kinetic power can now be written as a function of the abrasive speed ratio, abrasive mass loading, and the initial waterjet velocity

$$P_{abrasive} = \frac{1}{2}\dot{m}_a \Psi_a^2 v_w^2 \tag{10}$$

The hydraulic power of the waterjet is defined as

$$P_{hydraulic} = \eta_{pump} P_{pump=} \frac{1}{2} \dot{m}_w v_w^2 \tag{11}$$

Rearranging

$$P_{abrasive} = R\Psi_a^2 \eta_{pump} P_{pump} \tag{12}$$

The kinetic efficiency is the ratio of the abrasive kinetic output power to the overall electrical power consumption (input power) of the pump.

$$\eta_K = R\Psi_a^2 \eta_{pump} = \frac{a^2 R}{(R+b)^2} \eta_{pump} \tag{13}$$

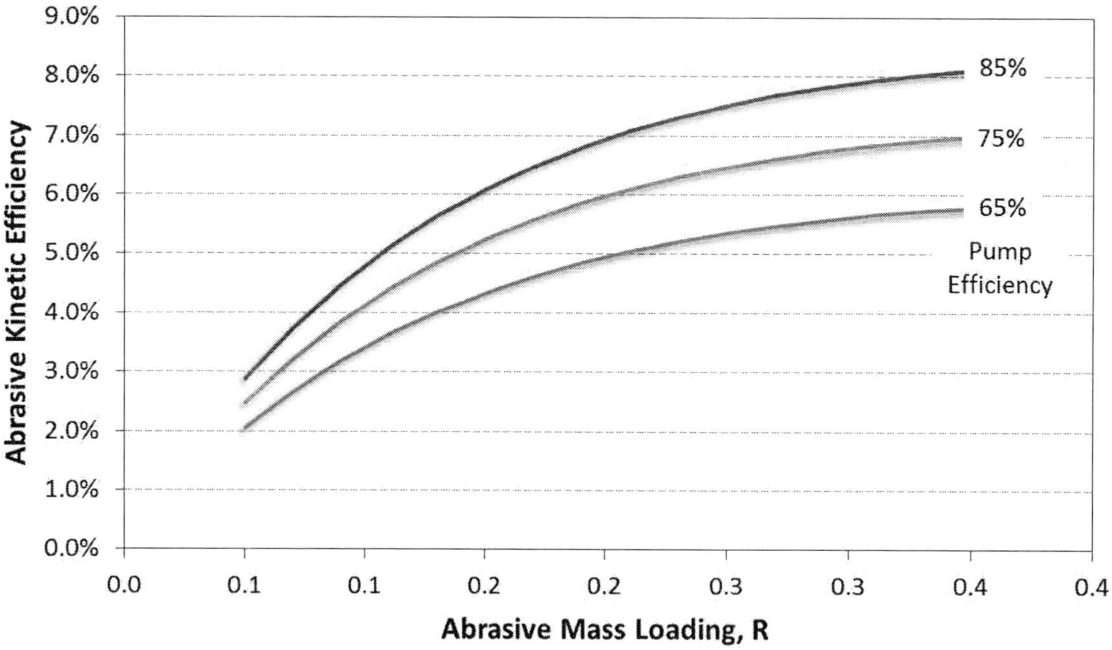

Figure 6. Abrasive kinetic efficiency for various abrasive mass loading ratios, R, and pump efficiencies.

It can be seen that the abrasive kinetic power exiting the nozzle is essentially a function of the abrasive mass loading, and efficiency and overall power consumption of the waterjet pump. Figure 6 shows the abrasive kinetic efficiency for different abrasive mass loadings and different pumping efficiencies. It can be seen in Figure 6 that there is a maximum kinetic efficiency to an abrasive waterjet. By taking the first derivative of Equation 13, the maximum efficiency occurs when the abrasive mass loading, R, is equal to the coefficient, b, or in this case when R=0.47.

Intuitively thinking, higher abrasive particle velocities that results from higher operating pressures should yield greater cutting power because the kinetic power is proportional to the particle velocity squared. But the increase in particle velocities are proportionally offset by a reduction in the mass flow rates of water and abrasives when maintaining constant power and abrasive mass loading. From equation 13 the overall kinetic efficiency is driven by the abrasive mass loading, and the overall efficiency of the pump.

4 EXPERIMENTAL CUTTING RESULTS

To illustrate how kinetic power dictates the overall cutting performance, a series of separation speed tests were conducted with a fixed electrical pump power rating. To keep the hydraulic power of the jet constant, different orifice sizes, and therefor water flow rates were used and the jet pressure was adjusted to keep the power constant. The orifice diameter for each data point is included in these plots so that the hydraulic power and jet pressure relate to each other for a given orifice diameter.

To compare how pump efficiency impacts the cutting performance, a standard 37kW (50 hp) pump rating was used in these experiments. Direct drive pumps that are typically 85% efficient will have 31 kW hydraulic power available at the nozzle, and a 65% efficient hydraulic intensifier pump will have 24 kW of hydraulic power available to their cutting nozzle. The cutting test results presented here compare these two types of pumps.

4.1 Effect of Pressure with constant Abrasive Load Ratio

Equation 13 shows that the abrasive kinetic power is proportional to the pump efficiency when the abrasive load ratio, R, is held constant. To experimentally verify this relationship, a series of cutting tests was conducted with a constant abrasive mass loading of 14% while the jet pressure was varied from 225MPa to 570MPa. Because the waterjet mass flow rate changed with each orifice diameter and jet pressure, the abrasive mass flow rate was adjusting accordingly to maintain a constant abrasive mass loading.

It can be seen in Figure 7 and Figure 8 that the separation speeds are greater when the hydraulic power available to the nozzle is greater. In this case, the higher efficient pump, 31kW at 85%, delivered more hydraulic power to the nozzle than the lower efficient pump, 24kW at 65%. These figures also shows that there appears to be an optimal operating pressure at around 300 MPa where the separation speed is maximized and that separation speed begins to decrease as the operating pressure continued to increase. This optimal cutting pressure appeared at the same pressure for both of the 6061 Aluminum and A36 mild steel tests and with both the 31kW and 24kW tests.

The critical observation to note with these cutting results is that the proportional change in the cutting separation speed test for the 31kW and 24kW follows the proportional changes in the cutting efficiency, which agrees well with the abrasive kinetic power (Equation 13) being a function of the pumping efficiency for fixed abrasive mass loadings.

An explanation for the slightly decreasing cutting performance as the operating pressure continued to increase may be due to a greater amount of particle fragmentation within the mixing tube. The kinetic energy of individual particles decreases with fragmentation, and thus as less energy to remove material when it impacts the workpiece.

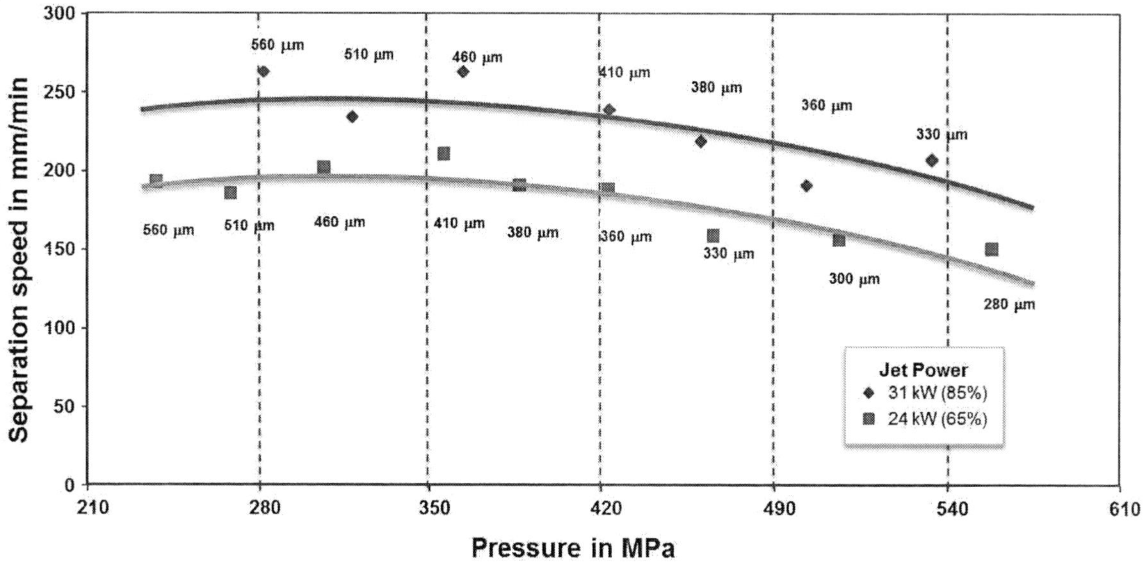

Figure 7: Separation speed for 25.4mm mild steel (A36) with fixed abrasive load (14%) with 37kW pump systems

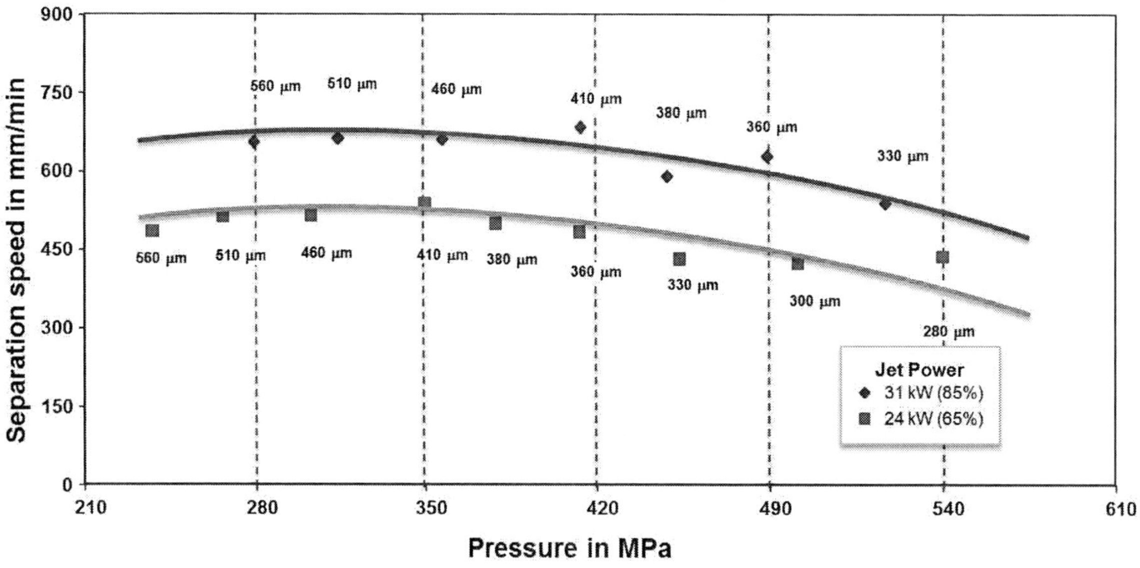

Figure 8: Separation speed for 25.4mm aluminum (6061) with fixed abrasive load (14%) with 37kW pump systems

4.2 Effect of Pressure with constant Abrasive Feed rate

The next series of tests that was conducted held the abrasive flow rate constant at 544 g/min while the jet pressure was varied from 225 MPa to 570 MPa. As the jet pressure as increased, the orifice diameter was reduced to maintain a constant hydraulic power at the nozzle.

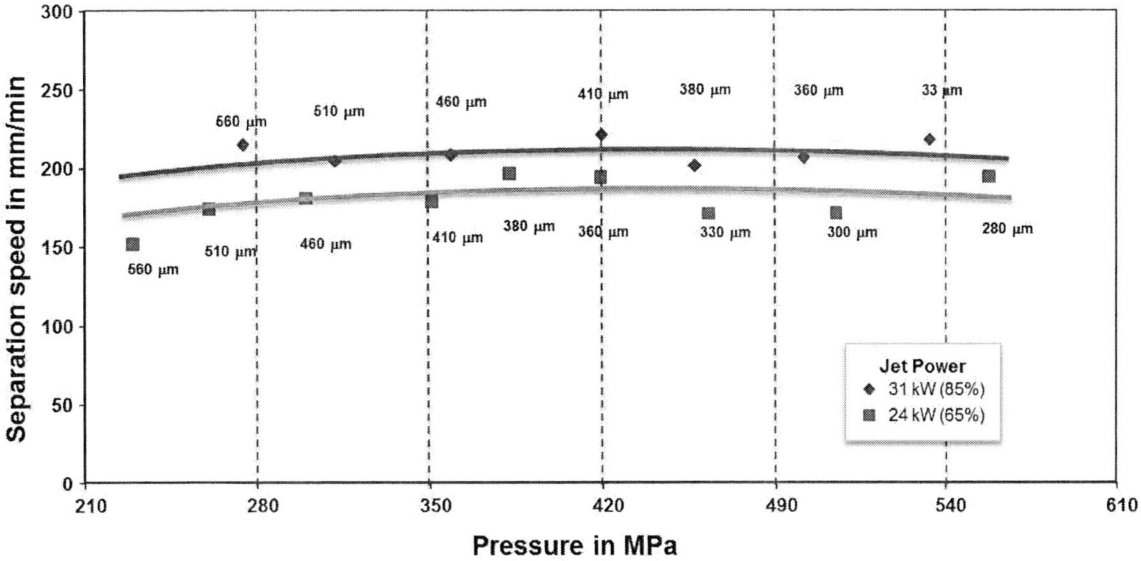

Figure 9: Separation speed for 25.4mm mild steel (A36) with fixed abrasive flow rate (544g/min) with 37kW pump systems

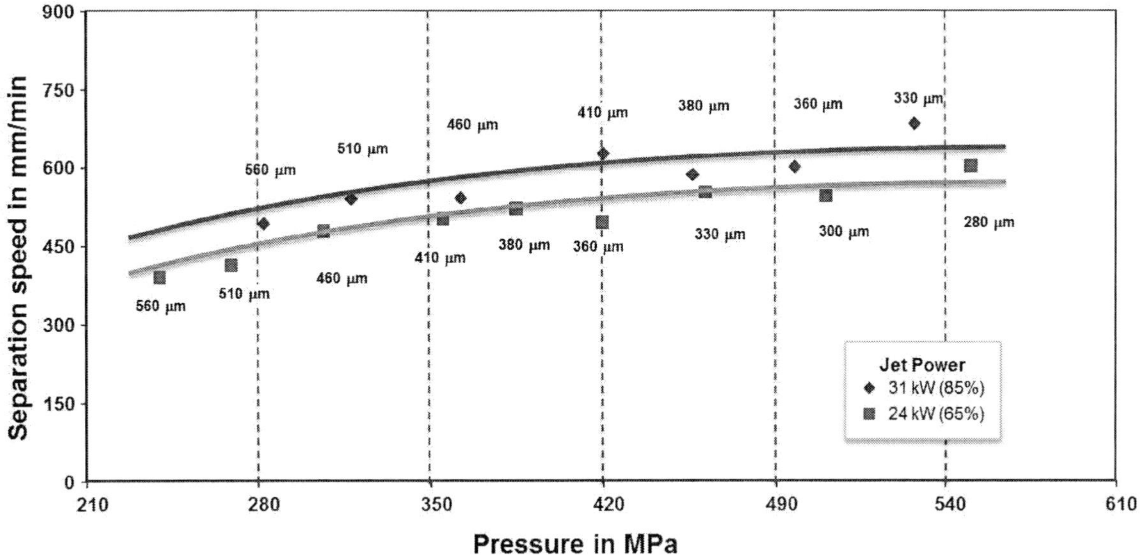

Figure 10: Separation speed for 25.4mm aluminum (6061) with fixed abrasive flow rate (544g/min) with 37kW pump systems

Figure 11 shows the data that was presented in Figure 10 as a function of the abrasive mass ratio, R. It can be seen that the separation speed as a function of the abrasive mass ratio follows the same trend as shown in Figure 6 and the abrasive kinetic efficiency Equation 13. Because the separation speed is following the same general trend, separation speeds could be modeled using the same type of relationship, where the *a* and *b* terms would be material dependent constants.

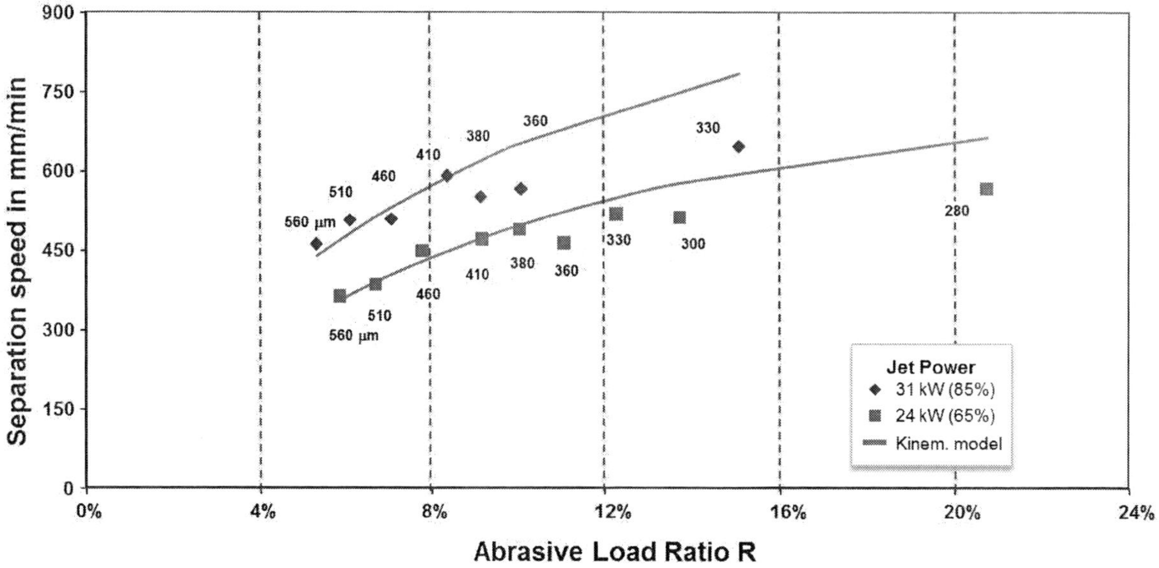

Figure 11: Separation speed for 25.4mm aluminum (6061) with fixed abrasive flow rate (544g/min) displayed over Abrasive Load Ratio, R.

4.3 Specific Cutting Efficiency Index

Some interesting results were observed when dividing the actual separation cutting speeds, v_S, by the abrasive kinetic power efficiency, η_K, to create the specific cutting efficiency index, η_C

$$\eta_C = \frac{v_S}{\eta_K} \tag{14}$$

Figure 12 and Figure 13 it can clearly be seen that the specific cutting efficiency index (SCEI) are identical for both the 31kW and 24kW tests for the various operating pressures. Because both plots lay on top of one another, this behavior appears to be independent of the hydraulic power at the nozzle (pump efficiency) but more on the overall electrical power draw of the pumps. The other interesting observation that as the jet pressure increased, the SCEI continually decreased for both the aluminum and the steel and that the steel decreased at a greater rate than the aluminum.

The exact reason for the reduction of the SCEI as a function of pressure is most likely due to the increased fragmentation of the abrasive particles within the mixing tube as the jet pressure increases, as discussed earlier. For both the aluminum and steel cutting tests the fragmentation ratios would be the same. Since steel requires more energy to remove material, than aluminum, it would be less sensitive to impacts from smaller particle impacts (lower particle kinetic energy). Further research area is needed to be able to model this behavior.

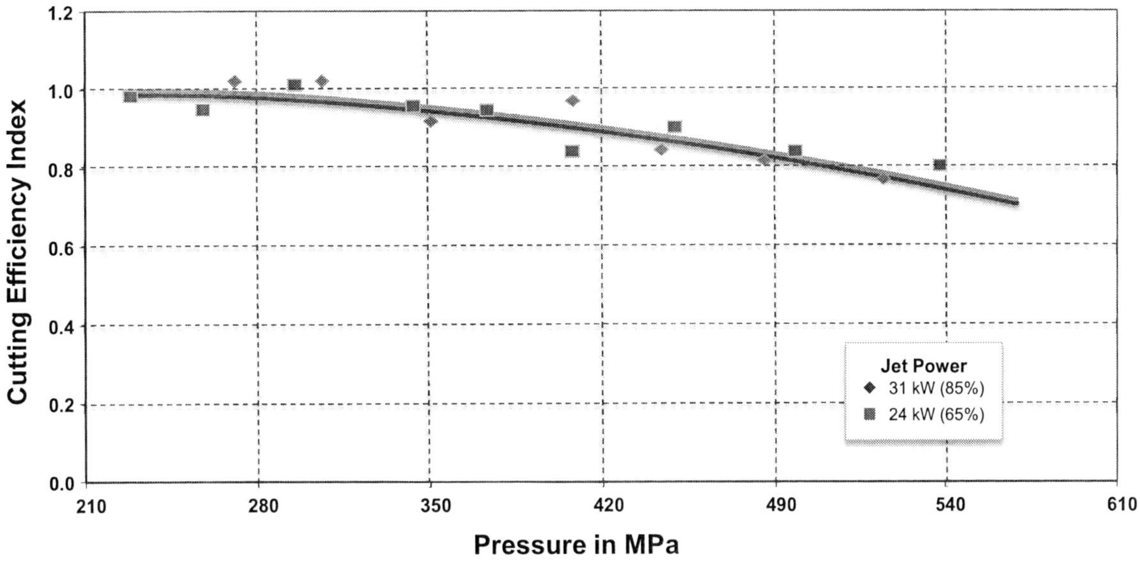

Figure 12: Cutting Efficiency Index for 25.4mm aluminum (6061) with fixed abrasive flow rate (544g/min) displayed over Pressure

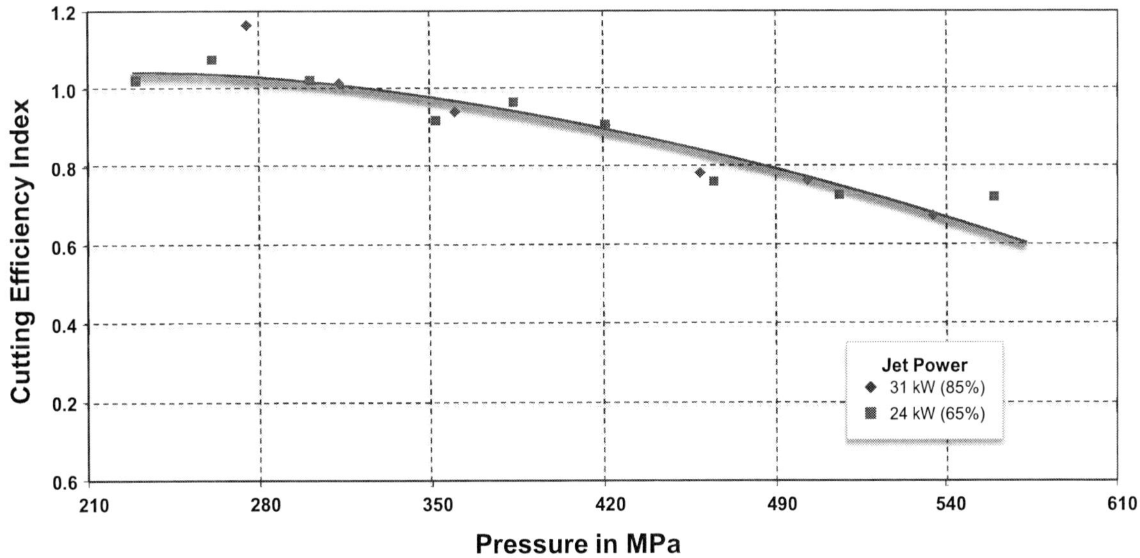

Figure 13: Cutting Efficiency Index for 25.4mm mild steel (A36) with fixed abrasive flow rate (544g/min) displayed over Pressure

5 DISCUSSION

The most important parameter in a waterjet cutting system is the hydraulic power that is being delivered to the waterjet nozzle because it is this power that will be converted into the cutting jet's abrasive kinetic power which does all of the actual cutting work. Though studying individual parameters is useful in understanding their relative effects on the cutting performance, but ultimately it is the overall hydraulic power that affects the overall cutting performance, not the individual parameter by itself.

The kinetic power of the abrasive particles was shown to be a simple function of the abrasive mass flow rate, an abrasive speed factor (which is also a function of the abrasive mass loading), the power consumption of the pump, and its efficiency. This relationship was shown to be independent of the operating pressure but more on the actual power consumption the pump. It was also shown that the velocity of the water exiting the nozzle is a function of the orifice diameter and power consumption of the pump. Generally waterjet velocity is shown to be a function of the upstream jet pressure, but in reality it is the orifice diameter that controls the jet velocity. For a fixed pump power rating, the efficiency of the pump has a significant impact on the resulting kinetic power of the abrasives exiting the nozzle since it limits the available hydraulic power being delivered to the nozzle.

A series of cutting tests were conducted by varying the waterjet pressure from 225 MPa to 570 MPa, while maintaining a constant abrasive load ratio of 14%. During these tests the orifice diameters were changed so that the overall power consumption of the pump was constant for all of the tests. The results showed a general trend of the cutting speed decreasing as the cutting pressure increased. The important observation from the cutting results was that the relative cutting speeds, in both the aluminum and steel tests, were proportionally identical to the proportional change in the pumping efficiency.

According to results published from other researchers, it is speculated that one of the reasons for a reduction in cutting speed as the jet pressure increases is due to particle fragmentation within the nozzle. Perec [11] showed that for 80 and 120 mesh garnet being accelerated through a waterjet nozzle operating at 390 MPa, that over 40% of the original garnet exiting the mixing tube has fragmented to less than 80μm when the original garnet had less than 1% by mass that was less than 80μm. The fragmentation rate was shown to be independent of the abrasive mass ratio between 15 and 25%. Ohlsen [12] showed that the average fragmented particle sizes exiting the nozzle decreased almost linearly with increasing operating pressure. Kulekci [13] showed that the cutting penetration depth rapidly drops towards zero as abrasive grains got smaller than 80 um in both aluminum and steel. Kulekeci also showed that for tests using abrasives whose average grain size ranged between 80 and 500 μm, their penetration depths remained within +/- 20% of a nominal penetration depth of 80 mesh garnet (~230μm average particle diameter). When looking at the totality of these researchers, it appears that as the jet pressure increases the particle fragmentation rate increases and this leads to an increase in smaller abrasive particles within the kerf. Since smaller particles are less efficient at removing material, this is resulting in a lower volumetric material removal rate within the kerf. Though the effect of particle fragmentation was not the focus of this paper, it does indicate that future research is needed to fully understand the impact of particle fragmentation on cutting performance to help further improve nozzle designs and abrasive cutting models.

A second series of cutting tests were conducted by holding the abrasive mass flow rate constant at 544 g/min while varying the operating pressure from 225 MPa to 570 MPa and varying the orifice diameters so that the overall power consumption of the pump was constant for all of the tests. Here the general results showed that for aluminum, there was a small increase in the separation speed as the jet pressure increase, but was relatively flat for the steel cutting results. The separation speeds were consistently higher with the higher efficient pumps. When the separation speed was plotted as a function of the abrasive load ratio, it showed that separation speeds followed the same trends as modeled from the abrasive kinetic power equation (a function of pump power, pump efficiency, abrasive mass loading, and abrasive speed factor). Again, the cutting results are driven by the abrasive mass flow rate, power consumption and the efficiency of the pump.

When the separation speed is divided by the efficiency of converting the power consumption to abrasive kinetic power, a specific cutting efficiency index is created. The interesting thing to note is that the curves for both 65% and 85% efficient pumps lay on top of one another, and that the cutting index continually declines as the operating pressure rises.

6 CONCLUSIONS

For any given waterjet cutting technology, the maximum material removal rates occur when the pumps are operating at their maximum output capacity. When the input power consumption is held nearly constant the actual cutting results shows that separation speeds tends to slow down as the operating pressure increases (within the normal operating pressure ranges of cutting systems). When all parameter ratios are held constant, actual cutting separation speeds increase proportionally with pumping efficiency since more hydraulic power is available at the nozzle when the input power consumption is held constant. This clearly shows that pumping efficiency plays a significant role in comparing cutting performance. Even though the average velocity of the abrasive particles decreased (abrasive kinetic energy) as the abrasive mass loading increased its abrasive kinetic power increased with the increase in the abrasive mass flow rate. A greater increase in the abrasive kinetic power resulted in an increase in actual separation cutting speeds.

The general trend in all of these cutting results shows that when the overall power consumption is held constant the available hydraulic power delivered to the waterjet nozzle that is converted into the abrasive kinetic power of the jet determines the actual separation cutting speeds. Greater hydraulic power results in greater cutting speeds. And for constant power consumption higher operating pressures tends to result in lower cutting speeds.

7 ACKNOWLEDGEMENTS

The authors wish to thank Ryan Boehm from OMAX Corporation for his tireless and dedicated work in running and collecting the abrasive particle velocity data.

8 REFERENCES

[1] Liu, H.-T., Schubert E., and McNiel, D. (2011) “Micro AWJ technology for meso-micro machining,” *Proc. of 2011 WJTA-IMCA Con.*, Houston, September 19-21.

[2] Westkämper, Engelbert; Gottwald, Bernhard; Henning, Axel: Intelligent means of process control during the high pressure water jet cutting. In: IEEE Industrial Electronics

Society u. a: IECON '98 - Vol. 4: Proceedings of the 24th Annual Conference of the IEEE Industrial Electronics Society, August 31 - September 4, 1998. pp. 2361-2365

[3] Henning, Axel: Computer Aided Manufacturing for Three-Dimensional Abrasive Water Jet Machining. In: Hashish, Mohamed (Hrsg.); Water Jet Technology Association: American Waterjet Conference <9, 1997, Dearborn> -1997, pp. 729-742

[4] Liu, H.-T., Miles, P., Hibbard, C. and Cooksey, N. (1999) "Measurements of Water-Droplets and Abrasive Speeds in Waterjets and Abrasive Waterjets," *Proc. 10th Ame. Waterjet Conference*, Houston, Texas, August 14-17.

[5] Veenhuizen, Scott, Operating Efficiency of Crankshaft Drive Pumps, Proceedings of the 6th Pacific Rim International Conference on Water Jet Technology, Sydney, Australia, October 9-11, 2000: pp 249-252.

[6] Henning, A.: Modellierung der Schnittgeometrie beim Schneiden mit dem Wasserabrasivstrahl, Dissertation, Universitaet Stuttgart, ISBN 978-3-939890-28-7, Jost Jetter Verlag, Heinsheim

[7] Westkämper, E.; Henning, A.; Radons, G.; Friedrich, R.; Ditzinger, T. (2000): "Cutting Edge Quality through Process Modeling of the Abrasive Waterjet" In: Teti, Roberto (Hrsg.); CIRP; Proceedings of 2nd CIRP International, Seminar, June 21-23, 2000, pp. 179-188

[8] Goletti, M., Monno M., Dal Lago S., (2010) "Pressure Signal Comparison in WJ/AWJ Intensifiers," *20th International Converence on Water Jetting*, BHR Group, Graz, Austria, 20 - 22 October 20, pp. 233-245.

[9] Henning, A., Liu, H.T., Olsen, C., (2010) "Economic and Technical Efficiency of High Performance Abrasive Waterjet Cutting," Proceedings of the 18th International Conference on Pressure Vessels, PVP2010, ASME, 18-22 July, Bellevue, Washington.

[10] Herbig, S., Trieb, F., (1999) "Calculation of the Efficiency Rate of High Pressure Pumps," Proc. 10th Ame. Waterjet Conference, Houston, Texas, August 14-17.

[11] Perec, A., (2011) "Abrasive Grain Breakage Process During the High Pressure Waterjet Formation," Proc. of 2011 WJTA-IMCA Con., Houston, September 19-21.

[12] Ohlsen, J., (1997) "Recycling von Festoffen beim Wasserabrasivstrahlverfahren," VDI Fortschritt-Berichte, Reihe 15, Nr. 175.

[13] Kulekci, M., K., (2002) "Processes and Apparatus Developments in Industrial Waterjet Applications," International Journal of Machine Tools and Manufacturing, Vol. 42, pp. 1297-1306.

9 NOMENCLATURE

C_D	Coefficient of Discharge
d_m	Diameter of the mixing tube
d_o	Orifice diameter
l_m	Length of the mixing tube
$\dot{m}_a$	Abrasive mass flow rate
$\dot{m}_{air}$	Air mass flow rate

$\dot{m}_W$	Waterjet mass flow rate
p	Water operating pressure
P_{pump}	Overall pump power consumption
$P_{hydraulic}$	Hydraulic power of the jet
$P_{abrasive}$	Kinetic power of particles
R	Abrasive load (mass) ratio
v_a	Particle velocity in mixing tube
v_W	Water velocity at orifice, in mixing tube
v_S	Cutting separation speed
η_C	Specific cutting efficiency index
η_K	Abrasive kinetic efficiency
η_{pump}	Pump efficiency
ρ_W	Density of water
Ψ_a	Abrasive speed ratio
a, b	Constants for abrasive speed ratio
B, K	Water compressibility coefficients

2015 WJTA-IMCA Conference and Expo
November 2-4 • New Orleans, Louisiana

Paper

COMPUTATIONAL FLUID DYNAMICS USED IN THE DESIGN OF WATERBLAST TOOLING

J. Schneider
StoneAge, Inc.
Durango, Colorado, U.S.A.

ABSTRACT

Advancements in computer technology in the last decade have made analysis and simulation programs more capable and commercially available. SolidWorks Flow Simulation is a computational fluid dynamics software package designed to analyze fluid flow. Modeling allows for parts and systems to be optimized virtually and reduces prototype manufacturing; this saves time and money during product development.

Analysis using SolidWorks Flow Simulation was performed on models of physical test setups. The program's input parameters and output variables were refined and selected to best reproduce the physical performance testing of waterjets. Flow rates ranging between 40L/min (10.5gpm) and 160L/min (42gpm) at a pressure of 69MPa (10,000psi) were tested. The simulation results were then compared with empirical data to determine if computational fluid dynamics performed by SolidWorks Flow Simulation could be effectively utilized in the design of waterblast tooling.

Organized and Sponsored by WJTA®-IMCA®

1. INTRODUCTION

In the waterblast industry it is commonly necessary for the design of a part or system to be optimized for fluid flow characteristics and manufacturability using computer analysis. The best options for machined features, feeder tube selections or even entire flow path designs can be assessed earlier in the design, or in a more comprehensive way.

SolidWorks Flow Simulation was the software package used for this paper to analyze how water flows through the 3D models created in the base CAD program, SolidWorks. Several simulations were set up and run to create and test a validation model for the results described in the paper The Effect of Upstream Feeder Tube Diameter and Length on Waterjet Performance in Cleaning Applications by Doug Wright, published in the 2015 WJTA proceedings.

2. SIMULATION CONFIGURATIONS AND SETUP

The analysis performed in this paper was comprised of different combinations of three separate tube diameters, in three lengths, with three different orifice sizes. The tube inside diameters were 7.9mm (.312"), 20.7mm (.815") and 27.9mm (1.10"). The three lengths were 102mm (4"), 305mm (12"), and 914mm (36"). The three orifice diameters were 1.6mm (.063"), 2.3mm (.090") and 3.2mm (.125").

The simulation parameters were held constant across all simulations. The analysis was internal using liquid water at 20°C (68°F) with a .4μm (16μin) surface roughness. The turbulence model used was Intensity and Length with the default parameters set at 2% and 1.05mm (.0413"). A static pressure inlet boundary condition of 69MPa (10,000psi) was used in conjunction with an outlet volume flow boundary condition with respect to each orifice size, set at 40L/min (10.5gpm), 83L/min (22gpm) and 160L/min (42gpm), as shown in Figure 1. The code in program was not written to calculate the mixing of two media, such as a waterjet shooting into air. A water chamber was added after the orifice to allow the mesh to fully resolve the waterjet and to locate the outlet flow boundary condition.

3. MESH AND RESULTS ANALYSIS

3.1 Mesh

The mesh settings and strategy had an overwhelming influence on the results produced. Several mesh refinements were made on the initial models in an effort to produce the most consistent results across all the configurations before the final studies were solved.

Control planes were set up to center the mesh on the axes of the inlet and the feeder tube/ orifice, shown in Figure 1. Centering the mesh reduced the number of cells needed to resolve the features and allowed the mesh to better represent the symmetrical nature of the tubes and orifice. A courser global mesh was used to analyze the inlet, head, and feeder tube, shown in Figure 2. The final mesh produced a result of 12 cells across a given diameter and was tailored to resolve the bulk of the flow.

Three local meshes were used to resolve the transition into the orifice, the orifice itself and the region immediately following the orifice. Local meshes allow for specific refinement in key areas of interest or concern, usually known by the designer, which may not be resolved as well when left up to the program's general mesh refinement process. A meshing technique was used utilizing solid parts inserted into the flow regime to provide a volume around which the program could specifically refine the mesh. These solids were disabled, or set so they did not change or influence the flow path. The finer local mesh and the mesh solids are shown in Figure 3. The mesh created by these local settings had an elevated level of curvature refinement and produced 24-28 cells across the flow path to better resolve the bulk of the fluid flow and account for the influence of the significantly changing geometry.

Numerical values in a cross section of the flow were calculated by taking the values from each cell and averaging them. When the number of cells across the diameter was increased it produced a result which was less influenced by the zero-slip wall conditions. The actual value produced was directly related to the cell count until convergence was neared by having a sufficient number of mesh cells. An adaptive mesh, which refined the mesh during the calculation in areas with a high gradient was also explored. However, with the results correlating to the number of mesh cells, an adaptive mesh was not used in order to eliminate any additional source of error due to a difference in number of cells between the studies. Each model had between 150,000 and 250,000 mesh cells with approximately 60% being fluid cells and the majority of the remaining cells being partial cells.

3.2 Results Analysis

Two styles of results were produced by the program, numerical and pictorial. Average values taken from the cross section of the fluid at the end of the orifice were used for the numerical results. These results are graphed in Figures 5, 7, 9, 10, and 11. The physical performance test results are graphed in Figures 4, 6, and 8. Images depicting the flow in cross sections parallel to and perpendicular to the orifice axis can be found in Figures 12 through 23.

3.2.1 Physical Performance Test Results

In the paper The Effect of Upstream Feeder Tube Diameter and Length on Waterjet Performance in Cleaning Applications by Doug Wright, an improvement in the physical cutting performance of a waterjet was found by increasing either the length of the feeder tube and/or the diameter of the feeder tube in order to achieve more ideal upstream conditions. One set of results shown in Figures 4, 6, and 8, illustrate a percent performance improvement in terms of the ratio of the tube inside diameter divided by the orifice diameter for three separate lengths of tube. Another data trend indicated the optimal tube diameter shifted toward a larger tube inside diameter as the length was increased.

3.2.2 Numerical Results

As part of the setup for a simulation, goals were created. Goals are used as convergence criteria and help tailor the equations being solved for each cell in the mesh to produce the desired output. Surface goals of average volume flow rate and average total pressure were created for the cross section of the flow where the waterjet exits the orifice. In Figures 5, 7, and 9, the ratio of the tube inside diameter divided by the orifice diameter for three separate length of tube were plotted against the total pressure at the orifice.

Total Pressure at the orifice was found to be the output parameter which most closely followed the physical test results for each setup; however, the trend of the curves did not fully reproduce the physical performance curves. Total pressure is the static pressure plus the dynamic pressure. Dynamic pressure is the density of the fluid times one half the velocity squared. In varying the number of mesh cells across the orifice, some studies which had as few as 14 cells showed a better correlation to the physical results. This suggests the zero-slip wall condition and the boundary layer formation have a significant influence on the jet quality, and increasing the number of mesh cells may damp out the realistic magnitude of this influence. In the manufacturing of a nozzle the surface finish in the orifice has a measurable impact on the waterjet performance, which influences the boundary layer formation and supports this correlation.

Other parameters were explored including waterjet velocity and turbulence intensity; these results are graphed in Figures 10 and 11 for the 305mm (12”) length tube. The average velocity of the waterjet as it exited the orifice did not trend with the physical results. With a larger diameter feeder tube, the velocity in the feeder tube is slower because velocity is directly driven by the diameter and flow rate. This might have been what influenced the waterjet and decreased the exit velocity. In the physical tests, the larger tube diameters generally increased the waterjet performance. Further investigation is needed to determine if these results show an inverse effect correlating a lower velocity waterjet as an indicator of jet quality. The average turbulence intensity of the cross section of the flow at the orifice also does not fully represent the physical data. Figure 11 shows the smaller diameter tubes had less turbulence than the larger diameters; however, the larger tubes produced better performing waterjets. In nozzle design and testing, a more cohesive jet has better cutting performance. Assuming the energy of a cohesive waterjet is more focused, it would be expected to have a lower turbulence intensity, which is not represented by this data set.

3.2.3 Image Results

Turbulence intensity provided the strongest visual representation of how the water flowed through the system. These plots overlaid with velocity streamlines in a cross section parallel to the flow are shown in Figures 12 through 17. The water started out smooth and uniform, gained turbulence as it changed direction, and then again smoothed out as it traveled the remaining length of the feeder tube. These plots show that the larger diameter tubes took more length to smooth out the turbulence than the smaller diameter tubes. The large diameters had a less constricting effect at the entry into the feeder tube and allowed a recirculation zone to form. This supports the physical results which indicated a minimum tube length to tube inside diameter ratio must be maintained and is suggestive of the physical data that showed the optimum inside diameter trending larger with increased length.

Figures 18 through 23 show several cross sections perpendicular to the flow, with a turbulence intensity plot overlaid by velocity vectors. These plots show how the water moved in a direction other than axially down the tube. A shift in characteristics happened as the length was increased. A dual recirculation effect could be seen where the water from the cavity in the head entered the feeder tube from each side while it was rotating in opposite directions. Over the length of the longer tubes the dual recirculation dampened out and the flow combined into one rotation direction before it reached the orifice. In the shorter lengths and smaller diameter tubes the flow reached the orifice before it had time to unify. The opposed rotation can be seen in the first cross section on the left side of Figures 18, 19 and 21 for the 7.9mm (.312") inside diameter tube in the 102mm (4") and 305mm (12") long tubes, and also in the 27.9mm (1.1") inside diameter, 102mm (4") tube. In the physical performance graph of Figure 4, the performance reached 60% by the time the rotation was unified. This dynamic influence may be a contributing effect to the optimal tube inside diameter to orifice diameter ratio being 5-7 for a 3.2mm (.125") orifice size, while maintaining a minimum length to tube inside diameter ratio.

3.2.3 Flow Simulation Capabilities

The results obtained in these studies were achieved by attempting several different approaches and using the one which fit best. A further investigation would lead to utilizing more advanced meshing techniques, researching and optimizing the turbulence model parameters, and a more strategic use of goals to leverage their interdependence to produce the desired results. Significant attention would need to be given to setting up every model in a proper, unified method due to the difference in comparative results being 1% to 3% of the input value.

The physical performance of a waterjet is determined by many variables which can be heavily influenced by slight changes. These studies have shown that of the parameters explored, including pressure drop and waterjet velocity, waterjet performance is not dependent on just a single output parameter.

4. CONCLUSION

SolidWorks Flow Simulation provides a variety of unique ways to extract data. Numerical data can be used to calculate pressure drop and velocities, and the imagery can answer questions about geometrical features, help explain effects present in test data and build intuition about the fluid dynamics of a system. Simulation studies looking for physical cutting performance results for waterjet are pushing the limits and require a deep knowledge of the program. While the results produced were not able to reliably predict the physical test results, they did follow the trend and shed light on the fluid dynamics involved in a feeder tube and orifice system. Flow Simulation is a viable tool to aid in the design of waterblast tooling but is not a replacement for physical performance testing at this point in time.

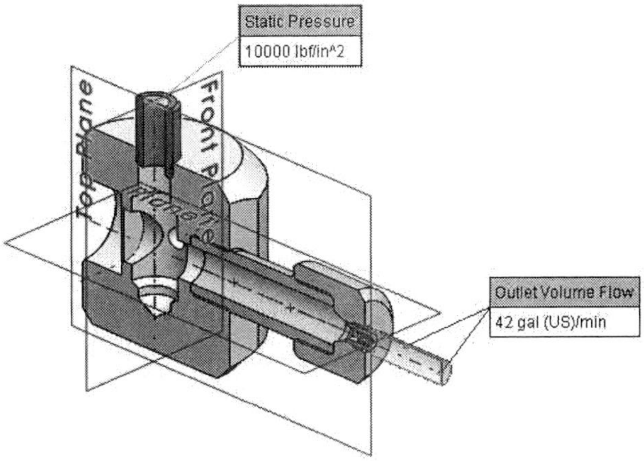

Figure 1: Shown are Mesh Control Planes and the Model Boundary Conditions of 69MPa (10,000psi) and 160L/min (42gpm) for the 20.7mm (.815") Inside Diameter, 102mm (4") Long Tube with 3.2mm (.125") Orifice

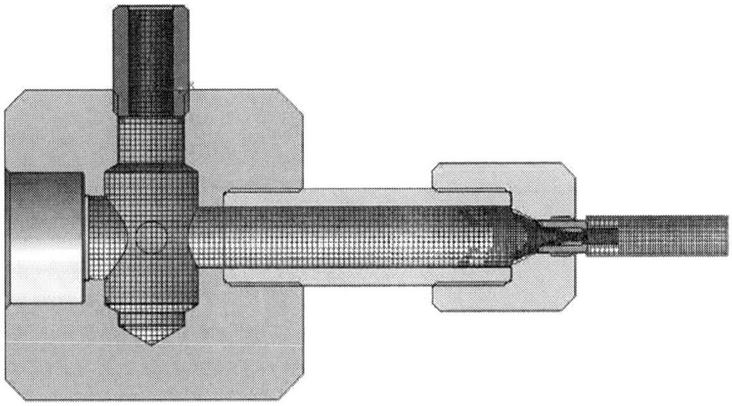

Figure 2: Global Mesh - Showing 20.7mm (.815") Inside Diameter, 102mm (4") Long Tube with 3.2mm (.125") Orifice

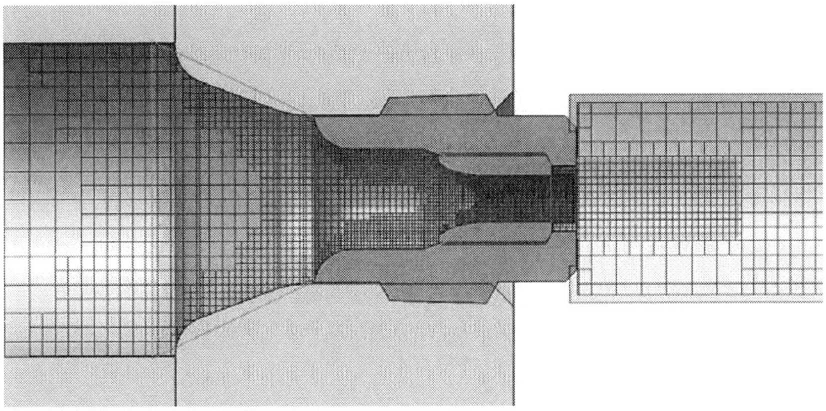

Figure 3: Local Mesh Refinement at the Orifice using Mesh Solids - 20.7mm (.815") Inside Diameter Tube with 3.2mm (.125") Orifice

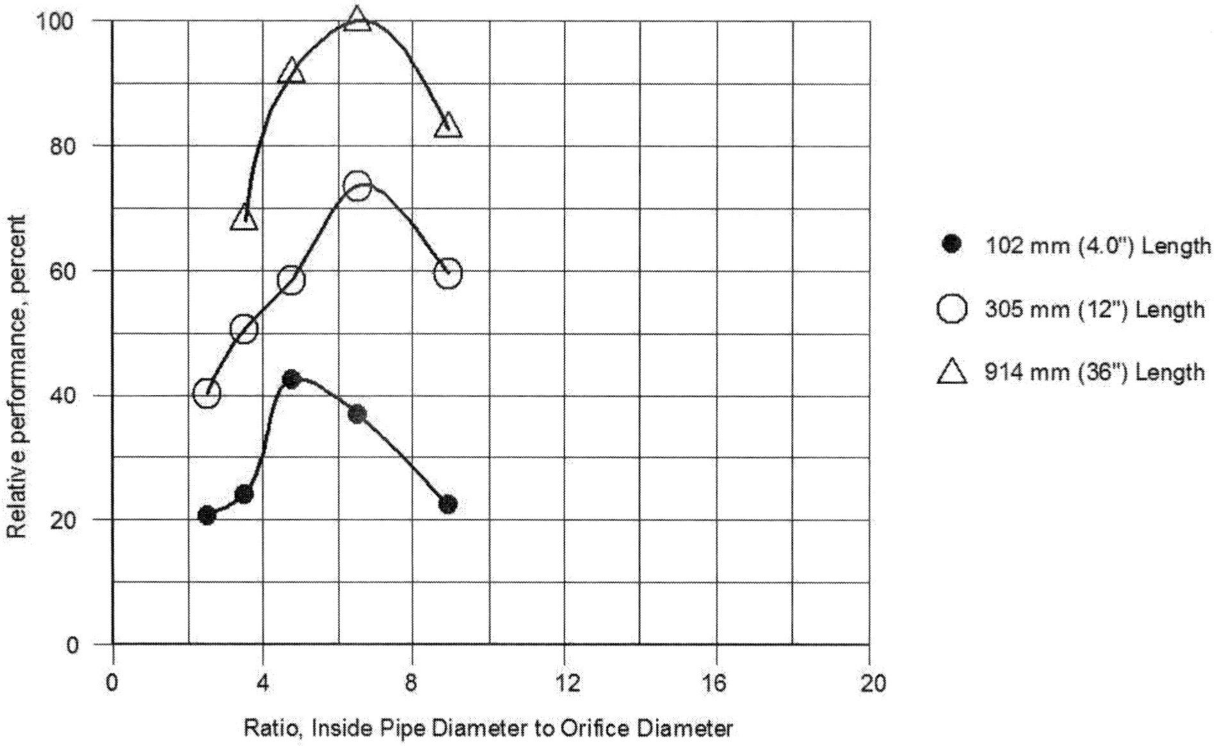

Figure 4: Relative Performance with Increasing Feeder Tube Inside Diameter at 3.2 mm (.125") Orifice Size

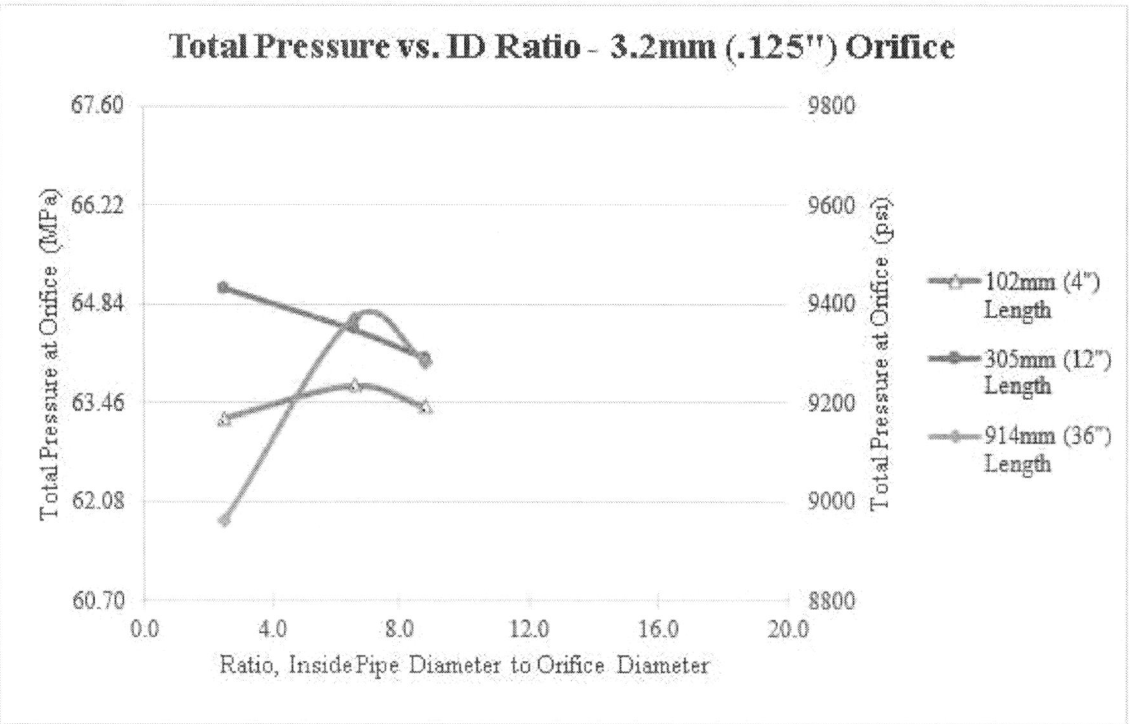

Figure 5: Simulation Results using Total Pressure as an Indicator of Relative Performance for a 3.2mm (.125") Orifice Size

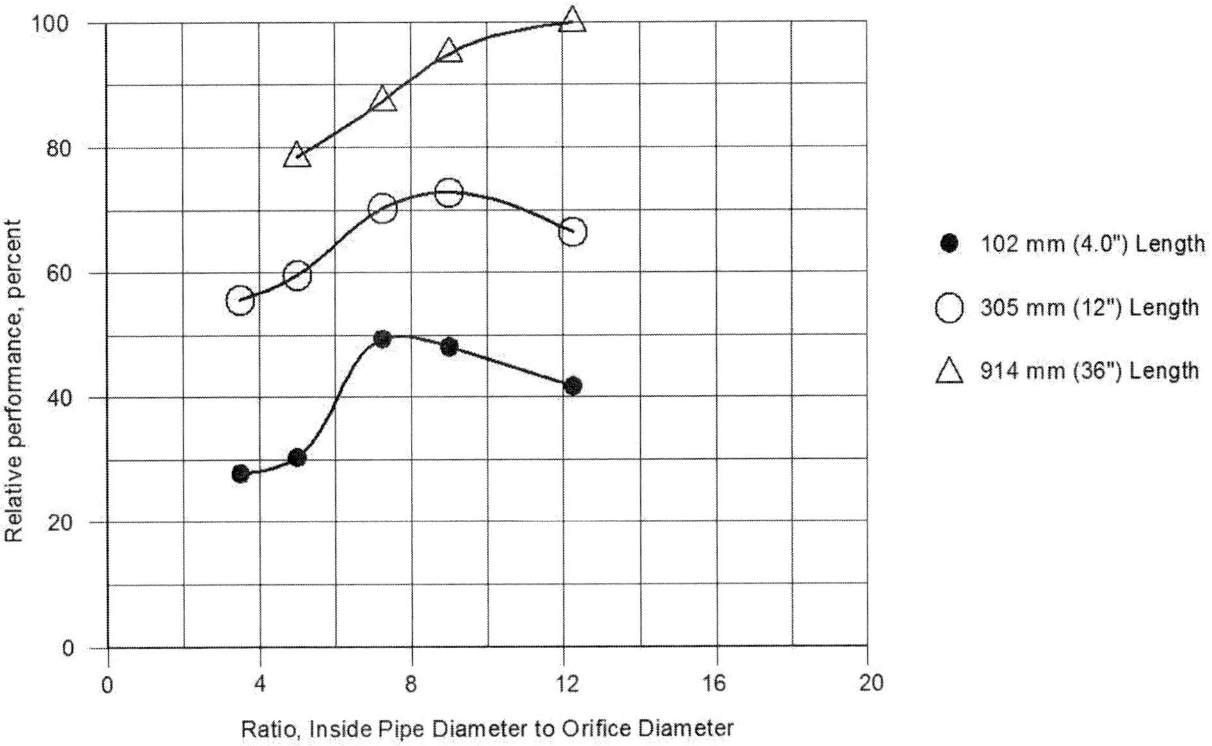

Figure 6: Relative Performance with Increasing Feeder Tube Inside Diameter at.2.3 mm (.090") Orifice Size

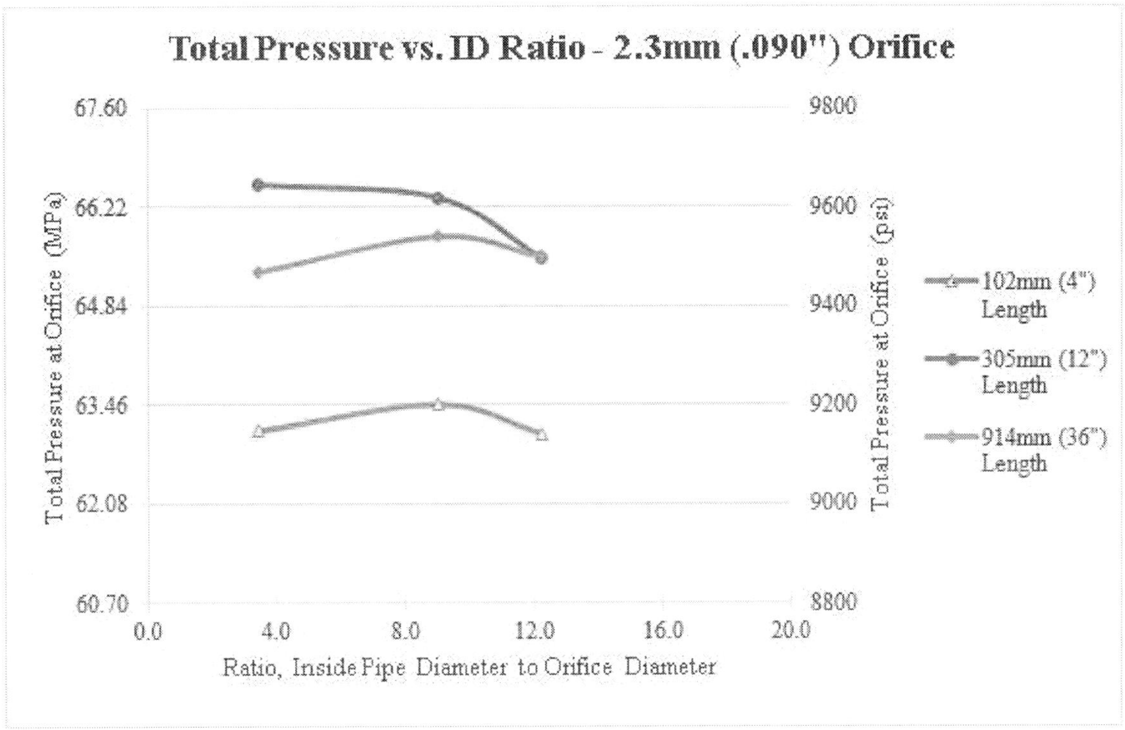

Figure 7: Simulation Results using Total Pressure as an Indicator of Relative Performance for a 2.3mm (.090") Orifice Size

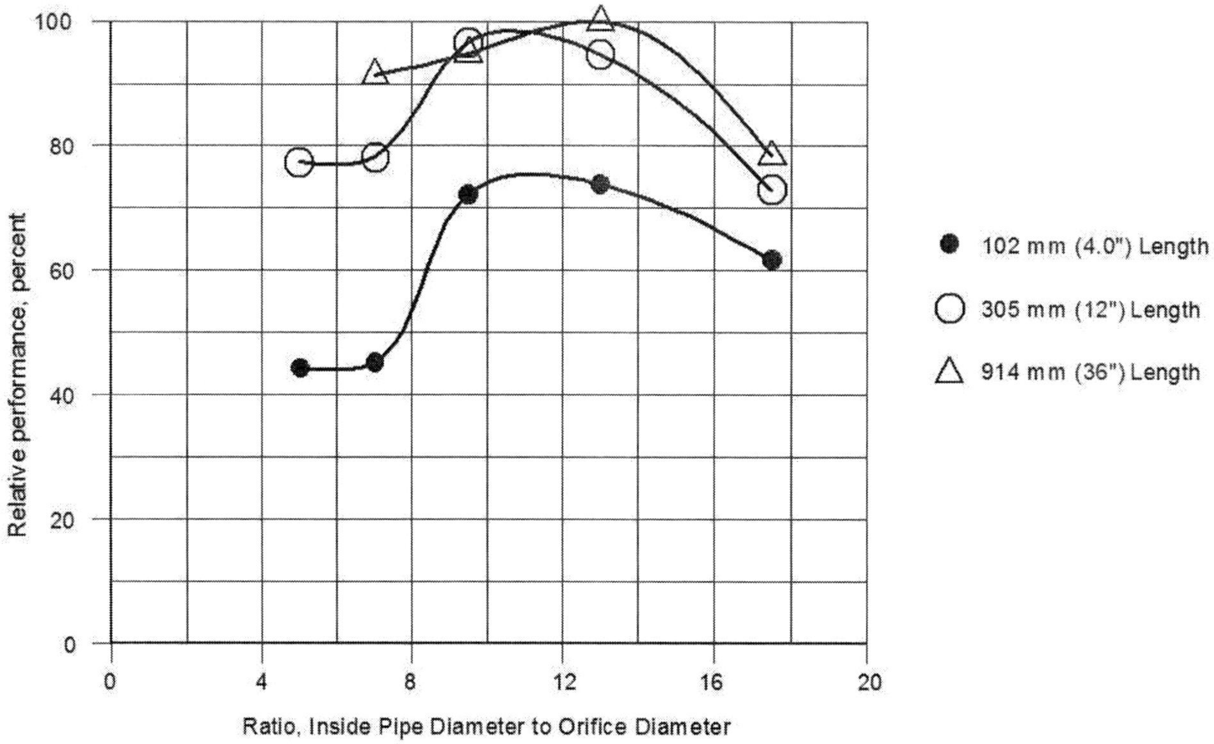

Figure 8: Relative Performance with Increasing Feeder Tube Inside Diameter at.1.6 mm (.063") Orifice Size

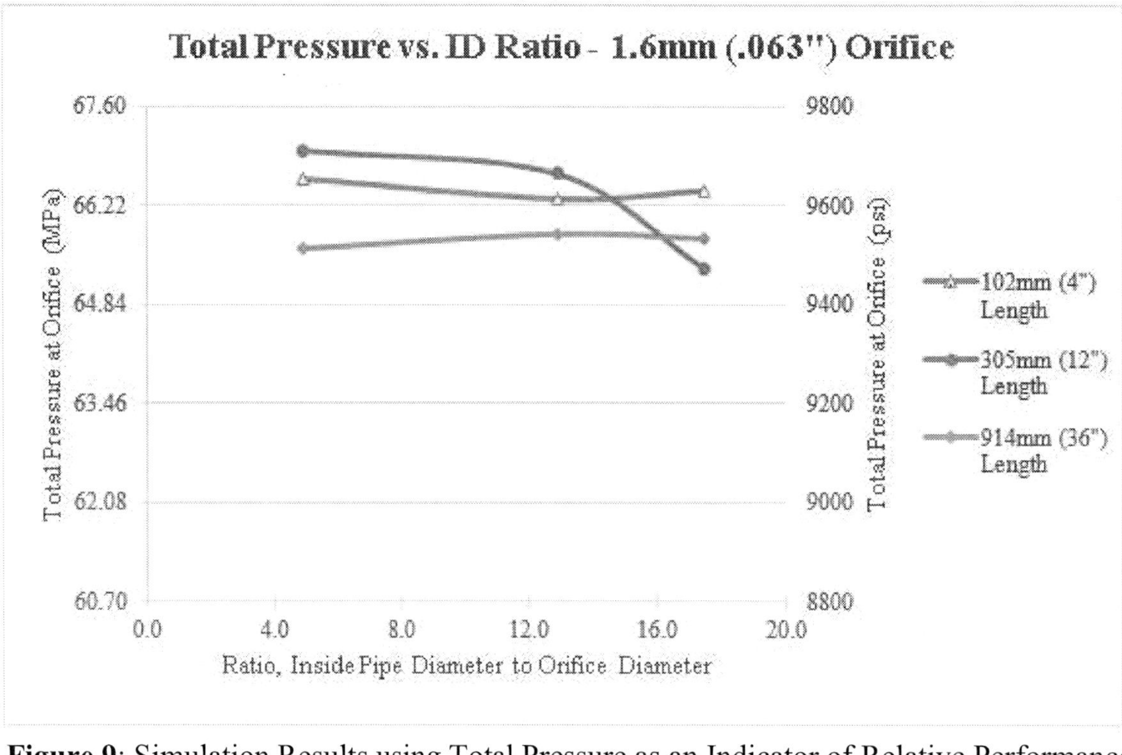

Figure 9: Simulation Results using Total Pressure as an Indicator of Relative Performance for a 1.6mm (.063") Orifice Size

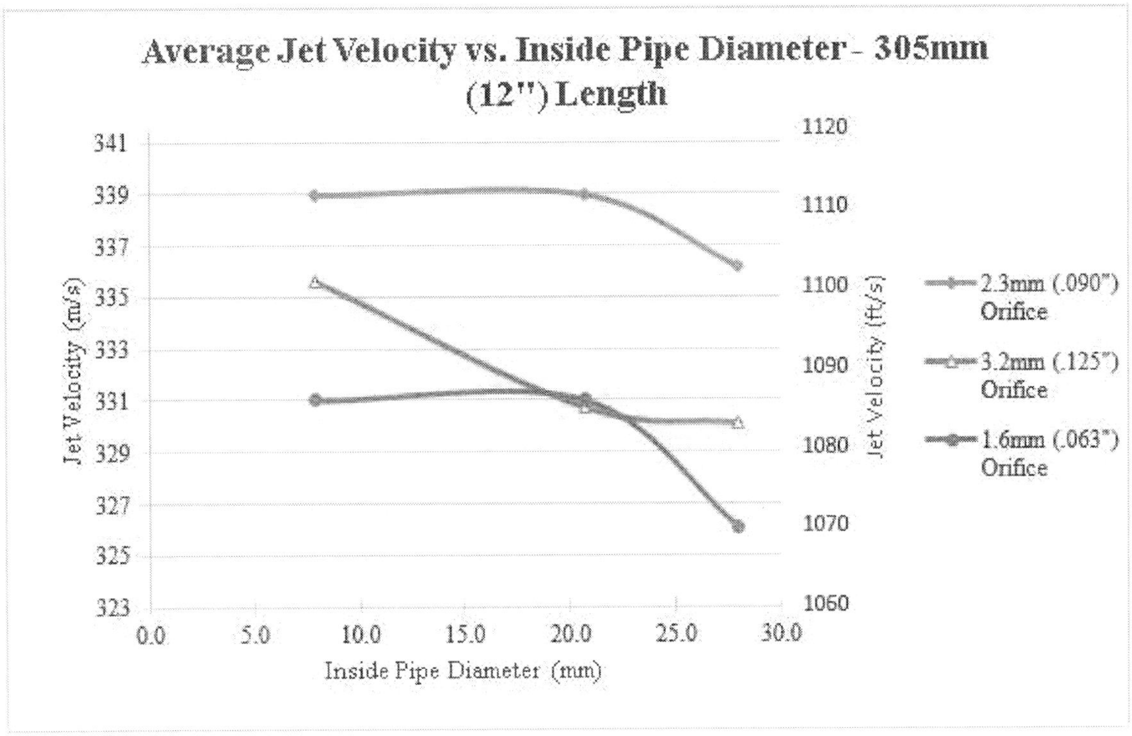

Figure 10: Simulation Results Correlating the Velocity of the waterjet to the Inside Tube Diameter for each Orifice Diameter

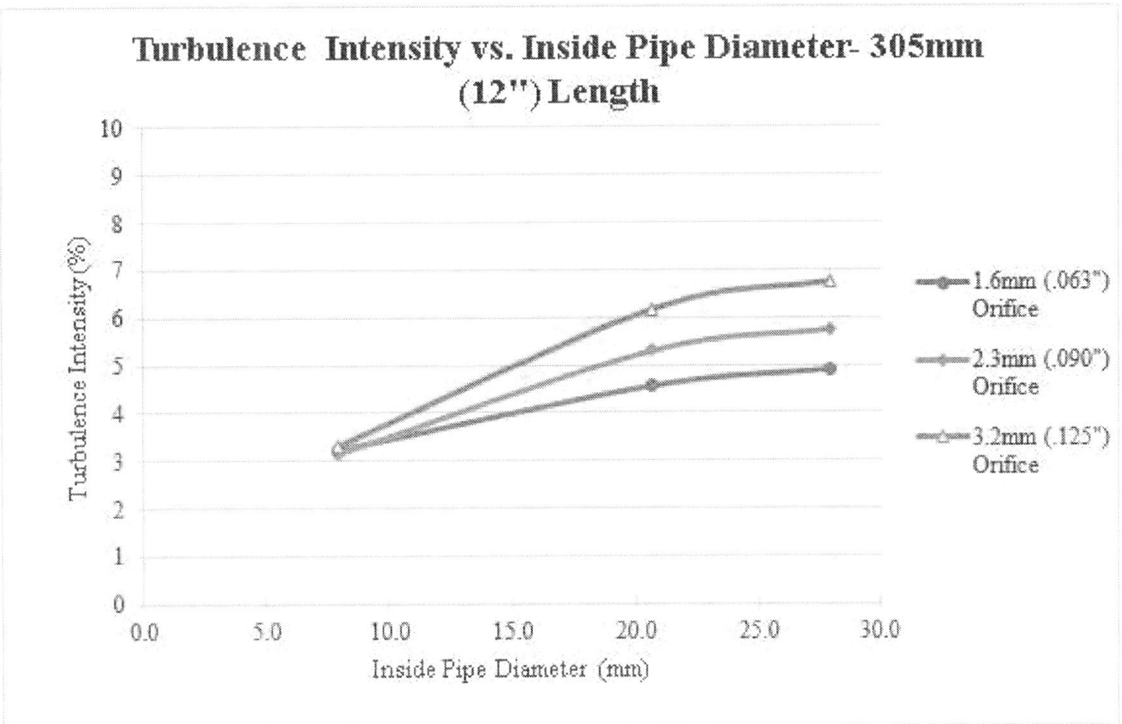

Figure 11: Simulation Results Correlating the Percent Turbulence Intensity of the waterjet to the Inside Tube Diameter for each Orifice Diameter

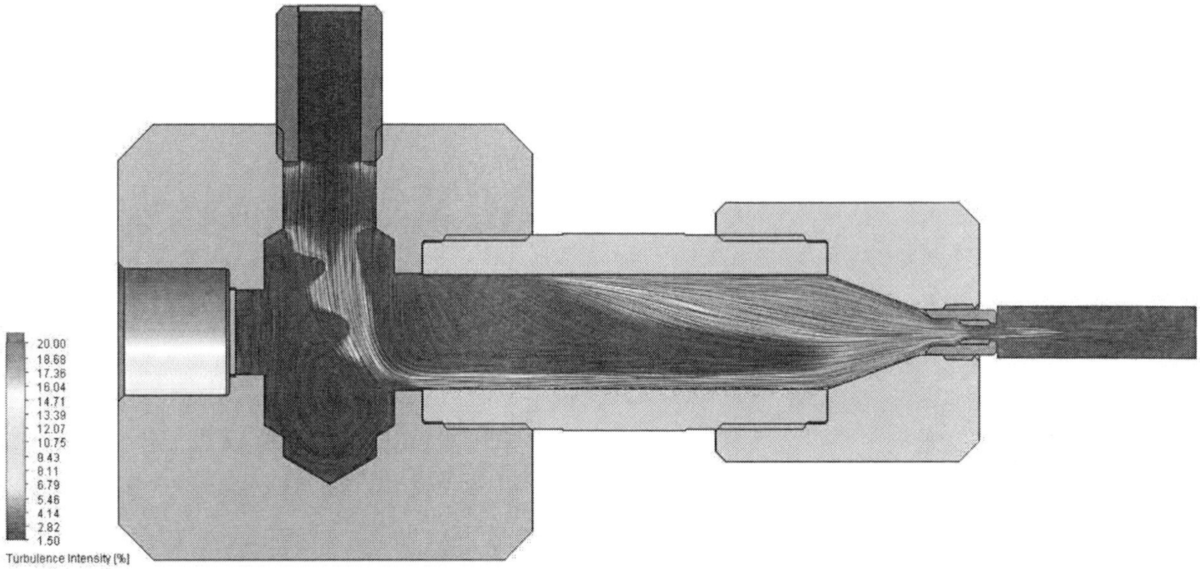

Figure 12: Turbulence Intensity with an Overlay of Velocity Streamlines - Showing 27.9mm (1.1") Inside Diameter, 102mm (4") Long Tube with 3.2mm (.125") Orifice

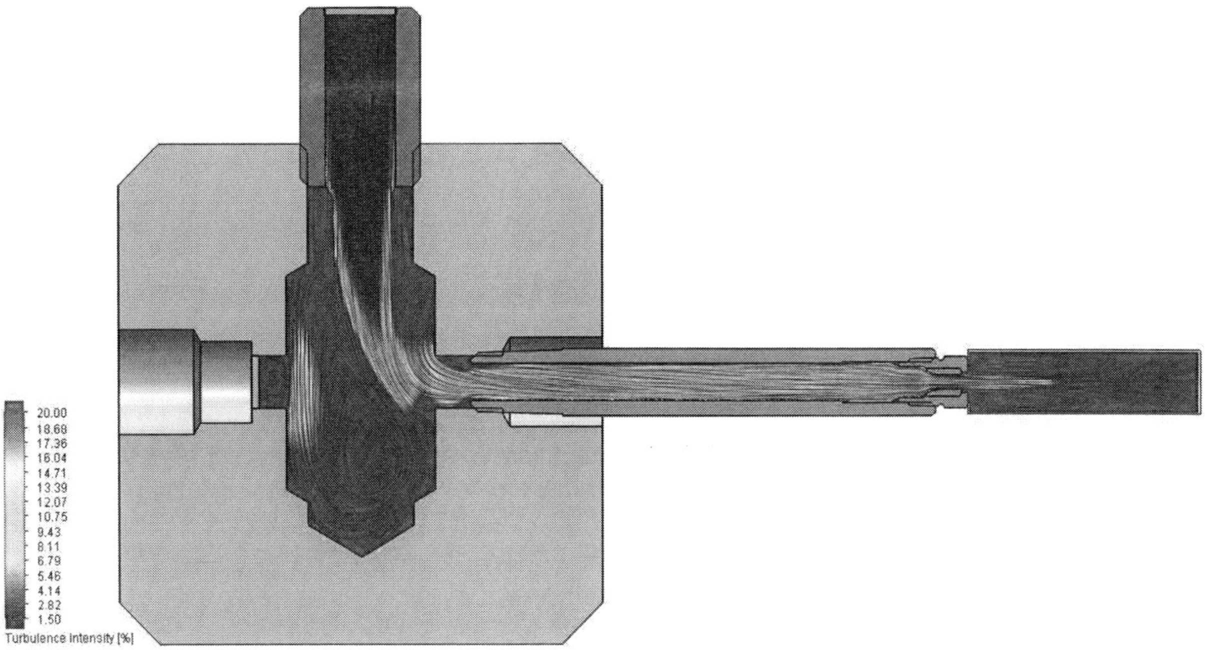

Figure 13: Turbulence Intensity with an Overlay of Velocity Streamlines - Showing 7.9mm (.312") Inside Diameter, 102mm (4") Long Tube with 3.2mm (.125") Orifice

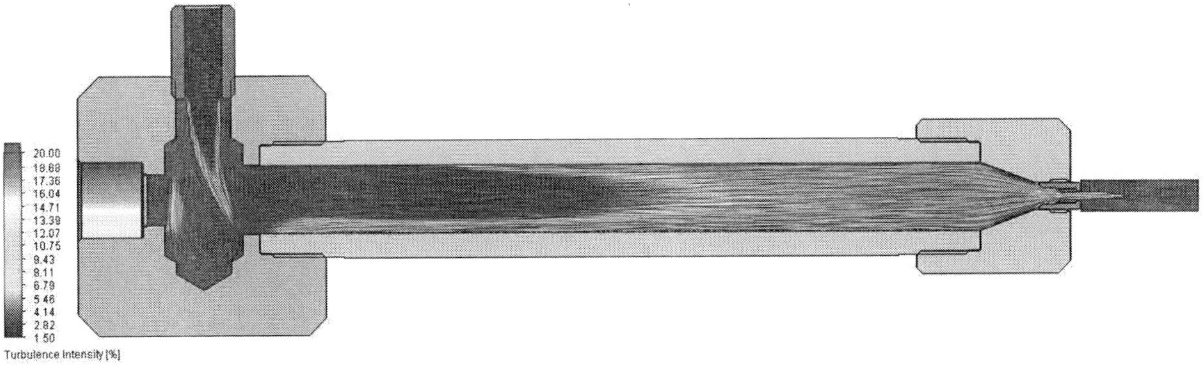

Figure 14: Turbulence Intensity with an Overlay of Velocity Streamlines - Showing 27.9mm (1.1") Inside Diameter, 305mm (12") Long Tube with 3.2mm (.125") Orifice

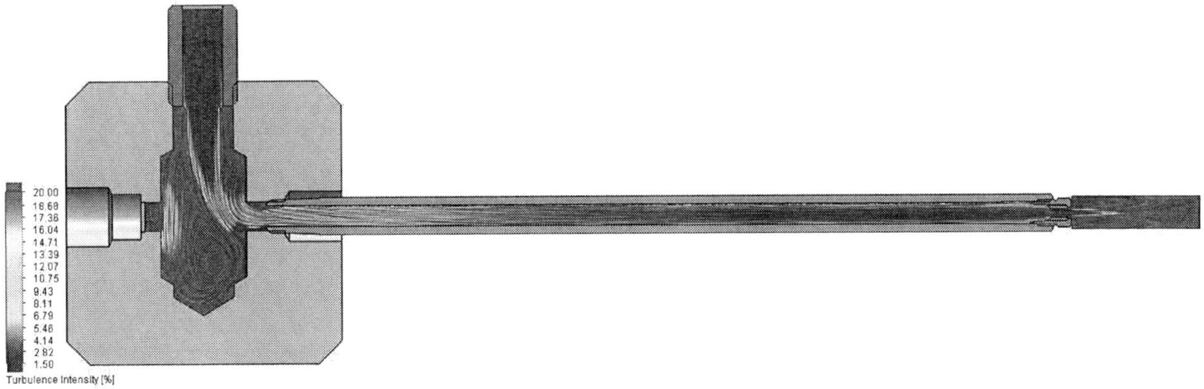

Figure 15: Turbulence Intensity with an Overlay of Velocity Streamlines - Showing 7.9mm (.312") Inside Diameter, 305mm (12") Long Tube with 3.2mm (.125") Orifice

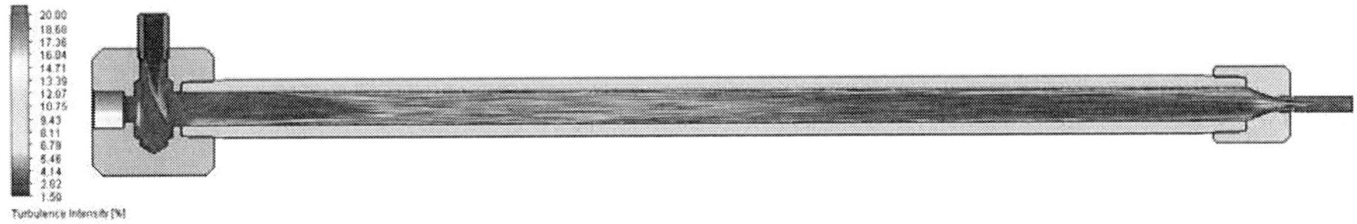

Figure 16: Turbulence Intensity with an Overlay of Velocity Streamlines - Showing 27.9mm (1.1") Inside Diameter, 914mm (36") Long Tube with 3.2mm (.125") Orifice

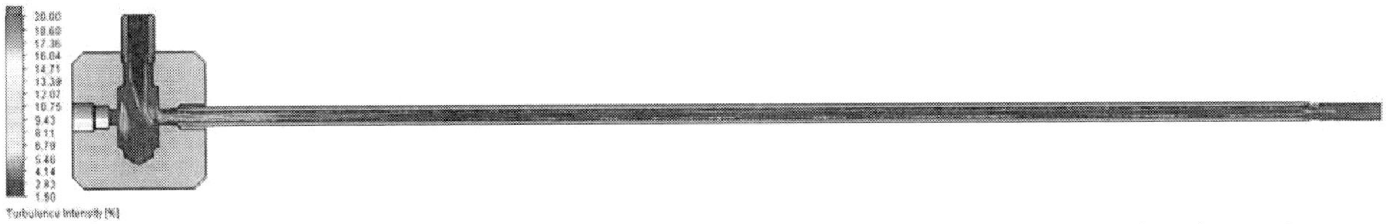

Figure 17: Turbulence Intensity with an Overlay of Velocity Streamlines - Showing 7.9mm (.312") Inside Diameter, 914mm (36") Long Tube with 3.2mm (.125") Orifice

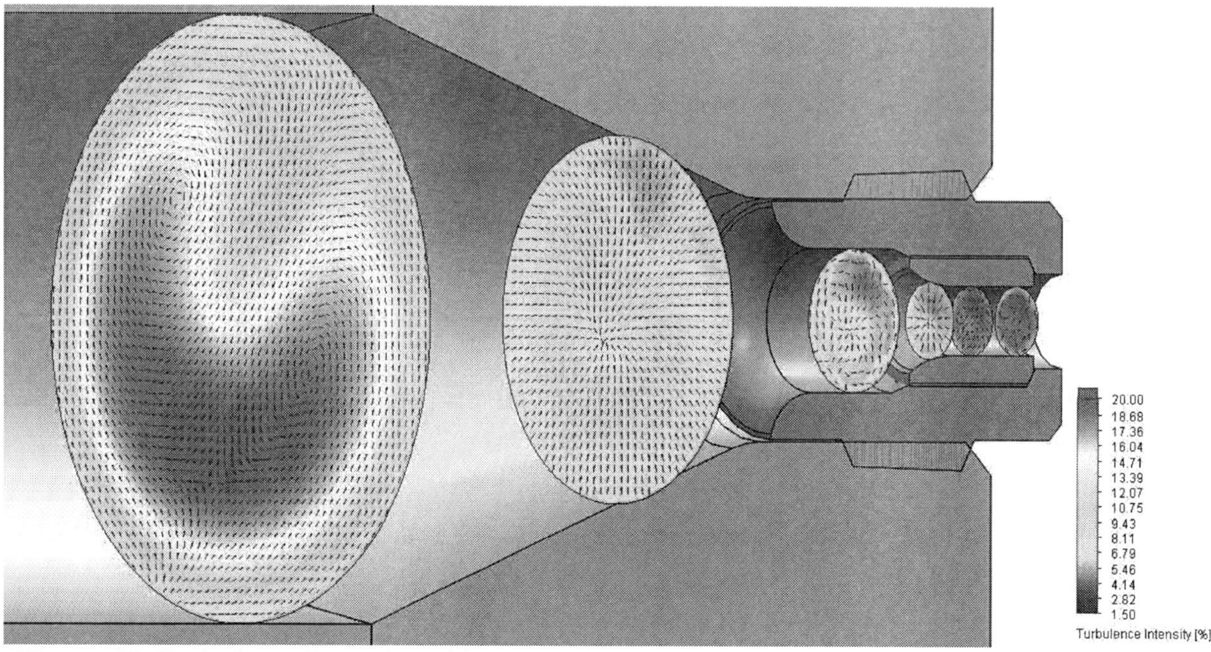

Figure 18: Turbulence Intensity with an Overlay of Velocity Vectors - Showing 27.9mm (1.1") Inside Diameter, 102mm (4") Long Tube with 3.2mm (.125") Orifice

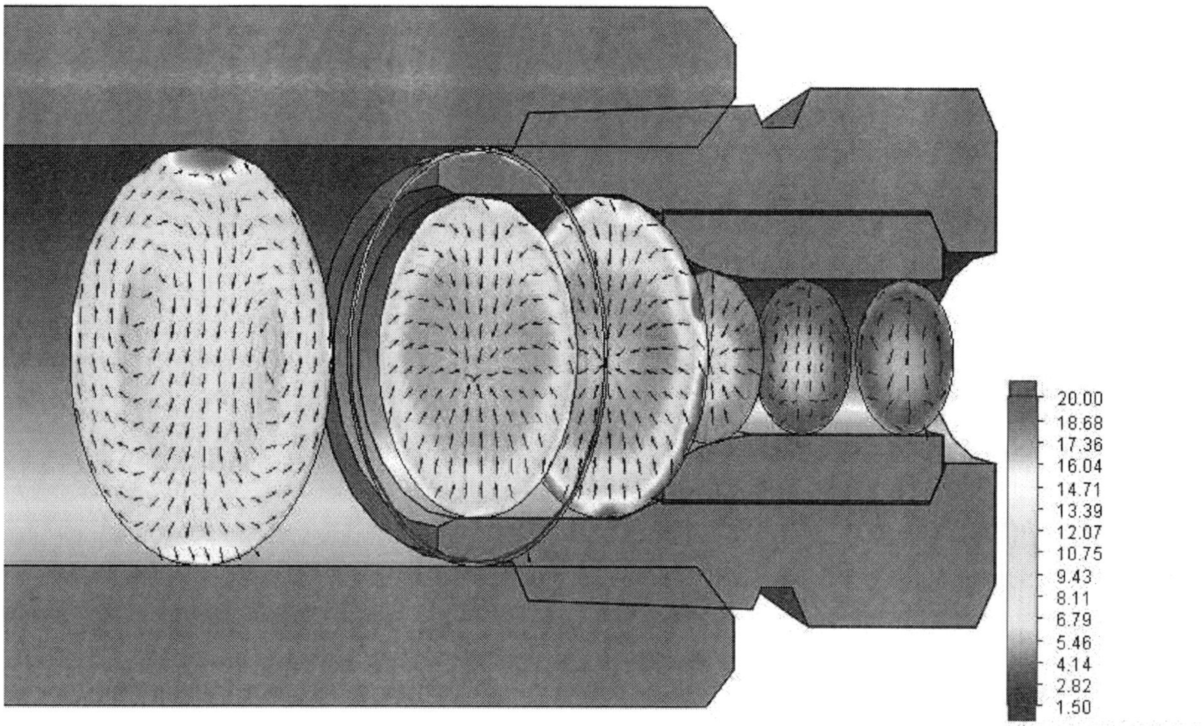

Figure 19: Turbulence Intensity with an Overlay of Velocity Vectors - Showing 7.9mm (.312") Inside Diameter, 102mm (4") Long Tube with 3.2mm (.125") Orifice

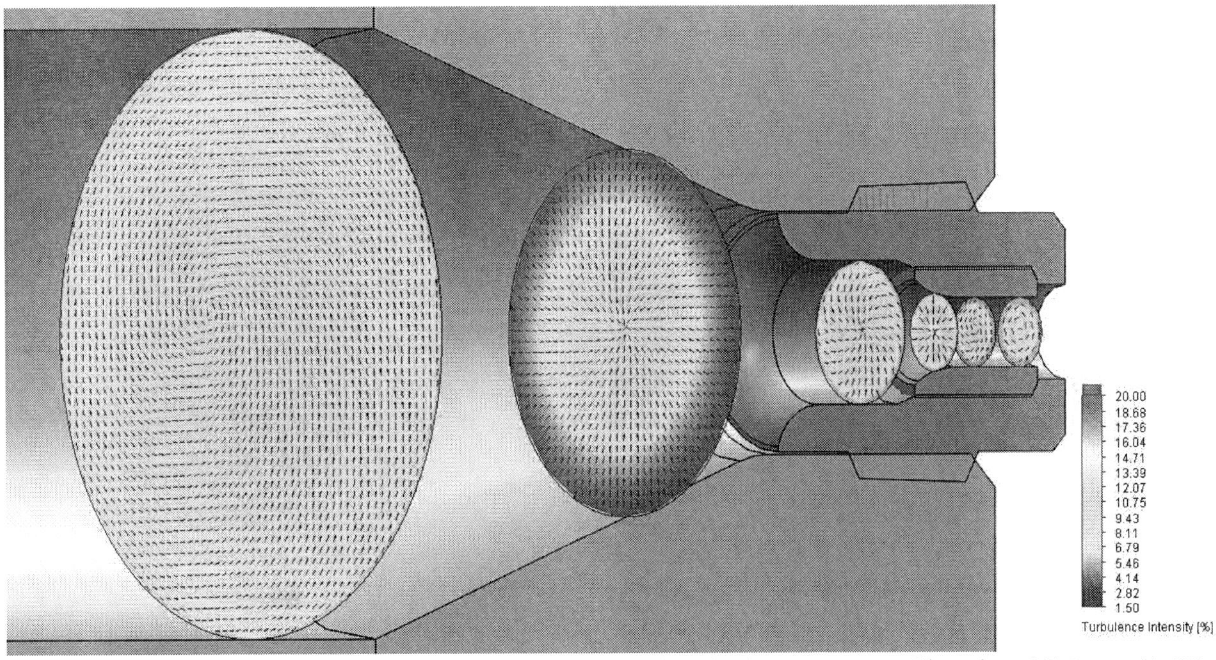

Figure 20: Turbulence Intensity with an Overlay of Velocity Vectors - Showing 27.9mm (1.1") Inside Diameter, 305mm (12") Long Tube with 3.2mm (.125") Orifice

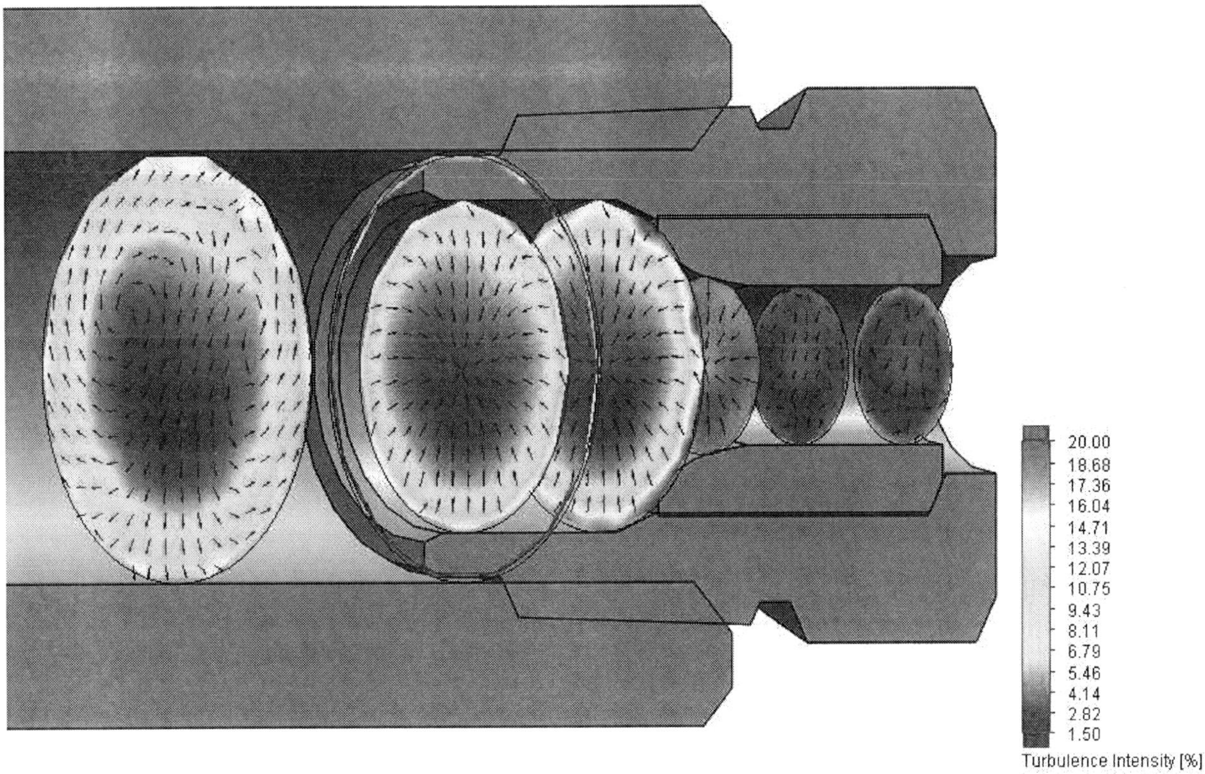

Figure 21: Turbulence Intensity with an Overlay of Velocity Streamlines - Showing 7.9mm (.312") Inside Diameter, 305mm (12") Long Tube with 3.2mm (.125") Orifice

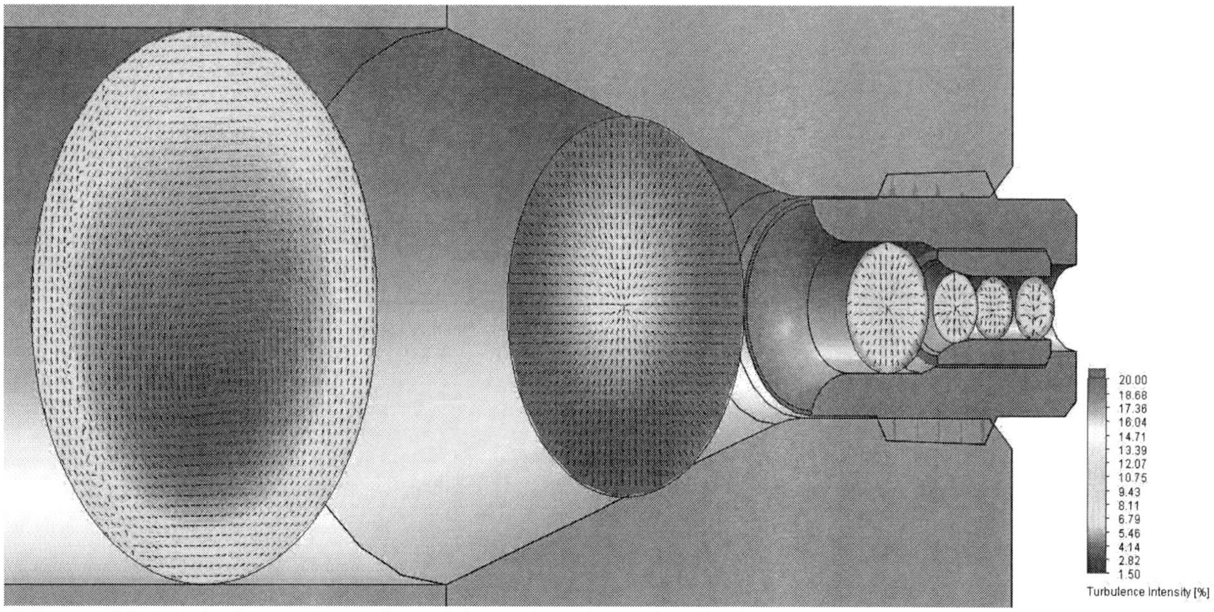

Figure 22: Turbulence Intensity with an Overlay of Velocity Vectors - Showing 27.9mm (1.1") Inside Diameter, 914mm (36") Long Tube with 3.2mm (.125") Orifice

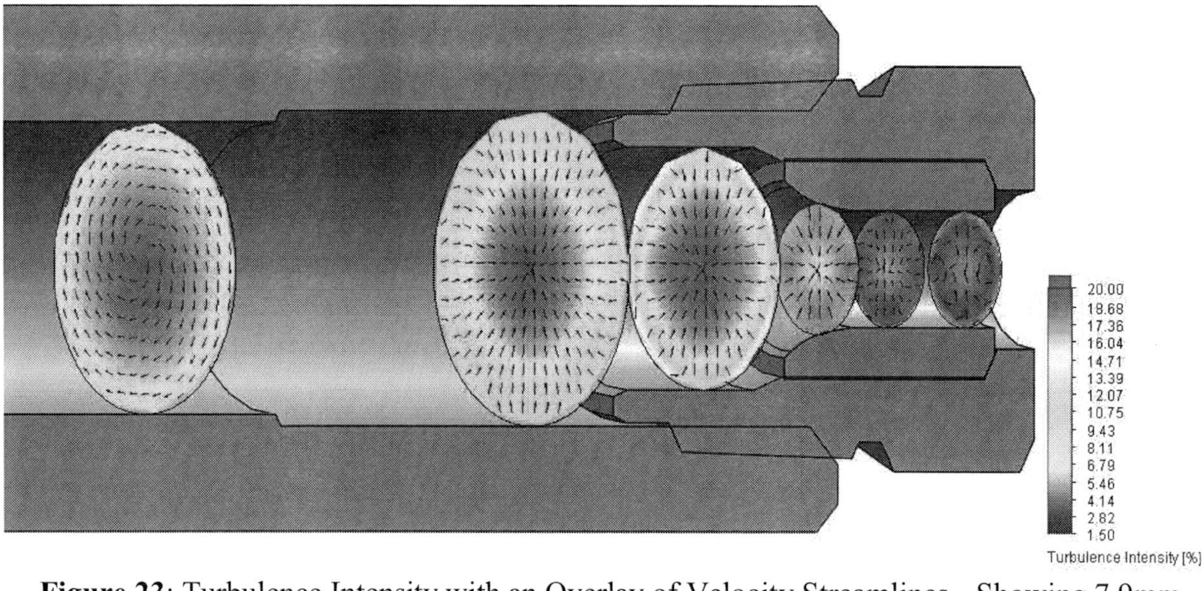

Figure 23: Turbulence Intensity with an Overlay of Velocity Streamlines - Showing 7.9mm (.312") Inside Diameter, 914mm (36") Long Tube with 3.2mm (.125") Orifice

2015 WJTA-IMCA Conference and Expo
November 2-4 • New Orleans, Louisiana

Paper

DESIGN OF EXPERIMENTS BASED WATERBLAST PROCESS MODELING AND OPTIMIZATION USED IN EQUIPMENT DESIGN AND SETUP FOR SPECIFIC APPLICATIONS

M. Kelly, C. Tricou
The Applied Research Laboratory at The Pennsylvania State University
State College, Pennsylvania, U.S.A.

K. Gromes
Terydon, Inc..
Navarre, Ohio, U.S.A.

ABSTRACT

There are many process parameters available for adjustment with waterblasting equipment. Understanding the significance of these parameters, their relative contribution, and interactions can yield order-of-magnitude increases in productivity. Modeling of these processes can also open up new applications for waterblasting that would otherwise be prohibitively expensive or otherwise unsuitable. This presentation will focus on applying Design of Experiments (DOE) based methodologies to model a dual-head Epi-Centric Waterjet Deck Blaster. Real-world results will be shown for extending waterblasting for removal of hard coatings from sensitive substrates and optimal setup of equipment for common applications.

1. INTRODUCTION TO EXPERIMENTAL DESIGN

Design of Experiments (DOE) is a scientific approach to experimentation. The goal of the experimentation is to optimize, maximize, or minimize one or more process outputs (Responses). Researchers choose those input variables (Factors) and ranges which may be controlled for the process of interest. The DOE approach to experimental design enables researchers to extract the maximum amount of information from a minimal number of trials. There are several types of experimental Designs researchers can utilize, having such uninformative names as: D-Optimal, Box-Behnken, Plackett-Burman, Taguchi OA, and others. All of these Designs have specific strengths and weaknesses. The general goal is to use a Design having a minimum number of trials to generate a mathematical function (Model) describing the process. The DOE process will inform researchers as to which factors, the order of those factors, interaction effects, and statistical significance of the model for each Response of interest.

A well-designed, or robust DOE experiment will:

- Aid in the selection and isolation of important variables
- Minimize the required number of experiments needed to yield meaningful results
- Maximize the amount of information extracted from the experiments
- Minimize the cost of product development and process control
- Provide statistical information on model accuracy and repeatability
- Characterize or account for random variability (noise) of process being modeled

1.1 Setting Range of Process Parameters

To better discern the relationship between machine or process input parameters (Factors) and outputs (Responses), it is helpful to widen the range over which experiments are performed. Rather than operating within a narrow range of pressures, flow rates, standoff distance, etc., expand the ranges beyond those typically seen in a 'normal' production environment. This strategy allows process cause-and-effect relationships to stand out against normal variability associated with measurement error, non-controlled environmental factors, normal process variability, etc. The series of graphs shown in

Figure 1 illustrate the benefit of expanding the range of a single Factor. The points plotted in Figure 1 were generated using a known polynomial function (continuous orange line). This polynomial is analogous to a mathematical function that precisely describes a particular process. To these results are added a random number between -5 and 5 to provide an artificial source of variability, or 'noise' to point within range evaluated (blue points). This 'artificial noise' represents effects of imprecise machine settings, effects of extraneous uncontrolled variables such as temperature and humidity, measurement error, etc. The best fit a polynomial function through the artificially-randomized points (black line in range of data points). Repeating this example and calculating a cumulative average to illustrate convergence over 100 iterations shown in

Figure 2. Curve fitting for 'Normal' range had an average R^2 of 0.492 ± 0.178, the wider range had an average R^2 of 0.794 ± 0.076, and the full process range had an average R^2 of 0.960 ± 0.012. Using 'normal' (narrow) process ranges, researchers would only be able to 'see' about 50% ($R^2 = 0.492$) of the underlying cause-and-effect relationship governing the process. Expanding it slightly beyond the normal range would enable researcher to 'see' almost 80% of the underlying relationship ($R^2 = 0.794$) between Responses (e.g., production rate) and Factors (e.g., machine and process settings used). Dramatically opening up the range, as has been done in the graph on the far right, allows researchers to understand and mathematically explain 96% (R2 = 0.960) of the variation seen in the Response variable (production process output of interest) resulting from various machine settings and other process parameter settings.

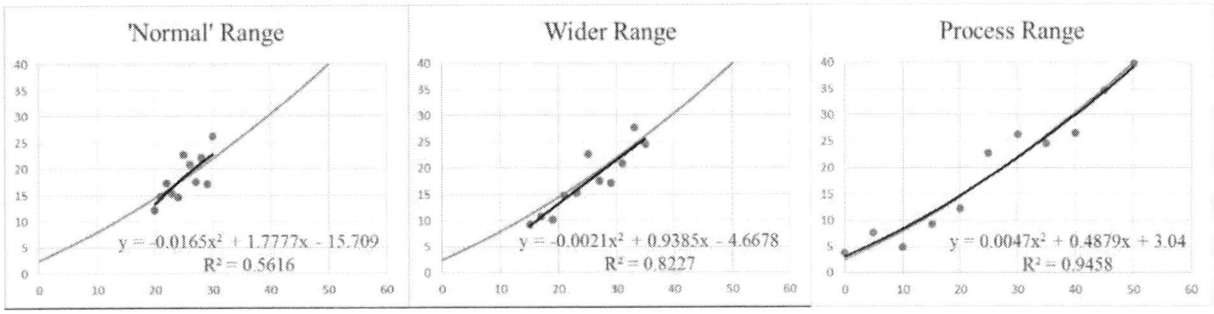

Figure 1. Effects of fitting curves to noisy data over different varible ranges. Know curve (orange) for generating ponts (blue) with random noise between limits and best fit polynomial (black). Best fit polynomial equation and fit shown for each.

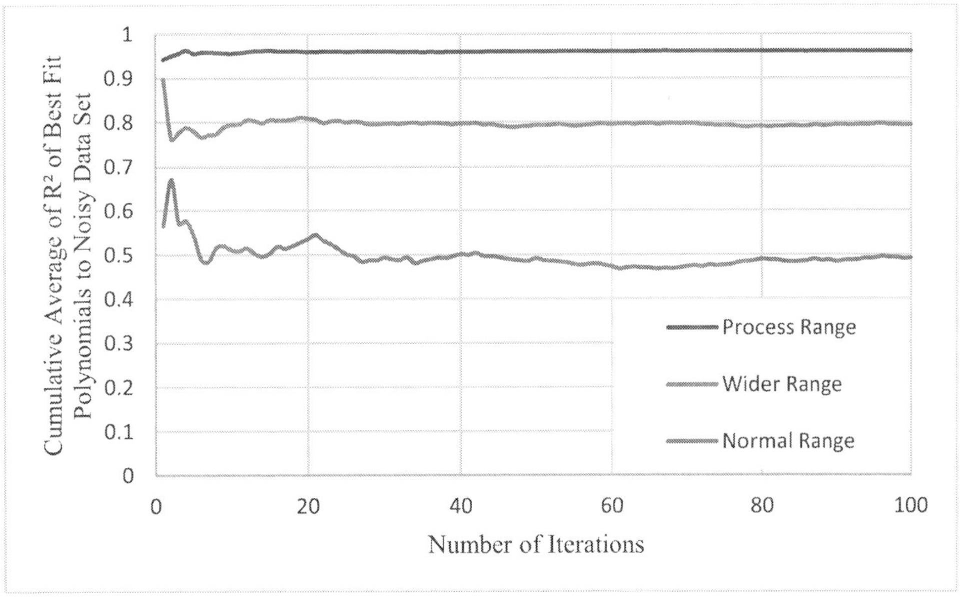

Figure 2. Convergence of best fit polynomial R2 to the three example ranges.

There are additional benefits associated with expanding the range beyond some (arbitrary) narrow range typically used in production as can be seen in Figure 3. An experimenter 'nibbling around the edges' of a typical production range (green region in Figure 3) might conclude that performance drops off if he or she moves too far away from their production 'comfort zone'. However, those researchers will not find dramatic improvements (red region in Figure 3) that can only be found by dramatically increasing the range of the variables studied. Even when a wide process range evaluated the region highlighted in red would be identified, however further experimentation may be required to generate an accurate model and find local optima.

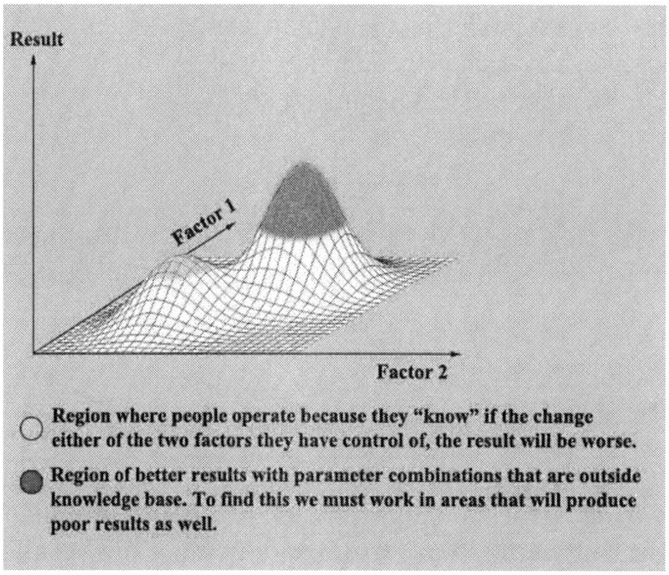

Figure 3. Local & global maxima & minima

1.2 Sizing an Experiment

Generating mathematical functions requires one data point more than the order of curvature. Any two points make a line (for which order=1). For a quadratic, or 2nd order polynomial, three points are required. One point to determine the endpoints of the curve, and one interior to the endpoints to create curvature. Clearly, using the minimum number of points in this way is statistically meaningless. The slope of a line created using only two points might provide information on a general trend, but the slope will vary wildly depending upon how much random error or noise is present, and the researcher would have no way to ascertain how much variation is present. To obtain useful, statistically-significant information, additional points must be included in the experimental design. Typically, quadratic designs (second order) are chosen as a starting point because they help to identify localized maxima (without large misrepresentation around points that are coincidently located on maxima or when extrapolated). An example of a quadratic (2nd order) single-Factor (one variable) Factor Response graph is shown in
Figure 1. The value of the factors appears on the x-axis and Responses on the y-axis. As above, 3 data points are required for fit and additional points are required for statistical significance. A second order polynomial best-fit to the data points will take the form of Equation 1.

$$y = Ax^2 + Bx + C$$

Equation 1. Basic form of second order polynomial for modeling Response from 1 Factor experiment

A quadratic design is needed to identify curvature in a process. This means a minimum of 3 levels (low, medium, high) must be investigated for each variable studied. To fully 'map-out' a design space (full factorial design) trials would need to be run for each variable at each level. The number of experiments that must be performed quickly becomes large and unwieldy, as shown in Figure 3. A 4-Factor full-factorial design (Figure 4C) would require 81 trials. Additional trials are needed to assess lack of fit and statistical accuracy. Pure replicates (exact duplicates of some of the trials) are used to assess process variability, while additional trials (with Factor levels located at points other than the low, medium, and high levels of each variable) would also be needed to assess model lack of fit. Full-factorial designs are commonly done out of ignorance or at the behest of regulatory requirements, where the whole parameter space is evaluated for each permutation of each selected level for each Factor - with replicates. This rapidly becomes overwhelming, as the number of trials increase by 3 to the exponent of number of factors plus replicated trials for evaluation of error and additional trials to assess lack of fit. However, systematic changes of multiple factors with each trial allows one to generate a mathematic matrix of equations of factors and responses which can be simultaneously solved. This dramatically reduces the number of trials required to generate functions relating factors to responses. For a 5-Factor D-Optimal DOE the number of trials required for a determinate matrix drops from 243 trials to 15. An additional 10 trials would be required (5 pure replicates, 5 for lack of fit) to determine statistical significance and lack of fit for a total of 25 experimental trials. The additional trials provide for statistical analysis of variance to determine how strongly each Factor contributes to a particular Response, correlation of the model to the data, and confidence level the models will predict an accurate Response. Continuous factors can be varied continuously over the Factor range(s). Ordinal factors are similar to continuous factors that have discrete set-points, and Categorical variables are discrete, non-numerical variables (e.g., apples & oranges, Jack & Jill, etc.) Categorical factors require all other factors to be varied for each category group. Ordinal factors must be evaluated at each level for each other Factor being evaluated.

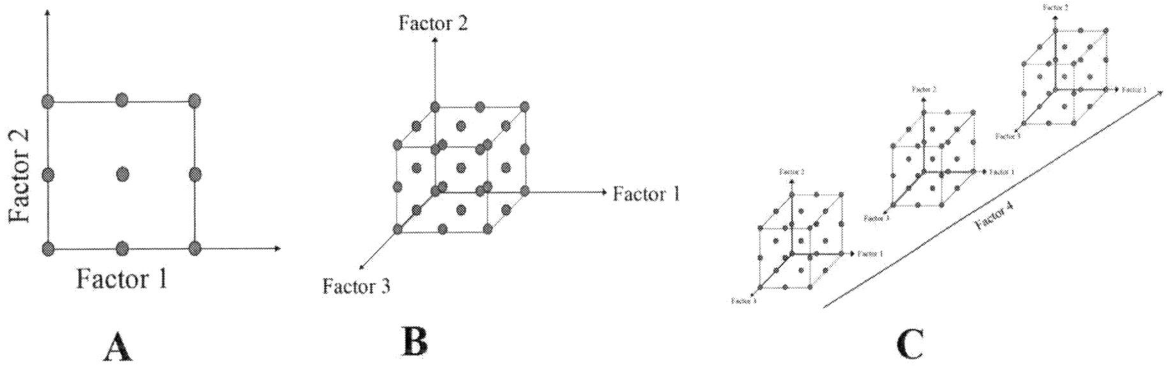

Figure 4. Experimental point reprenstation for 2 Factor (A), 3 Factor (B), and 4 Factor (C)

1.3 Challenges and Shortcomings of Design of Experiments Based Process Modeling

There are few shortcomings to DOE based strategies and techniques and most of these can be addressed with proper planning for robustness and quantification. DOE modeling should be avoided when a process may have true singularities discontinuities or kinks (sections of very high curvature) in the Response curve. This is rare, but electromagnetic fields and quantum effects are two possible examples. . Other shortcoming include: when the cost of each experiment is too high to replicate; when the cost of failed experiments is unacceptable (e.g., human trials); and when experiments have unique components that cannot be accurately quantified or reproduced (behavioral or complex societal factors).

The biggest challenge in using DOE is getting production personnel on-board with the Experimental Design process. Researchers on production lines are tolerated at best, and generally reviled when they start to turn knobs or otherwise mess with the equipment and process. The expertise and experience of production personnel is invaluable when scoping out and conducting experiments, but detrimental if they are allowed to alter subsequent trials based on individual / preliminary results. It is also difficult for production personnel who are unfamiliar with DOE to move out of their comfort zone with regard to expanding ranges of current processes. Production workers and managers are often confident in the process they currently use due to historical experience. Production personnel and product developers also need to understand and accept that obtaining 'bad results' in the course of the optimization is not a 'bad thing'. To uncover and understand the cause-and-effect relationships between input parameters and final surface quality (for example), it is necessary to have a wide range of results. Education and inclusion of production personnel (and others) in the experimental design process is crucial. As is ensuring reasonable expectations with regard to trial-to-trial results when it is time to perform the experiments.

1.4 Software

Full-feature packages allow for process managers with working knowledge of statistics to be successful with minimal training. There are a number of open-source software products that can be used in combination to obtain accurate results with more effort and knowledge of DOE processes and statistical analysis. JMP, Minitab, and Design-Expert are the most recognizable software packages for Multifactor DOE.

2. DOE APPLICATION TO WATERBLASTING

Waterblasting incorporates several complex and non-linear phenomena such as: water traveling through air; and material Response to water impacts. As such, the DOE approach is an excellent tool to understand how machine settings and other process parameters affect output parameter(s) of interest. The DOE method can be used to optimize equipment for specific applications, inform equipment/component design, or evaluate efficacy of new equipment claims.

3. EXAMPLE OF DOE FOR SELECTIVE COATING REMOVAL

ARL was contracted to develop a tool and process to remove durable anti-slip coatings from an elastomeric substrate. For discussion purposes the title 'Rubber' with a following letter designation differentiate the substrates evaluated. Rubber "A" has a relatively firm durometer but can still be gouged or cut with sharp metal tools. Rubber "B" is ~20% softer and is more easily damaged. Both materials have thin seams with a compatible seam filler that is softer than either material. In some areas, the anti-slip coatings overlie patches of much firmer material that cannot be detected until the anti-slip coating is removed. Currently, the only approved process for coating removal is manual hand-sanding using an orbital air sander with 36 grit or finer abrasive. Current reported production rates average ~1 man hour per 0.93 m^2 (1 ft^2) removed, with typical job-size between 74 m^2 and 111 m^2 (800 ft^2 and 1,200 ft^2). The slowness of the current removal method has resulted in schedule delays and interruption of other work. Based on previous experiments for optimizing destruction of the same rubber, it was hypothesized that specialized waterblasting equipment could be optimized to remove the nonskid coating without causing damage to the rubber and to do so at a substantially improved production rate.

3.1 Equipment

The Epi-Jet, produced by Terrydon Inc. shown in Figure 5, was used to test the above hypothesis.

Figure 5. Epi-jet prototyper equipment (Left), View of underside showing nozzle carriers (Right)

Key features of the Epi-Jet that make it ideally suitable for the application are:

- The nozzle carriers are mounted at a uniform distance from the axis of rotation and maintain consistent angle if incidence to surface
- Controllable rotation of nozzle carriers
- Rotation of large central plate eliminates spray-bar stalling on substrate
- Ability to set the overlap of subsequent passes of rotating nozzle carriers
- Adjustable standoff distance between nozzle and substrate

An example of the path traced (epi-centric motion) is shown in Figure 6.

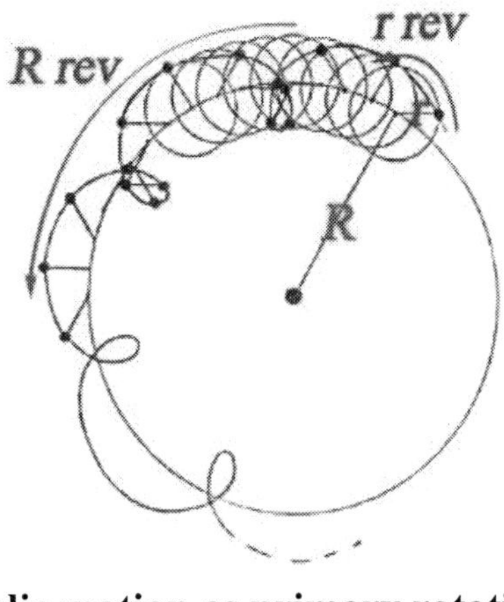

Figure 6. Epi-cyclic motion as primary rotation rate increases

In Figure 6, the rotation speed of the prime centroid (large disc) is increasing with time. The amount of energy delivered to a surface by a waterjet depends upon pressure, flow rate, dwell time, and extent to which tracks overlap. To model this, ARL created a single-jet representation of the planetary-motion jetting pattern using JAVA. The JAVA-based software creates an energy-density map, similar to a heat map, of the process. To better understand the energy intensity map, envision marbles streaming out of a tube as it moves over a grid of cups. The more marbles that end up in a cup, the more intense the energy imparted to that region. The software provides a visual map of the relative energy intensity experienced by the surface. The user inputs rotation speed of the large plate, rotation speed of the small spray nozzles, the radius of the large plate, radius of the nozzle carriers, etc. ARL used this tool to rapidly explore potential designs and provide as uniform surface energy intensity as possible. The program can model both single symmetric bar head and 2-planet epicyclical motion. The nozzle carriers and are evenly spaced around the perimeter of a circle. Input terms are as follows:

R:	Radius of large disk
R rev:	Rate at which nozzle carriers 'orbit' around the centroid of the large disk
r:	Radius of small-head / nozzle-carrier
r rev:	small-head / nozzle-carrier rotation rate
x rate:	Forward translation of the system (left-to-right motion on the screen)
hole count:	Number of nozzles in each nozzle carrier
marbles/sec:	Rate term representing how fast 'marbles' are dropped onto the grid
x length:	Left-to-right display size of grid
y length:	Top-to-bottom display size
x grid count:	Number of cups in a row
y grid count:	Number of cups in a column
time inc:	Time increment

Extending the basic example shown in Figure 6, two small-head nozzle carriers with two jets per nozzle carrier rotate about a central axis (large plate) as the entire system moves from left to right at constant speed is dynamic motion basis for epi-centric water blasting as shown in Figure 7. While this type of pattern appears complicated at first, the basic principles of machine tool path determination apply. There needs to be a minimum number of tooth paths (jets) per inch based on tool diameter. For water jets there can be spacing between adjacent jet paths as direct impact of a jet on a surface is not needed to effect coating removal. The tool head (nozzle carrier) overlap is kept at some known percentage (per head-pass) to produce uniform surface coverage. For common machining this is analogous to a mill-facing operation where a facing to meet surface finish requires the head to split the previous pass by some fraction of the facing-head diameter.

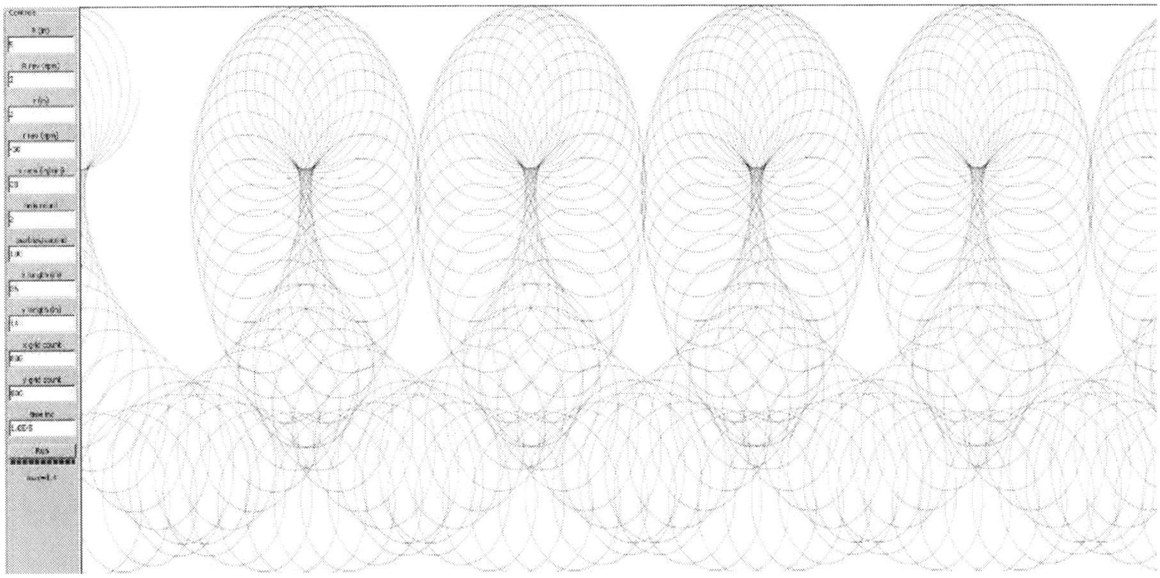

Figure 7. Exagerated epi-cyclic motion example by increasing ration of translation speed to rotation rates

The additional degree of freedom of the epi-cyclic motion results in a fundamentally different jetting pattern from conventional deck-blasters. The jetting patterns of a common deck-blaster and epi-cyclic blast system appear in Figure 8.

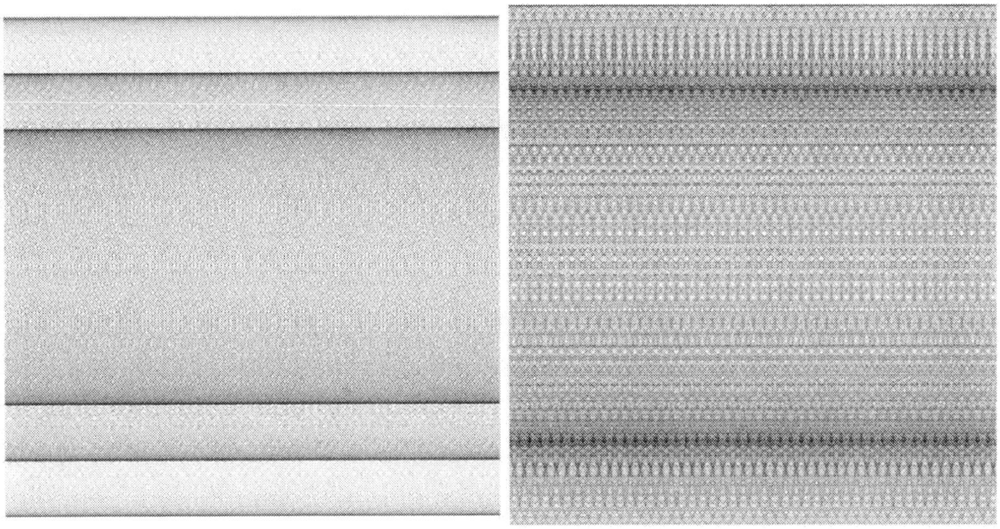

Figure 8. Jetting patterns of conventional 6-jet deck blaster (left) and 4-jet epi-jet (right)

In Figure 8, areas where the jet has passed multiple times or dwell appear. The darker a spot, the greater energy intensity / dwell time that spot has received. Regardless of rotation rate, a deck-blaster with a single axis of rotation will always have high-energy-intensity path lines at distances from center where individual jets are located (shown as the dark horizontal lines on left side of Figure 8). The epi-cyclic blasting pattern (right side of Figure 8) has two higher-energy paths appearing as slightly diffuse darker bands near top and bottom of graphic. These more diffuse higher-energy tracks have less than half the maximum intensity of a typical deck blaster.

3.2 Conducting the Experiments and Process Optimization

A 2nd order, 4-Factor 'D-Optimal' experimental design was chosen for both Rubber 'A' and Rubber 'B' experiments. This resulted in a total of 25 trials for each series of experiments. The 2nd order design will identify curvature and interaction effects within the design space.

Pressure, nozzle size, and flow through the nozzle are coupled. Any two of these variables can be used to calculate the third. Pressure and flow-per-nozzle were chosen as factors for the design, making it easier to limit total flow to the 6 gpm limit imposed by the pump. Nozzles are supplied in integer multiples of 0.1 mm or 0.001". Flow-charts were used to identify the nozzle size providing the 'closest match' in inches to the target flow for a given pressure, and the flow-per-nozzle value in the design was replaced with the theoretical flow through that nozzle.

As mentioned previously, the DOE process generates results outside of what would be considered 'acceptable', and damage to the rubber during trials was expected. Consequently, all trials were performed on rubber scheduled for removal.

Removal Percentage and Post-Blast Surface Condition were the Response (output) parameters of interest. It is difficult to accurately estimate coating removal percentage visually. High-quality photographs and specialized software can be used to accurately estimate removal percentage, but high-quality photographs were not immediately available to ARL personnel.

ARL consulted with Puget Sound Naval Shipyard and Intermediate Maintenance Facility (PSNSY&IMF) rubber and paint engineering authorities to rate the test patches for acceptable condition of rubber and acceptable condition for applying a new coating. ARL then created a combined parameter (rating between -5 and +5) that incorporated relative removal efficiency and post-blast rubber surface condition. At a rating of +5 no coating was removed. At a ranking of zero (0) no coating remained and the rubber was completely unscathed. At a rating of less than zero (0) damage occurred to the rubber ranging from very minor cosmetic scratches to major erosion/gouging. While subjective in nature, the results are valid as long as the rating or ranking system is consistent.

For all trials, the surface condition of the pattern center and pattern edges differed due to the fact that the energy density of the pattern Center is lower than the energy density of the pattern Edge. To account for this, the Edge and Center conditions were rated separately for each trial, representing two distinct and independent outputs.

Rubber "A"

Experiment Design

Factors studied and their ranges are shown in Table 1.

	Range Minimum		Range Maximum	
Factor 1: Pressure	1034 bar	(15.0 Ksi)	2068 bar	(30.0 Ksi)
Factor 2: Flow	2.16 l/min	(0.57 gpm)	5.68 l/min	(1.50 gpm)
Factor 3: Standoff*	0.00 mm	(0.00 in)	25.40 mm	(1.00 in)
Factor 4: Head Rotation	1200 rpm		2000 (rpm)	

*(distance between orifice and surface is an additional 1.344 inches to the standoff)

Table 1. Experimental Design Space for Rubber A

Trial run sheet with actual/correct values and evaluated responses is shown in Table 2.

Conducting and Evaluation of DOE Trials

Trials for Rubber A were conducted at PSNSY&IMF on a US Navy asset in areas where rubber was to be removed for maintenance. Total Area for testing was limited to < 3.7 m^2 (40 ft^2). Aluminum sheet metal was used as a mask to frame 457 mm (18") wide by 203 mm (8") long test patches for each trial.

Results

	Factor 1		Factor 2		Factor 3		Factor 4	Response 1	Response 2
Run	Pressure		Flow		Standoff		Rotation	Edge Rating	Center Rating
	bar	(Ksi)	l/min	(gpm)	mm	(in)	(rpm)	(- damage, 0 best, + paint remains)	
1	2068	(30.0)	2.16	(0.57)	19.05	(0.75)	2000	-0.2	0.0
2	1034	(15.0)	5.64	(1.49)	12.19	(0.48)	1200	4.0	2.5
3	2068	(30.0)	2.16	(0.57)	0.00	(0.00)	1200	-4.0	0.5
4	1034	(15.0)	5.64	(1.49)	12.19	(0.48)	1200	4.5	5.0
5	2068	(30.0)	2.16	(0.57)	19.05	(0.75)	2000	0.0	1.0
6	1034	(15.0)	5.64	(1.49)	25.15	(0.99)	1200	3.0	3.0
7	1207	(17.5)	3.48	(0.92)	19.05	(0.75)	1204	3.0	2.5
8	1841	(26.7)	5.30	(1.40)	11.68	(0.46)	1630	-1.0	-0.5
9	1712	(24.8)	4.66	(1.23)	19.05	(0.75)	1200	-2.5	1.5
10	2068	(30.0)	2.16	(0.57)	0.00	(0.00)	1200	-1.0	0.5
11	1407	(20.4)	2.35	(0.62)	13.21	(0.52)	1200	4.0	5.0
12	2068	(30.0)	5.64	(1.49)	5.08	(0.20)	1200	-2.0	-1.0
13	1432	(20.8)	5.60	(1.48)	26.16	(1.03)	1636	2.0	3.0
14	2068	(30.0)	5.64	(1.49)	0.00	(0.00)	2015	-1.5	0.5
15	1241	(18.0)	3.56	(0.94)	10.41	(0.41)	1672	3.0	3.0
16	1034	(15.0)	3.97	(1.05)	0.00	(0.00)	1200	3.0	3.0
17	2068	(30.0)	2.88	(0.76)	25.15	(0.99)	1200	-3.0	-1.5
18	2068	(30.0)	5.64	(1.49)	0.00	(0.00)	2000	-3.0	-1.0
19	1407	(20.4)	5.56	(1.47)	0.00	(0.00)	1582	2.0	2.0
20	1034	(15.0)	2.31	(0.61)	0.00	(0.00)	2000	4.0	5.0
21	1034	(15.0)	2.31	(0.61)	25.15	(0.99)	1620	5.0	5.0
22	2068	(30.0)	3.67	(0.97)	9.91	(0.39)	1690	-2.0	-1.0
23	1412	(20.5)	4.20	(1.11)	25.91	(1.02)	2000	2.5	4.0
24	1034	(15.0)	2.31	(0.61)	0.00	(0.00)	2000	5.0	5.0
25	2068	(30.0)	5.64	(1.49)	25.40	(1.00)	1848	-0.5	0.5

Table 2. Design of Experiments table of of experiments and responses for Rubber A

Test patch ratings for Edge and Center were the two Responses for this experiment and equations were generated with a best fit regression.

$$\begin{aligned} Rubber\ A\ Edge\ Rating &= 15.7895 - 0.4382 * Pressure(ksi) - 12.5503 * Standoff(in) + 0.0005 \\ &* Head\ Rotation\ (rpm) + 0.0025 * Standoff(in) * Head\ Rotation(rpm) \\ &+ 5.7139 * Flow\ per\ Nozzel(gpm) * Flow\ per\ Nozzel(gpm) - 3.2484 \\ &* Standoff(in) * Standoff(in) \end{aligned}$$

Equation 2. Best fit regression equation of model to relate inspector rating of pattern edge to process parameters of experimental equipment for Rubber A

$$\begin{aligned} Rubber\ A\ Center\ Rating &= 5.8785 + .06418 * Pressure(ksi) - 13.4912 * Flow\ per\ Nozzel(gpm) \\ &- 0.0276 * Pressure(ksi) * Pressure(ksi) + 5.8168 \\ &* Flow\ per\ Nozzel(gpm) * Flow\ per\ Nozzel(gpm) \end{aligned}$$

Equation 3. Best fit regression equation of model to relate inspector rating of pattern center to process parameters of experimental equipment for Rubber B

All terms are statistically significant (95% confidence level). The resulting models had F-values of less than 0.0001. The F-value provides a measure of model significance, and can be interpreted as the likelihood that correlations between factors and responses are due to chance alone. For an F-value of 0.0001 there is less than one chance in 10,000 that the correlation between model inputs and the resulting surface condition are due to chance. The Edge Ranking model explained ~90% of the variation seen in the data (Adjusted R^2 = 0.9008). The Center Ranking model explained ~84% of the variation seen in the data (Adjusted R^2 = 0.8404).

These models are shown graphically in Figure 9. In Figure 9 the full ranges of Pressure and Flow per Nozzle appear on the primary axes at fixed values of the other 2 factors. Thus, these represent a 'slice' of the design space. Graphs are color shaded with the same Response scale with green being in the coating completely removed range and rubber undamaged.

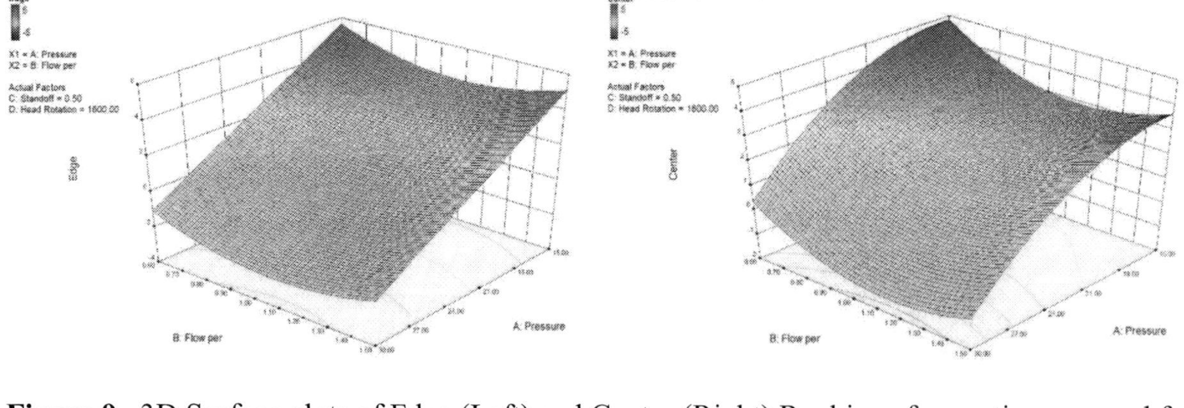

Figure 9. 3D Surface plots of Edge (Left) and Center (Right) Rankings for coating removal from Rubber A.

It should be noted that additional Responses can be measured (power consumption, water consumption, run cost, etc.) and models for these quantities calculated in order to further optimize the process by minimizing cost or water consumption. There are a number of strategies for weighted simultaneous solving to achieve the most acceptable aggregate of responses. A simple method for limited number of responses is to graphically solve by overlaying contour plots as shown in Figure 10. Inside the design space, maximum Flow per Nozzle, 1896 bar (27.5 ksi) Pressure, 20.4 mm (0.805") Standoff, and maximum Rotation provided a perfect rating for both responses is found by the intersection of the blue and black zero level contour lines. This trial was run as a validation patch for the same PSNSY&IMF engineers. All nonskid was removed. Only trace amount of primer and construction paint markings remained. There was no visible scarring or cutting damage to Rubber A. PSNSY&IMF engineers adjudged the surface to be undamaged and "paint-ready". These process parameters on this test equipment produce an instantaneous removal rate of 16.7 m^2/hour (180 ft^2/hour). Because factors needed to be set to maximums to achieve an optimum results, additional improvements in the responses may be achieved by expanding the design space further. To further optimize an additional set of trials should be conducted to augment the design. Time, equipment limitations and services available at test site prevented augmentation of design. Using the mathematical model and extrapolating (accuracy of any model degrades rapidly with extrapolation) a solution can be found at a considerably lower Flow per Nozzle as shown in the intersection of the red and black zero-level contour lines in Figure 10. This result was validated as acceptable at a later date on a mockup.

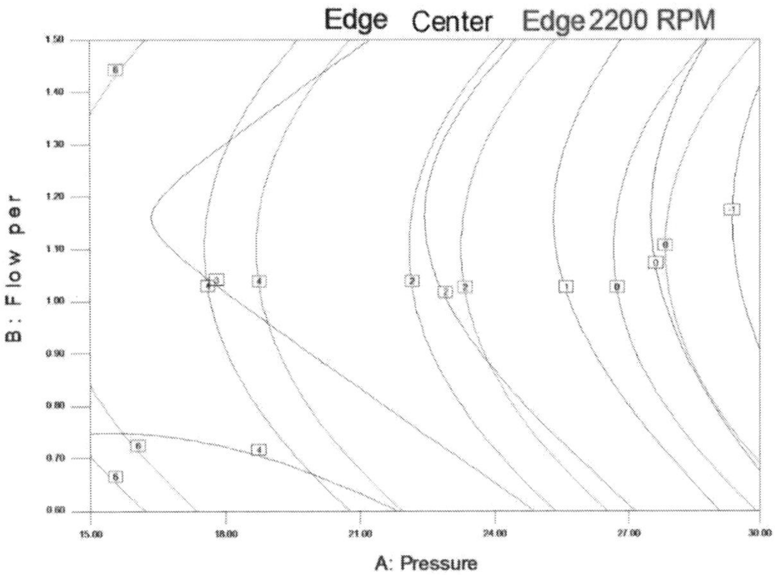

Figure 10. Overlay of 3 contour plots for Pressure and Flow per Nozzle. Standoff is fixed at 0.805" inches, Edge Rating, Center Rating, and Edge Rating with head rpm increased from 2000 to an extrapolated 2200 rpm.

Process Optimization Rubber "B"

Experiment Design

The ranges of the process space were modified based on results obtained during testing of Rubber 'A'. The factors and ranges selected for the Rubber 'B' Trials are shown in Table 3.

	Range Minimum		Range Maximum	
Factor 1: Pressure	1344 bar	(19.5 Ksi)	2103 bar	(30.5 Ksi)
Factor 2: Flow	2.16 l/min	(0.57 gpm)	5.94 l/min	(1.57 gpm)
Factor 3: Standoff*	0.00 mm	(0.00 in)	23.88 mm	(0.94 in)
Factor 4: Head Rotation	1700 rpm		3000 (rpm)	

*(distance between orifice and surface is an additional 1.344 inches to the standoff)

Table 3. Experimental Design Space for Rubber B

Trial run sheet with actual/correct values and evaluated responses is shown in Table 4. Additional trials were conducted due to errors in setup and unplanned variations in vacuum supply.

Conducting and Evaluation of DOE Trials for Rubber 'B'

Trials for Rubber B were conducted at Pearl Harbor Naval Shipyard and Intermediate Maintenance Facility (PHNSY&IMF) on a US Navy asset in areas where rubber was to be removed for maintenance. Area for testing was limited to < 2.79 m2 (30 ft2). Aluminum sheet metal was used as a mask to frame 457 mm (18") wide by 152 mm (6") long test patches for each trial. Difficulties in maintaining adequate vacuum service to the job site existed and some trials were run with little or no vacuum adding additional noise to results. The range of ratings differed compared to experiments performed on Rubber 'A' due to more severe damage observed on the softer rubber.

Results

	Factor 1		Factor 2		Factor 3		Factor 4	Response 1	Response 2
Run	Pressure		Flow		Standoff		Rotation	Edge Rating	Center Rating
	bar	(Ksi)	l/min	(gpm)	mm	(in)	(rpm)	(- damage, 0 best, + paint remains)	
1	1820	(26.4)	2.35	(0.62)	9.65	(0.38)	2311	-0.3	1.5
2	1662	(24.1)	5.53	(1.46)	15.24	(0.60)	2610	-2.0	0.0
3	2075	(30.1)	2.16	(0.57)	22.61	(0.89)	3000	0.0	0.3
4	2034	(29.5)	5.56	(1.47)	17.78	(0.70)	2935	-6.0	-4.0
5	2027	(29.4)	5.56	(1.47)	17.78	(0.70)	2935	-6.0	-3.0
6	2103	(30.5)	5.68	(1.50)	0.00	(0.00)	1960	-10.0	-7.0
7	2075	(30.1)	2.88	(0.76)	0.00	(0.00)	1700	-1.5	-0.3
8	1379	(20.0)	2.35	(0.62)	0.00	(0.00)	1700	3.0	3.0
9	1386	(20.1)	2.35	(0.62)	0.00	(0.00)	3000	4.5	4.0
10	1386	(20.1)	2.35	(0.62)	23.11	(0.91)	1700	4.0	5.0
11	2048	(29.7)	5.07	(1.34)	18.29	(0.72)	1700	-7.0	-4.0
12	1441	(20.9)	4.24	(1.12)	0.00	(0.00)	2942	-5.0	-2.0
13	2062	(29.9)	5.11	(1.35)	0.00	(0.00)	2942	-4.0	-3.5
14	1958	(28.4)	4.96	(1.31)	14.48	(0.57)	1707	-6.0	-4.0
15	1344	(19.5)	4.13	(1.09)	10.16	(0.40)	2330	4.0	5.0
16	2062	(29.9)	3.26	(0.86)	19.05	(0.75)	1710	-5.0	-2.0
17	2082	(30.2)	3.29	(0.87)	0.00	(0.00)	2942	-3.0	-2.0
18	2082	(30.2)	3.29	(0.87)	0.00	(0.00)	2942	-4.0	-2.0
19	1965	(28.5)	4.50	(1.19)	10.41	(0.41)	2454	-1.5	-0.3
20	1800	(26.1)	4.28	(1.13)	22.61	(0.89)	2318	-2.0	0.0
21	1800	(26.1)	4.28	(1.13)	22.61	(0.89)	2318	-4.0	-2.0
22	1551	(22.5)	3.18	(0.84)	0.00	(0.00)	1700	-2.5	-1.0
23	1613	(23.4)	3.63	(0.96)	16.00	(0.63)	3000	-0.5	2.0
24	1613	(23.4)	5.94	(1.57)	0.00	(0.00)	2935	-3.0	2.0
25	1558	(22.6)	5.87	(1.55)	18.80	(0.74)	1700	-2.0	-0.5
26	1475	(21.4)	5.72	(1.51)	0.00	(0.00)	1700	-3.0	-1.0
27	1344	(19.5)	5.45	(1.44)	23.88	(0.94)	3000	1.0	1.0
28	1379	(20.0)	2.35	(0.62)	0.00	(0.00)	3000	0.0	0.3
29	1379	(20.0)	2.35	(0.62)	22.61	(0.89)	1700	0.5	1.0
30	2103	(30.5)	5.68	(1.50)	0.00	(0.00)	1960	-7.0	-10.0
31	1379	(20.0)	2.35	(0.62)	0.00	(0.00)	3000	3.0	2.5

Table 4. Design of Experiments table of of experiments and responses for Rubber B

$$Rubber\ B\ Edge\ Rating = 15.5377 - 0.4948 * Pressure(ksi) - 4.8048 * Flow\ per\ Nozzel(gpm)$$

Equation 4. Best fit regression equation of model to relate inspector rating of pattern edge to process parameters of experimental equipment for Rubber B

$$\begin{aligned} Rubber\ B\ Center\ Rating &= 6.3243 + .0834 * Pressure(ksi) + 8.8472 * Standoff(in) - 0.0021 \\ &* Head\ Rotation\ (rpm) + 3.1192 * Flow\ per\ Nozzel(gpm) - 0.5462 \\ &* Pressure(ksi) * Flow\ per\ Nozzel(gpm) + 0.0027 * Head\ Rotation\ (rpm) \\ &* Flow\ per\ Nozzel(gpm) - 9.48264 * Standoff(in) * Standoff(in) \end{aligned}$$

Equation 5. Best fit regression equation of model to relate inspector rating of pattern center to process parameters of experimental equipment for Rubber B

All terms appearing in the preceding equations are statistically significant at the 95% confidence level. Non-significant terms have been eliminated. Both models had F-values less than 0.0001. There is less than one chance in 10,000 that the correlation between model inputs and the resulting surface condition could be due to chance. The Rubber B model was able to explain ~71% of the variation seen in the data (Adjusted R2 = 0.7122). The Rubber 'B' Center model was able to explain ~76% of the variation seen in the data (Adjusted R^2 = 0. 7566). These models are shown graphically in Figure 11.

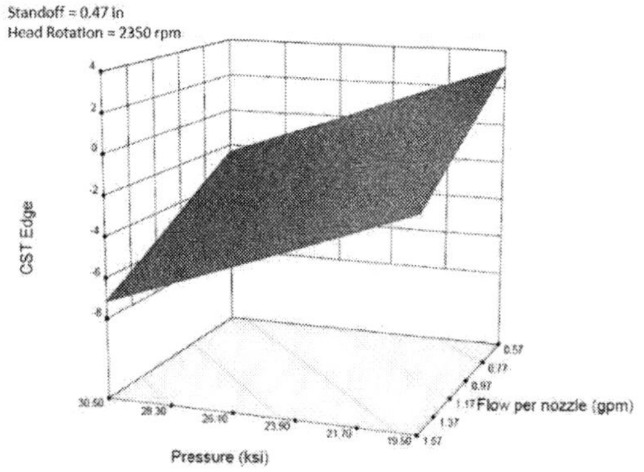

Figure 11. 3D Surface plots of Edge and Center Rating for Rubber B anti-skid coating removal

PHNSY&IMF asked ARL to optimize the process to leave the bottom-most layer of anti-skid coating intact, even though this would require secondary sanding to get to a paint-ready surface. After optimizing to achieve the desired condition, ARL performed a dozen validation trials. In some of these, ARL went over the same area 2-3 times to assess the effect of multiple passes. ARL consistently achieved the requested single-pass surface condition. With multiple passes nearly all anti-skid coating was removed without causing damage to the surface. The single-pass removal rate (leaving most of the lower layer of anti-skid coating) is ~16.7 m^2/hour (180 ft^2/hour). At three passes, the removal rate to a (nearly) paint-ready surface is ~ 5.6 m^2/hour (60 ft^2/hour). A process maps was generated to aid in equipment setup and use for shipyard workers, shown in Figure 12.

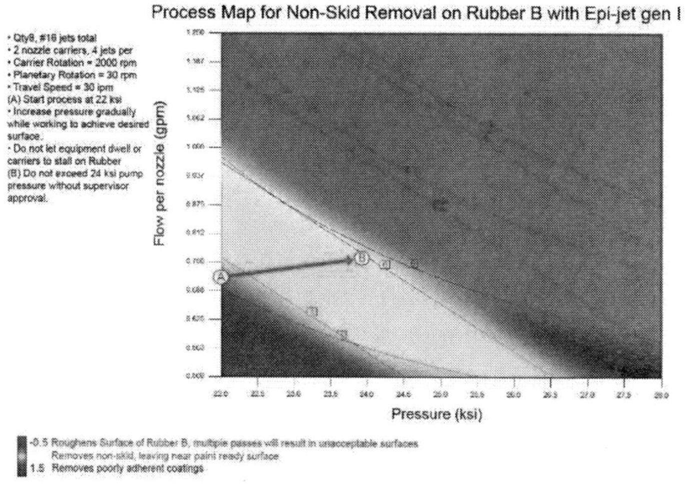

Figure 12. Process map for opperators removing anti-skid coatings from Rubber B

4. CONCLUSIONS

The Design of Experiments process can be a powerful tool to model the waterblasting process. The mathematical models produced can be used to optimize for applications that would otherwise be prohibitive to try. Models can also be used to identify process parameter ranges when results become unacceptable, which is very useful in establishing acceptable process windows. Additional responses on water usage, equipment wear, production rate, and operating costs can be used to predict job costs and size support equipment.

2015 WJTA-IMCA Conference and Expo
November 2-4 • New Orleans, Louisiana

Paper

UNDERWATER EFFECTIVITY CHARACTERIZATION OF WATERJET PERFORMANCE IN CLEANING APPLICATIONS

R. Doornbos
StoneAge, Inc.
Durango, Colorado, U.S.A.

ABSTRACT

A significant amount of research has been conducted in the past regarding the effectiveness of waterjet nozzles as well as the jet quality produced. This research is commonly conducted in an operating condition where the waterjet must travel through air to reach the target. However, many field applications exist where water jetting takes place in an underwater condition. These applications may occur in a body of water, piping system, or a vessel that fills during the cleaning process.

This paper investigates the effectiveness and performance of a waterjet nozzle in operation underwater, while comparing these results with those of a waterjet nozzle operating in air. In order to complete this testing, the variables of nozzle pressure, flow, standoff distance, and traverse velocity were compared on the same target material. Parameters for the investigation included various flow rates, orifice sizes, pressures from 69 MPa (10,000 psi) to 276 MPa (40,000 psi), standoff distances from 51 millimeters (2 inches) to 559 millimeters (22 inches), and traverse velocities of 0.61 m/sec (2 ft/sec) to 4.42 m/sec (14.5 ft/sec).

Machinable wax was used as a target material. The average depth of target material removed was measured to determine jet effectiveness and compared between the two different operation modes.

Organized and Sponsored by WJTA®-IMCA®

1. INTRODUCTION

Many waterjet cleaning applications may occur in a body of water, piping system, or vessel that fills during the cleaning process. Currently, limited research has been conducted to assess the actual effectiveness and performance of a waterjet traveling through water as opposed to air. Through the years, some have asserted that underwater cleaning can be more effective than common cleaning operations in air. The theory behind this view pertains to the phenomenon of cavitation. As a waterjet contacts a substrate underwater, the impact of the waterjet creates subsequent cavitation in the surrounding area further enhancing the effectiveness or power of the waterjet. The collapsing voids (cavitation) that implode near the surface of the substrate cause cyclic stress through repeated implosions; subsequently, increasing the material removal and effectiveness of the waterjet. An opposing view held by some, is that a submerged waterjet will show very limited effectiveness and performance or simply not function at all, based on the deterioration of the waterjet as it passes through water.

2. TESTING SETUPS

2.1 Effectivity Testing - Traversing Target and Stationary Lance

Figures 1 and 2 depict the testing setup used to evaluate and test underwater effectiveness. In the tests conducted, the machinable wax target was traversed in a rolling carriage across a rail system using a pneumatic cylinder. The cylinder was capable of providing a consistent traverse speed. Speed was measured by two magnetic pickups oriented in close proximity to the lance location. A lance with an inside diameter of 4.78 millimeters (0.188 inch) and length of 813 millimeters (32 inches) was used upstream of the waterjet orifice. Two types of nozzle orifices were adapted to fit the testing setup, Carbide Nozzles for the 69 MPa (10,000 psi) and the 138 MPa (20,000 psi) testing, and a Sapphire Nozzle for the 276 MPa (40,000 psi) testing.

2.2 Verification and Correlation Setup – Traversing Lance and Stationary Target

Typical waterjet cleaning is completed with the waterjet traversing across a target. Therefore, verification was required of the testing completed using the traversing target test setup described in Section 2.1 and shown in Figures 1 and 2. The correlation testing was completed by duplicating a previous data set while translating the lance in water. The lance translation setup is shown in Figure 3. The limitation of the lance translation test setup was that the maximum traverse speed was limited to 0.61 m/sec (2 ft/sec) based on the design of the test fixtures.

3. EFFECTIVITY TESTING

3.1 Baseline Testing in Air

After the initial test setup was assembled as shown in Figures 1 and 2 (without filling the tank with water), baseline testing was completed by translating the machinable wax target through the air and across the specified waterjet nozzle. Table 1 outlines the tests completed in this method varying pressure, flow, traverse velocity, and standoff distance. The purpose of this baseline testing was to obtain effectiveness and performance for the typical waterjet application. Different tests were completed to provide conclusive results based on the variables selected.

3.2 Underwater Comparison Testing

After the initial test setup was assembled as shown in Figures 1 and 2, underwater testing was completed by translating the machinable wax target through the water and across the specified waterjet nozzle. Table 2 outlines the tests completed in this method varying pressure, flow, traverse velocity, and standoff distance. The purpose of this underwater testing was to obtain effectiveness and performance for the waterjet in an underwater application, which allows for correlation to the baseline air testing. Once again, different tests were completed to provide conclusive results based on the variables selected.

3.3 Verification and Correlation Testing

As previously discussed, typical waterjet cleaning is completed with the waterjet traversing across a target. Both of the first two test setups utilized the traversing target test setup shown in Figures 1 and 2. The verification and correlation testing was completed by duplicating a previous data set as shown in Table 3, while translating the lance in water. The lance translation setup is shown in Figure 3.

3.4 Results

Based on the progression of the testing and the evidence provided by the initial results, some of the originally planned test steps outlined in Tables 1 through 3 were eliminated. This determination was made based on the initial evidence of limited to no material removal in the machinable wax target at specific standoff distances and / or speeds. When this occurred, the original test plan was altered to remove tests that included greater standoff distances and / or higher speeds while utilizing similar waterjet pressures and nozzle diameters. The progression of the Effectiveness Testing Results within this document identifies specific test pressures, nozzle sizes, and traverse speeds for each set of graphical results. The graphical results shown in Figures 4 through 8 directly relate to the Effectiveness Testing outlined in Tables 1 through 3, respectively.

The graphical results depicted in the Figures 4 through 8 indicate the relative performance data collected for the waterjet nozzle testing in both air and underwater. Each test was completed using the same waterjet nozzle with a constant pressure and flow at various standoff distances and two traverse speeds. The results are expressed in terms of nozzle diameters and percent of relative performance. The percent of relative performance was determined with the assumption that the largest amount of material removal for a specific set of test parameters would determine the 100 percent effective performance for comparison purposes.

Black solid lines within the graphical results indicate testing completed in air with a slow traverse speed. Red and Green solid lines indicate testing completed in water with a slow traverse speed, where Red results were acquired using a traversing target setup and Green were acquired using a traversing lance setup. Additionally in the graphical results, Black dashed lines indicate air testing with a fast traverse speed; whereas, Red dashed lines indicate water testing with a fast traverse speed.

No graphical results are provided for the 69 MPa (10,000 psi) with 1.07 millimeter (0.042 inch) waterjet nozzle testing based on the evidence found in the initial testing. A combination of the machinable wax threshold pressure (pressure at which the material can be removed), standoff distances selected, and the traverse speeds of the target did not allow for enough measureable data to provide a clear result.

The results show that in the testing completed, underwater material removal was generally more effective, approximately 10%, when compared to material removal when the waterjet passes through air at close distances. This phenomenon was true at very close standoff distances, typically within 100 nozzle diameters. On the other hand, this phenomenon does not hold true as the waterjet standoff distance was increased. The underwater test results show significantly reduced overall effectiveness as the standoff distance was increased beyond the 100 nozzle diameter threshold. These conclusions are true independent of the test pressure, flow, and nozzle diameter based on the testing completed. However, if the traverse speed of the waterjet was increased by a factor of six times the results were found to provide more variable results. More specific test for each test sequence can be found in Figures 4 through 8.

The data produced by holding all variables constant and simply varying the traverse speed of the target material were found to be very consistent as well. The graphical results for slow traverse speed testing versus increased traverse speed testing provided uniform and consistent results. This was true when comparing air to air as well as water to water curves. The reduction of effectiveness results of the air to air correlation were typically greater, ranging from 20% to 50% reduction. Whereas, the results of the water to water correlation were smaller, ranging from 5% to 20% reduction.

Finally, as shown in Figure 8, the test method used with a traversing target provided test results which did not significantly vary from those within the Verification and Correlation testing section from Table 3. The curves generated from the results of both testing methods closely match with little offset. One major difference found was that the target translation testing was 10% more effective at closest standoff distance when compared to the lance translation testing. This effect was most likely caused by loss of the waterjet power as it traversed through the water, by a lag in the waterjet, or a combination of both. The lag of the waterjet can be thought of as slightly bent or bowed waterjet stream creating a loss of power based on movement in multiple directions in the water (i.e. traversing through the water and moving through the water toward the target.)

4. CONCLUSIONS

Many conclusions can be drawn from the results and evidence provided by this Effectiveness and Performance Testing Program. The results of this testing show that underwater waterjet cleaning can be greatly effective within the 100 nozzle diameter range; conversely, underwater waterjet cleaning may be ineffective and very limited beyond the 300 nozzle diameter range. Additionally, the graphical trends show that the smaller and more cohesive the waterjet, the more effectiveness you will be able to achieve at a greater distance. This trend was true in both the air and underwater test conditions.

The effectiveness and performance testing completed in this test program was not intended to be able to produce underwater material removal rates for waterjet cleaning; however, it does give an indication that the closer cleaning distance to the substrate that can be achieved in the underwater condition, the more effective the cleaning will be. Furthermore, greater performance deterioration would be expected with larger standoff distances. Based on the results, if the cleaning distance can be reduced to within 100 nozzle diameters, waterjet effectiveness and performance can be expected to be approximately 10% more effective than when cleaning in air. This would authenticate the first theory outlined previously in this paper that the effectiveness of underwater waterjet applications is greater at distances closer than 100 nozzle diameters. However, this research does not pinpoint the exact mechanism or phenomenon (i.e. cavitation, etc.), which allows this to happen.

5. ACKNOWLEDGMENTS

Individuals:

Travis Watkins, StoneAge, Inc. - Test and Automation Engineer
Matt Hastey, StoneAge, Inc. - Sr. Product Engineer
Corey Waller, StoneAge, Inc. - Systems Engineer
Colton Andersen, StoneAge, Inc. - Product Engineer
Joe Schneider, StoneAge, Inc. - Product Engineer
Doug Wright, StoneAge, Inc. - Director of Research and Development

Organization:

StoneAge, Inc. - Durango, CO.

6. REFERENCES

None

7. NOMENCLATURE

None

Table 1. Baseline Testing Data Points – Target Translated in Air

Pressure [MPa (kpsi)] Note: +/- 7 MPa (1000 psi)	Calculated Flow [lpm (gpm)]	Nozzle Ø for Flow [mm (in)]	Traverse Velocity [m/sec (ft/sec)] Note: +/- 0.46 m/sec (1.5 ft/sec)	Standoff Distance [mm (in)] Note: +/- 2.54 mm (.1 in)	Standoff Distance [Nozzle Ø's]
69 (10)	18.0 (4.75)	1.07 (0.042)	0.61 (2)	559 (22)	524
				406 (16)	381
				203 (8)	190
				51 (2)	48
69 (10)	18.0 (4.75)	1.07 (0.042)	4.42 (14.5)	559 (22)	524
				406 (16)	381
				203 (8)	190
				51 (2)	48
138 (20)	11.3 (2.98)	0.71 (0.028)	0.61 (2)	559 (22)	786
				406 (16)	571
				203 (8)	286
				51 (2)	71
138 (20)	11.3 (2.98)	0.71 (0.028)	4.42 (14.5)	559 (22)	786
				406 (16)	571
				203 (8)	286
				51 (2)	71
138 (20)	25.4 (6.72)	1.07 (0.042)	0.61 (2)	559 (22)	524
				406 (16)	381
				203 (8)	190
				51 (2)	48
138 (20)	25.4 (6.72)	1.07 (0.042)	4.42 (14.5)	559 (22)	524
				406 (16)	381
				203 (8)	190
				51 (2)	48
138 (20)	57.2 (15.11)	1.60 (0.063)	0.61 (2)	559 (22)	349
				305 (12)	190
				83 (3.25)	52
				51 (2)	32
276 (40)	6.6 (1.74)	0.46 (0.018)	0.61 (2)	559 (22)	1222
				406 (16)	889
				203 (8)	444
				51 (2)	111
				25 (1)	56
276 (40)	6.6 (1.74)	0.46 (0.018)	4.42 (14.5)	559 (22)	1222
				406 (16)	889
				203 (8)	444
				51 (2)	111
				25 (1)	56

Table 2. Underwater Comparison Testing Data Points

Pressure [MPa (kpsi)] Note: +/- 7 MPa (1000 psi)	Calculated Flow [lpm (gpm)]	Nozzle Ø for Flow [mm (in)]	Traverse Velocity [m/sec (ft/sec)] Note: +/- 0.46 m/sec (1.5 ft/sec)	Standoff Distance [mm (in)] Note: +/- 2.54 mm (.1 in)	Standoff Distance [Nozzle Ø's]
69 (10)	18.0 (4.75)	1.07 (0.042)	0.61 (2)	559 (22)	524
				406 (16)	381
				203 (8)	190
				51 (2)	48
69 (10)	18.0 (4.75)	1.07 (0.042)	3.81 (12.5)	559 (22)	524
				406 (16)	381
				203 (8)	190
				51 (2)	48
138 (20)	11.3 (2.98)	0.71 (0.028)	0.61 (2)	559 (22)	786
				406 (16)	571
				203 (8)	286
				51 (2)	71
138 (20)	11.3 (2.98)	0.71 (0.028)	3.81 (12.5)	559 (22)	786
				406 (16)	571
				203 (8)	286
				51 (2)	71
138 (20)	25.4 (6.72)	1.07 (0.042)	0.61 (2)	406 (16)	381
				305 (12)	286
				203 (8)	190
				51 (2)	48
138 (20)	25.4 (6.72)	1.07 (0.042)	3.81 (12.5)	406 (16)	381
				305 (12)	286
				203 (8)	190
				51 (2)	48
138 (20)	57.2 (15.11)	1.60 (0.063)	0.61 (2)	559 (22)	349
				305 (12)	190
				83 (3.25)	52
				51 (2)	32
276 (40)	6.6 (1.74)	0.46 (0.018)	0.61 (2)	559 (22)	1222
				406 (16)	889
				203 (8)	444
				51 (2)	111
				25 (1)	56
276 (40)	6.6 (1.74)	0.46 (0.018)	3.81 (12.5)	559 (22)	1222
				406 (16)	889
				203 (8)	444
				51 (2)	111
				25 (1)	56

Table 3. Correlation Testing Data Points

Pressure [MPa (kpsi)] Note: +/- 7 MPa (1000 psi)	Calculated Flow [lpm (gpm)]	Nozzle Ø for Flow [mm (in)]	Traverse Velocity [m/sec (ft/sec)] Note: +/- 0.46 m/sec (1.5 ft/sec)	Standoff Distance [mm (in)] Note: +/- 2.54 mm (.1 in)	Standoff Distance [Nozzle Ø's]
138 (20)	25.4 (6.72)	1.07 (0.042)	0.61 (2)	305 (12)	286
				203 (8)	190
				152 (6)	143
				102 (4)	95
				51 (2)	48

Figure 1. Traversing Target Effectivity Testing – Top View

Figure 2. Traversing Target Effectivity Testing Setup – Front View

Figure 3. Verification and Correlation Setup – Traversing Lance

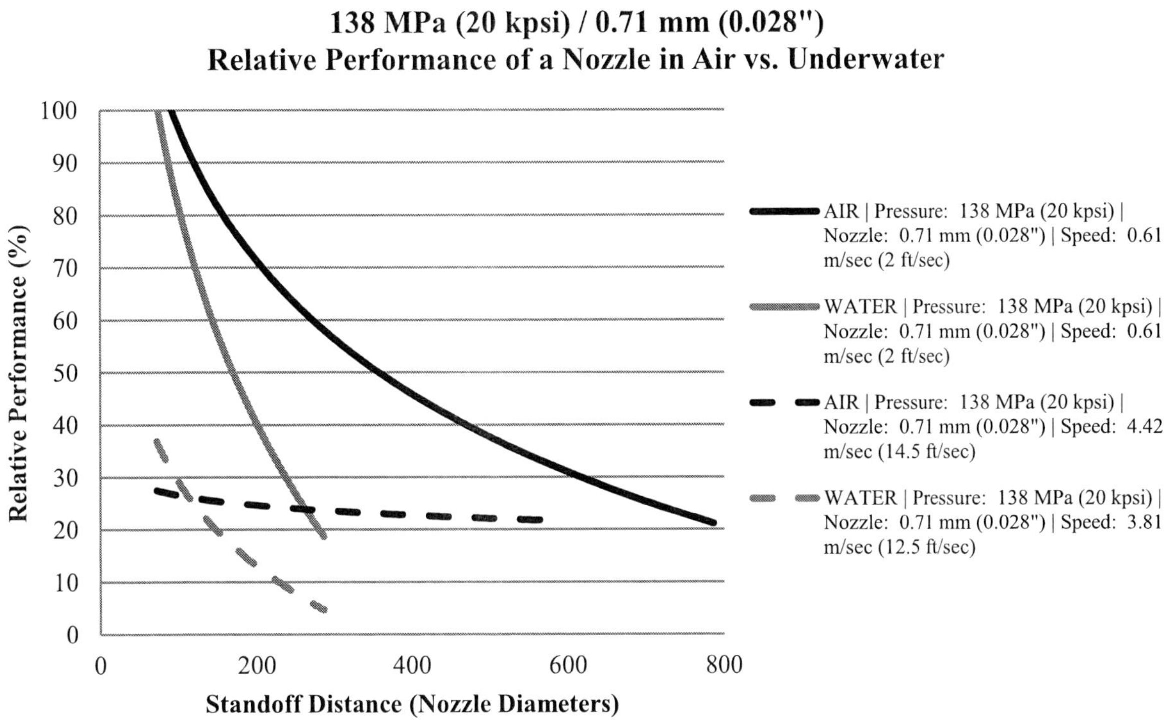

Figure 4. 138 MPa (20 kpsi) / 0.71 mm (0.028") Relative Performance

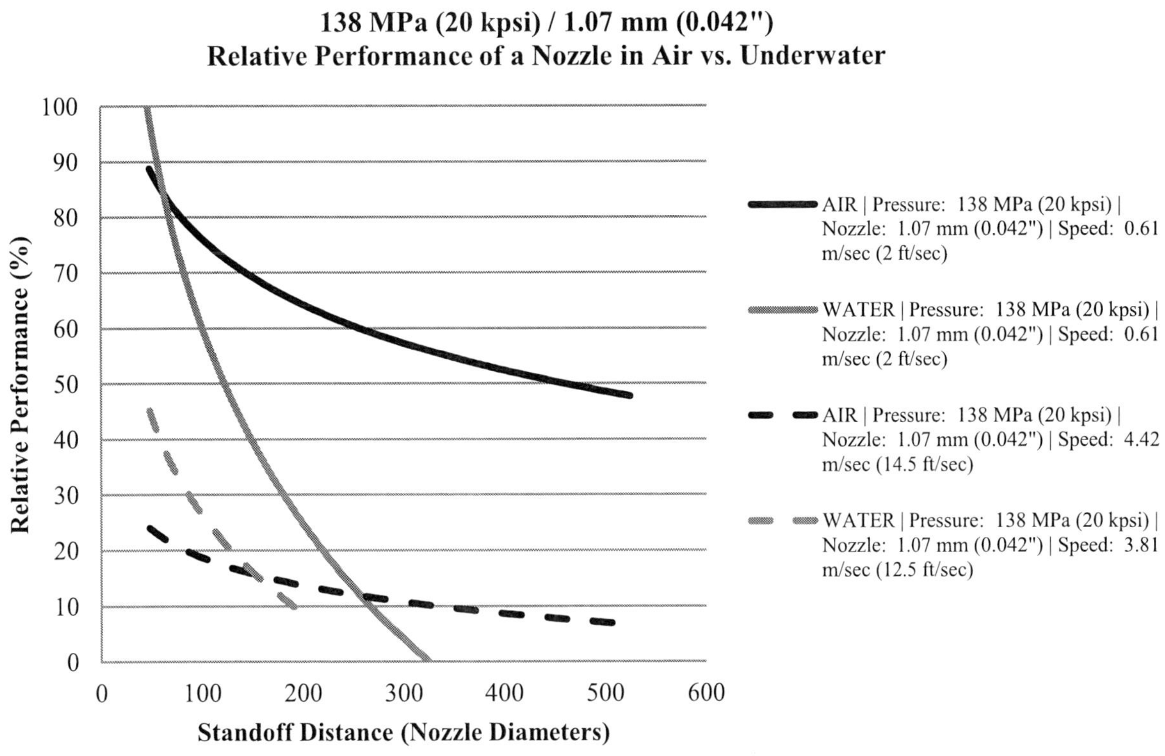

Figure 5. 138 MPa (20 kpsi) / 1.07 mm (0.042") Relative Performance

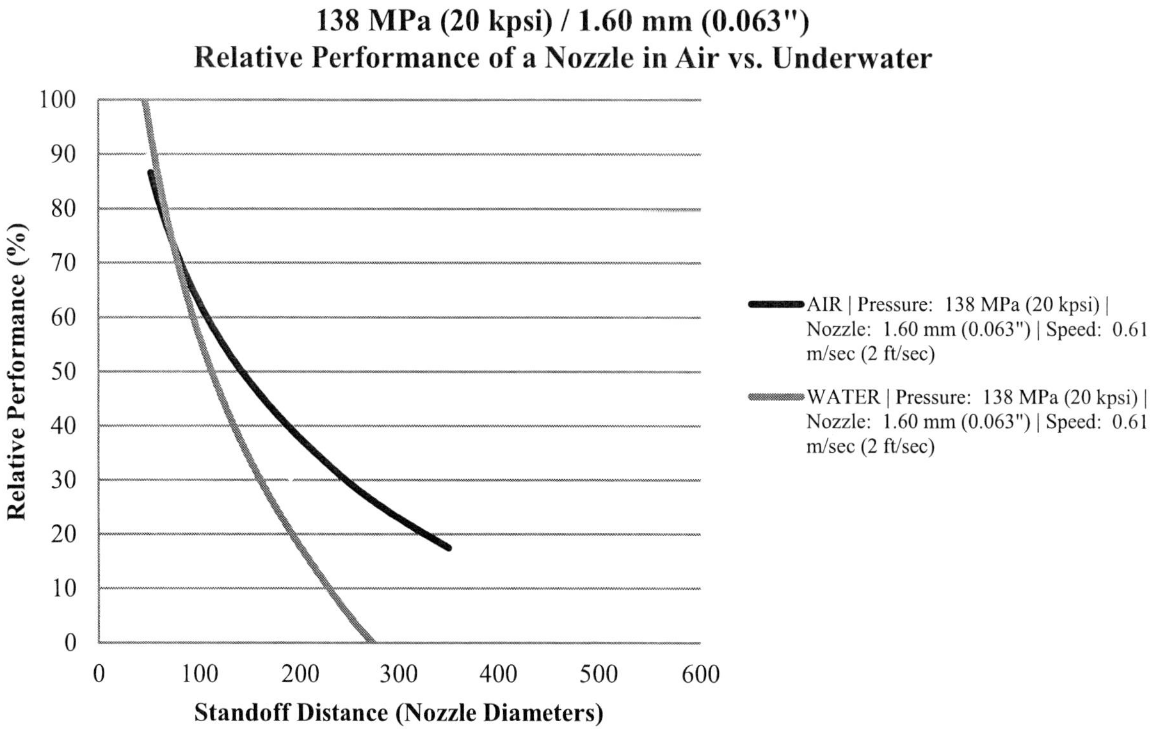

Figure 6. 138 MPa (20 kpsi) / 1.60 mm (0.063") Relative Performance

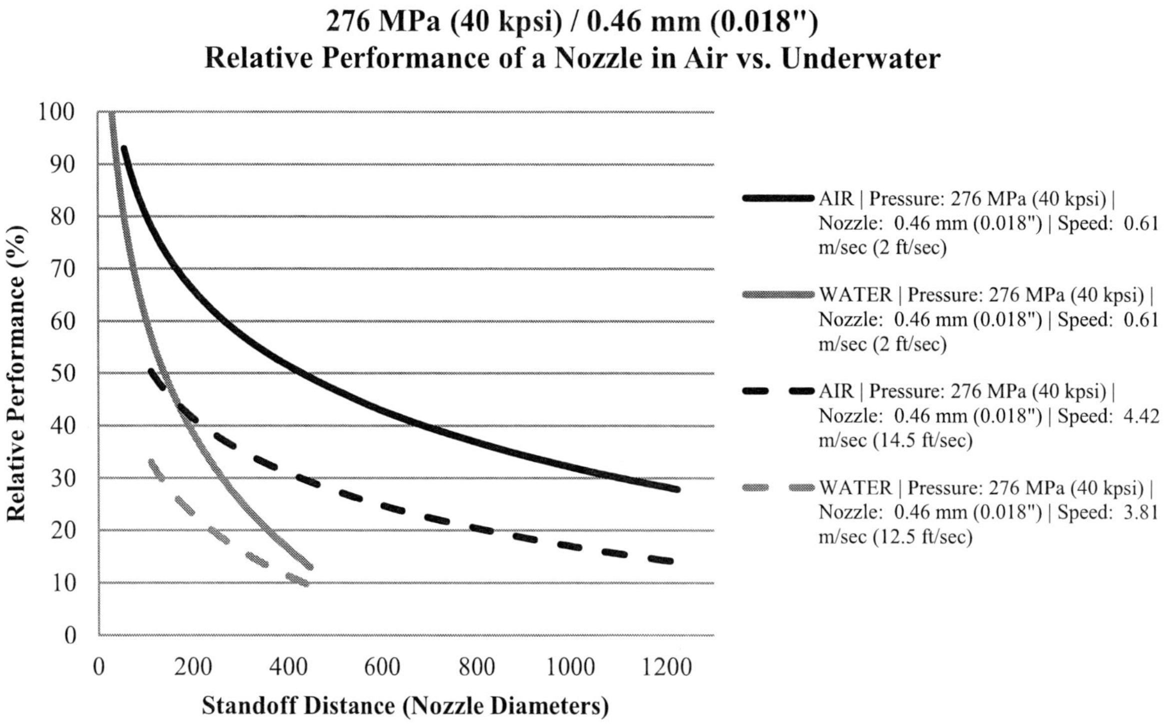

Figure 7. 276 MPa (40 kpsi) / 0.46 mm (0.018") Relative Performance

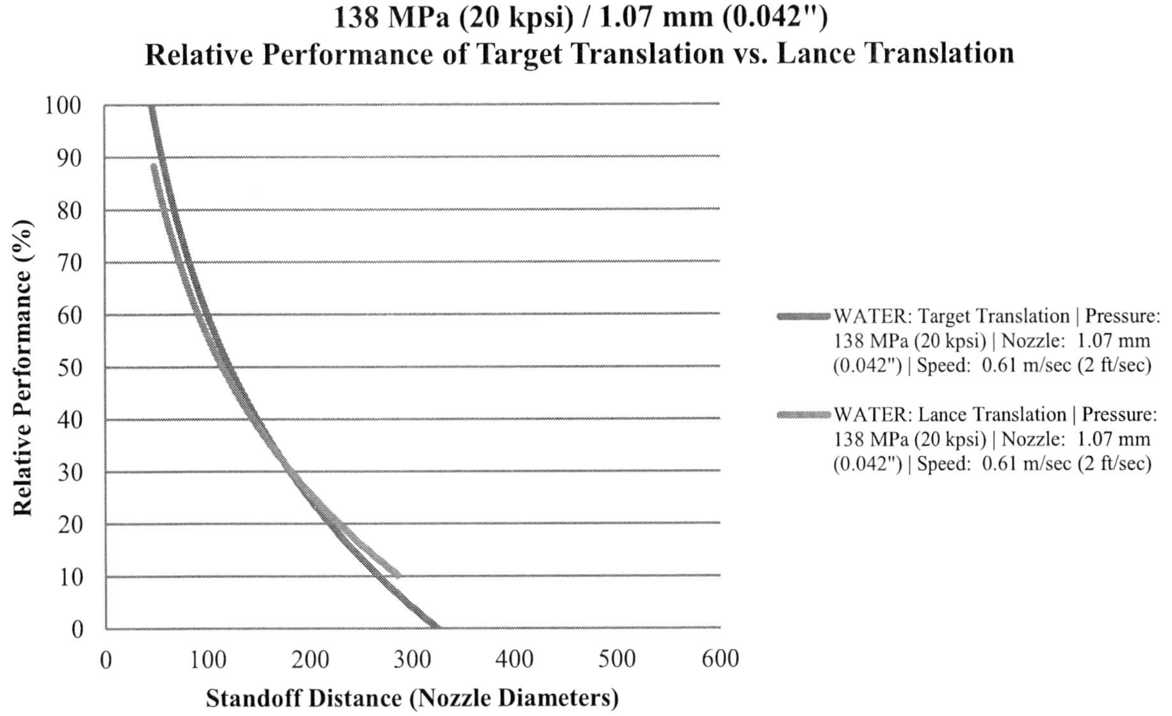

Figure 8. 138 MPa (20 kpsi) / 1.07 mm (0.042") Verification and Correlation

2015 WJTA-IMCA Conference and Expo
November 2-4 • New Orleans, Louisiana

Paper

THE EFFECT OF UPSTREAM FEEDER TUBE DIAMETER AND LENGTH ON WATERJET PERFORMANCE IN CLEANING APPLICATIONS

D. Wright
StoneAge, Inc.
Durango, Colorado, U.S.A.

ABSTRACT

The use of tubing to support and feed a waterjet nozzle is a common and easily achieved method of bringing the nozzle closer to the surfaces to be cleaned in large tanks, vessels and pipe lines. The use of such extensions brings the added benefit of improving the upstream conditions and producing a higher quality and further carrying jet stream. In a 1972 publication Hydrodynamics of High Pressure Fine Continuous Jets, Shavlovsky stated that the ideal inner diameter of this upstream tubing would be 8 to 10 times the diameter of the waterjet orifice, and the optimum length would be 40 to 50 times the inner diameter of the tubing. In today's waterjet cleaning, with high flow rates, there are practical limitations that make it difficult to achieve upstream diameter ratios only as great as 5 times the orifice diameter, and limitations on vessel opening size or internal clearances are more likely to determine the length that can be used.

The purpose of this research was to evaluate the effects on jet performance through testing the proportions of upstream tubing and orifice sizes commonly in use. The ratios of tubing length, tubing inside diameter and orifice diameter were varied to determine their relationships and combined effects on jet performance, and which relationships in combination are best optimized.

Organized and Sponsored by WJTA®-IMCA®

1. INTRODUCTION

The effectiveness and range of a high pressure waterjet can be as dependent on upstream conditions as the design of the nozzle orifice used to form the jet. Upstream conditions can vary from an orifice installed radially in a nozzle head for small pipe cleaning, to the orifice being placed axially in the end of a long straight pipe or tube, commonly called an extension arm, on a vessel cleaning tool such as a 3D or 2D swivel as illustrated in Figures 1 and 2. Within this range of upstream condition types, there exist relationships between the proportionate size of the orifice diameter relative to the inside diameter of the tube, and the length of the tube proportionate to the inside diameter of the tube.

In field applications, there is often a limitation to the maximum length of straight tube or extension arm that can be utilized due to internal obstructions of the vessel being cleaned. This also creates the need to have effective jet impact at ever greater standoff distances. While the orifice size is predetermined by the desired operating pressure and flow, and the maximum length of tubing behind the orifice is determined by physical constraints, it is still possible to select an inside diameter for the tubing feeding the orifice. The purpose of this testing was to determine the optimum relationship between the orifice size, tubing length and tubing inside diameter to produce the most effective jet impact.

2. TEST ARRANGEMENT

The test parameters consisted of tubing sections with inside diameters of 7.9 mm (.312"), 11.1 mm (.438"), 15.2 mm (.599"), 20.7 mm (.815") and 27.9 mm (1.10"). Nozzle holders were fabricated that allowed the use of the same carbide orifice insert for all tests fitted to the ends of the different tubing sections, as shown in Figure 3. Tubing lengths of 102, 203, 305, 610 and 914 mm (4, 8, 12, 24 and 36 in.) were utilized in each tubing size. Orifice diameters of 1.6, 2.3 and 3.2 mm (.063, .090 and .125 in.) were tested with each tubing combination. All tests were conducted at 69 MPa (10,000 psi), traversing the jet at 305 mm/second (12 in./second) across machinable wax samples placed at standoff distances of 100, 400 and 700 times each orifice diameter. The depth of cut produced in the wax was measured and averaged for the three standoff distances to provide a value for relative jet impact.

3. TEST RESULTS AND ANALYSIS

3.1 Feeder Tube Inside Diameter

The result of varying the inside diameter of the feeder tube at each tubing length is shown in Figures 4 through 7. The performance shown for each orifice size is relative to itself, not overall performance. For all lengths and orifice sizes, there are general improvements in performance with increasing inside diameter, up to an optimum before falling off. In the shorter length, an optimum occurs near the 15.2 mm (.599") inside diameter before performance deteriorates with increasing inside diameter. As the length is increased, the optimum shifts further to a larger inside diameter of 20.7 mm (.815"), indicating that a relationship exists between inside diameter and length. Relative to orifice size in the range tested, the curves generally show the same trends through the shorter lengths when expressed relative to inside diameter. At the longest length tested, the 2.3 mm (.090") orifice shows continued improvement at the 27.9 mm (1.10") inside diameter.

3.2 Feeder Tube Length and Inside Diameter

Figures 8-10 express the relationship of feeder tube length and inside diameter, showing the performance gains due to both parameters for a given orifice size. The horizontal axis in this chart series is displayed in terms of the ratio of the inside tube diameter to the orifice diameter.

The performance improvement gained by increasing length is greater with increasing orifice size, with a gain of 25% for the 1.6 mm (.063") orifice, 50% for the 2.3 mm (.090") orifice, and nearly 60% for the 3.2 mm (.125") orifice, showing a greater dependence on upstream conditions with larger orifice sizes and proportionally higher flow rates.

The average improvement due to increasing the inside diameter of the feeder tube across all lengths and orifice sizes was 24%. The greatest gains occurred within the 305 and 914 mm (12 and 36 in.) lengths and the 3.2 mm (.125") orifice size, each with a 32% improvement in relative performance.

The other trend visible in these relative curves is the shift in optimum inside diameter to larger bore sizes with increasing feeder tube length. For the 1.6 mm (.063") orifice size, the optimum stays within a 10 to 13 times ratio for inside tube diameter to orifice diameter; for the 2.3 mm (.090") orifice size the optimum occurs at a 7 to 9 times ratio for the shortest length and shifts to 9 to 12 times at the longest length. The 3.2 mm (.125") orifice shows an optimum at just less than 5 times this same ratio with the shortest length, shifting up to 6.5 times at the longest length tested. Continuing on this trend would predict that with longer allowable lengths, performance would improve further with larger inside diameters.

3.3 Feeder Tube Length Proportional to Inside Diameter

Feeder tube length can also be expressed as a relative proportion of the inside diameter, utilized in Figures 11 through 14, in combination with expressing the inside diameter as a relative proportion of orifice size. At lengths equivalent to a ratio of 12:1, the curves for the three orifice sizes show a trend toward a peak performance at nozzle diameter ratios in a range of 10 to 12 times the inside diameter. With an increase to ratios of 20:1 and 28:1, the small orifice size continues to show a peaking curve, while the curves for the two larger orifice sizes begin to trend more strongly upward and further toward higher inside diameter ratios. At a ratio of 40:1, all three curves are still trending upward and toward the higher inside diameter ratio, again indicating that with increasing allowable length, the optimum inside diameter ratio to orifice size would continue to increase beyond the optimums found in this range of testing.

4. CONCLUSIONS

The purpose of this testing was to determine the effect that the inside diameter of a tube feeding a waterjet nozzle has on jet quality, as well as to determine if there was a relationship between the inside diameter and the overall length of the tube. The results show a dependent relationship between the orifice size, the inside diameter, and the length of the feeder tube. Performance gains due to feeder tube length can be on the order of 50 to 60%, while performance gains due to optimization of the inside diameter of the feeder tube relative to the orifice size can be on the order of 20 to 30%. In shorter feeder tube lengths combined with larger orifice sizes, the optimum inside diameter may be as low as 5 to 7 times the orifice size, while increasing the length of the feeder tube shifts the optimum inside diameter into a range of 7 to 13 times the orifice size.

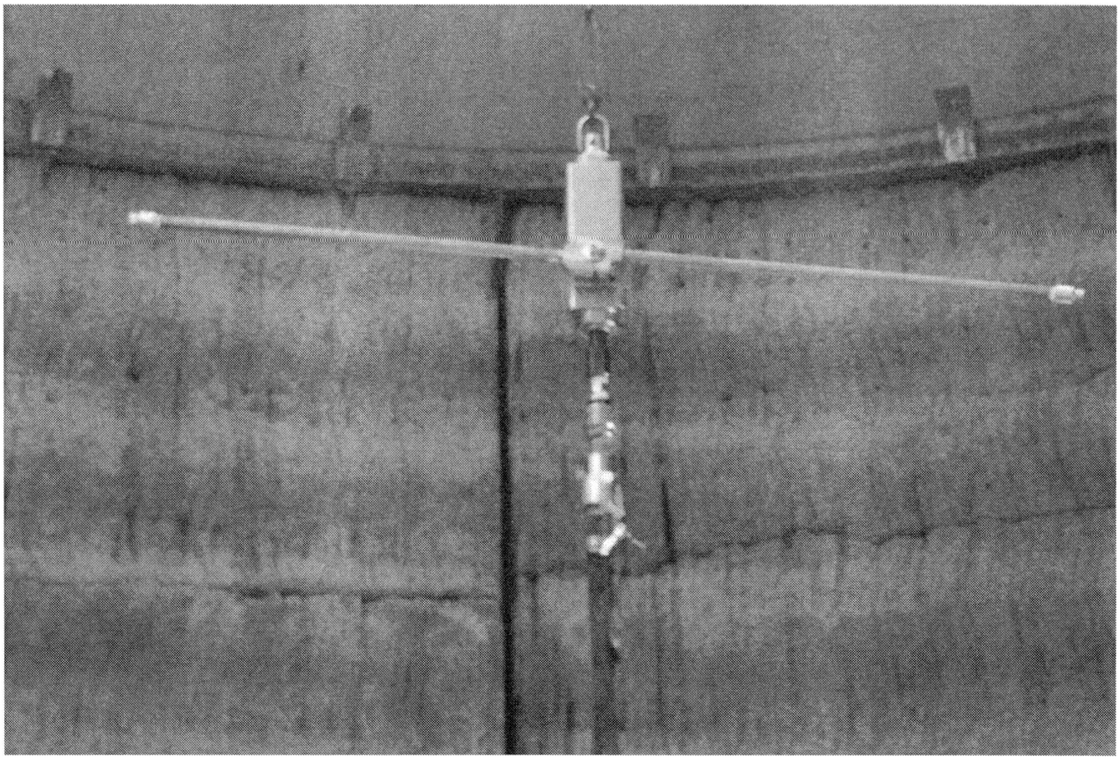

3D Tool with Long Extension Arms Used in Vessel Cleaning
Figure 1.

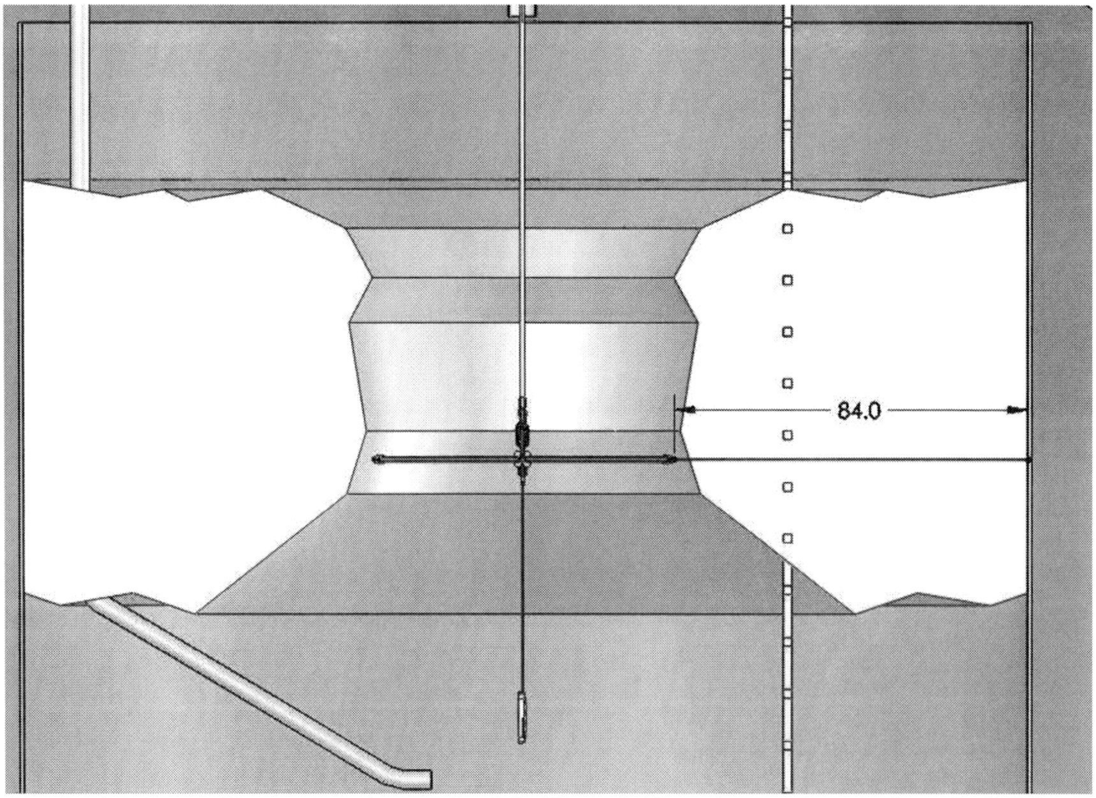

2D Tool with Long Extension Arms Used in Vessel Cleaning at Large Standoff Distances
Figure 2.

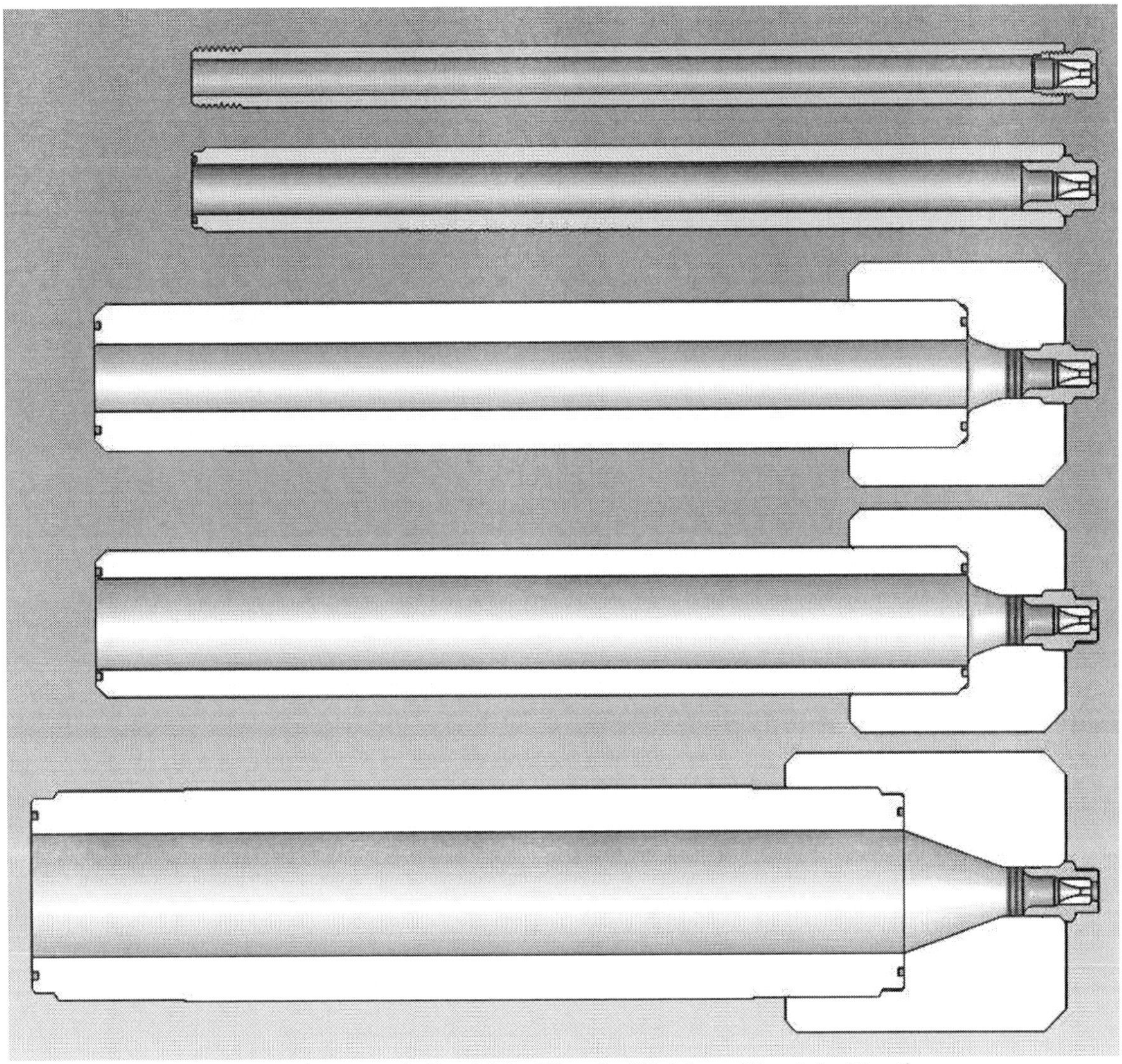

Feeder Pipes and Carbide Nozzle Holders Used in Testing
7.9 mm (.312"), 11.1 mm (.438"), 15.2 mm (.599"), 20.7 mm (.815"), 27.9 mm (1.10")
Figure 3.

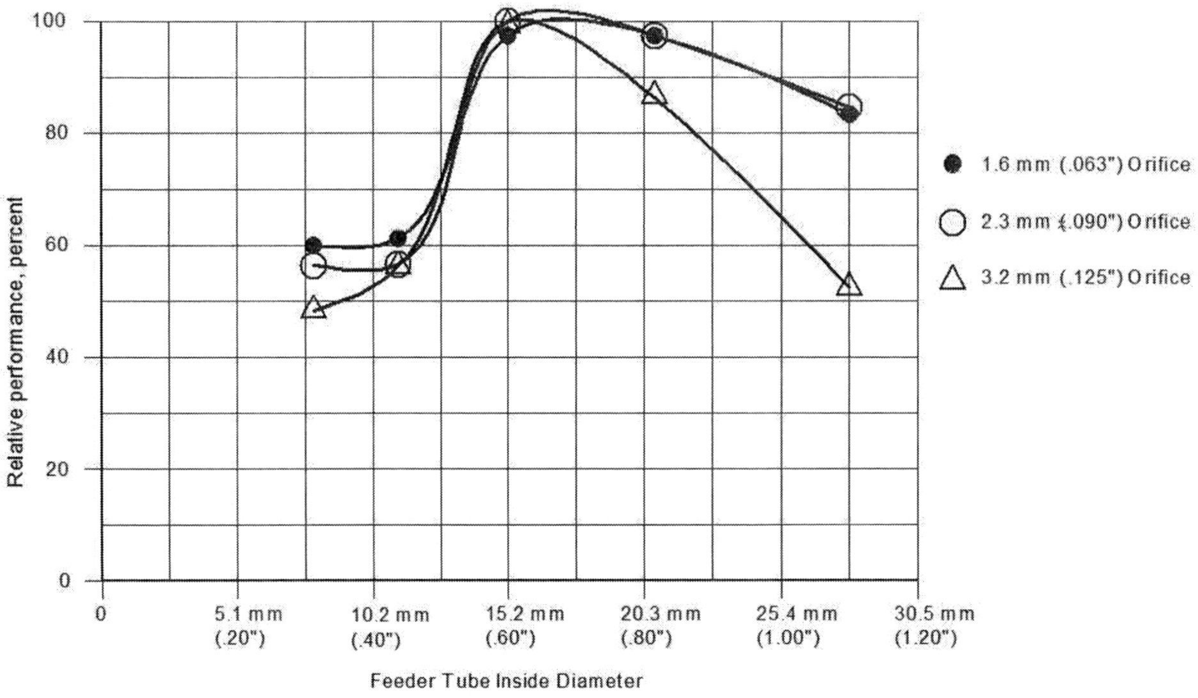

Relative Performance with Increasing Feeder Tube Inside Diameter at a Feeder Tube Length of 102 mm (4 in)
Figure 4.

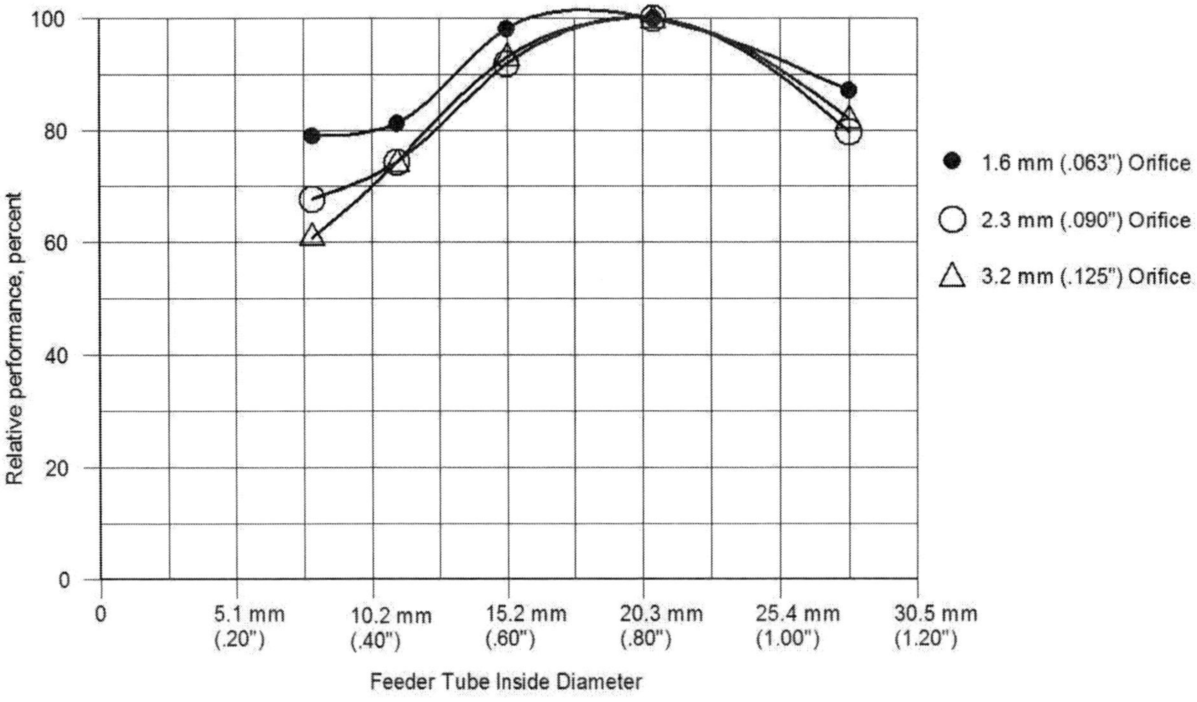

Relative Performance with Increasing Feeder Tube Inside Diameter at a Feeder Tube Length of 203 mm (8 in)
Figure 5.

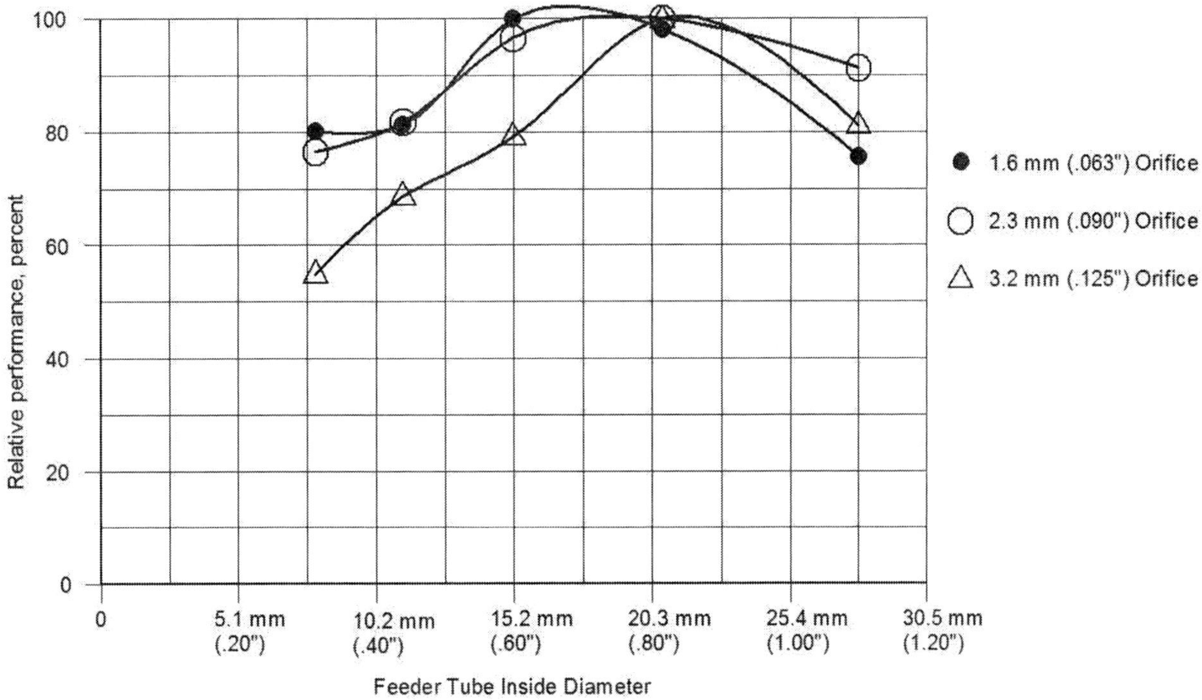

Relative Performance with Increasing Feeder Tube Inside Diameter at a Feeder Tube Length of 305 mm (12 in)

Figure 6.

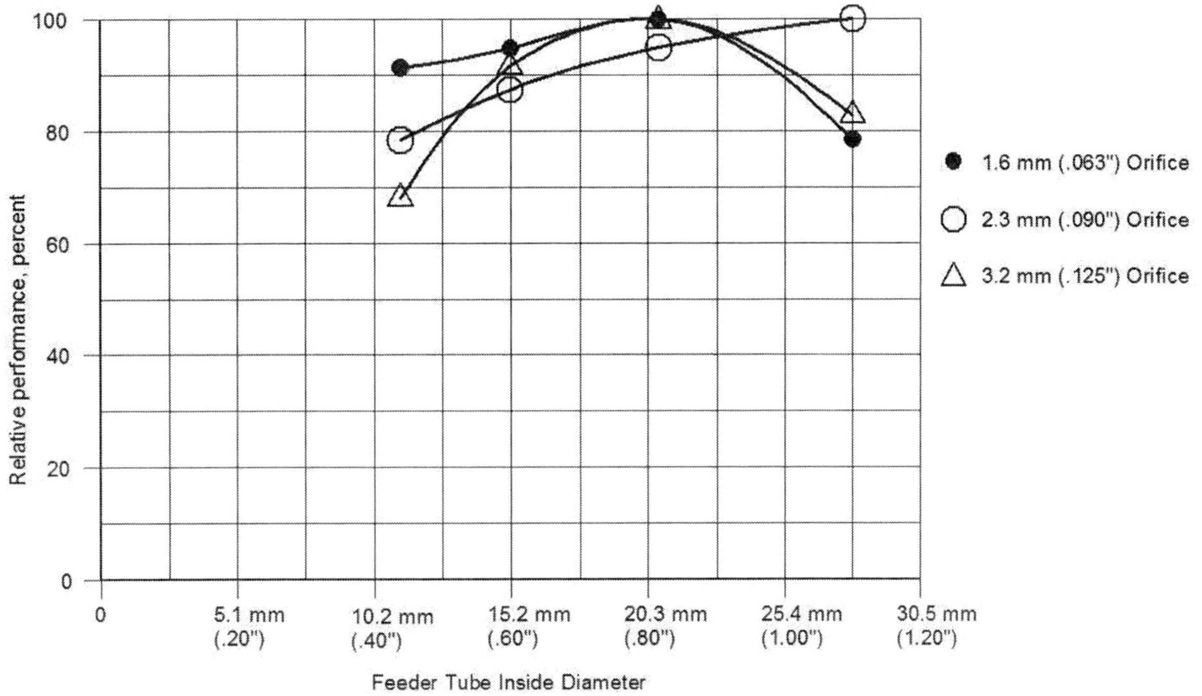

Relative Performance with Varying Feeder Tube Inside Diameter at a Feeder Tube Length of 914 mm (36 in)

Figure 7.

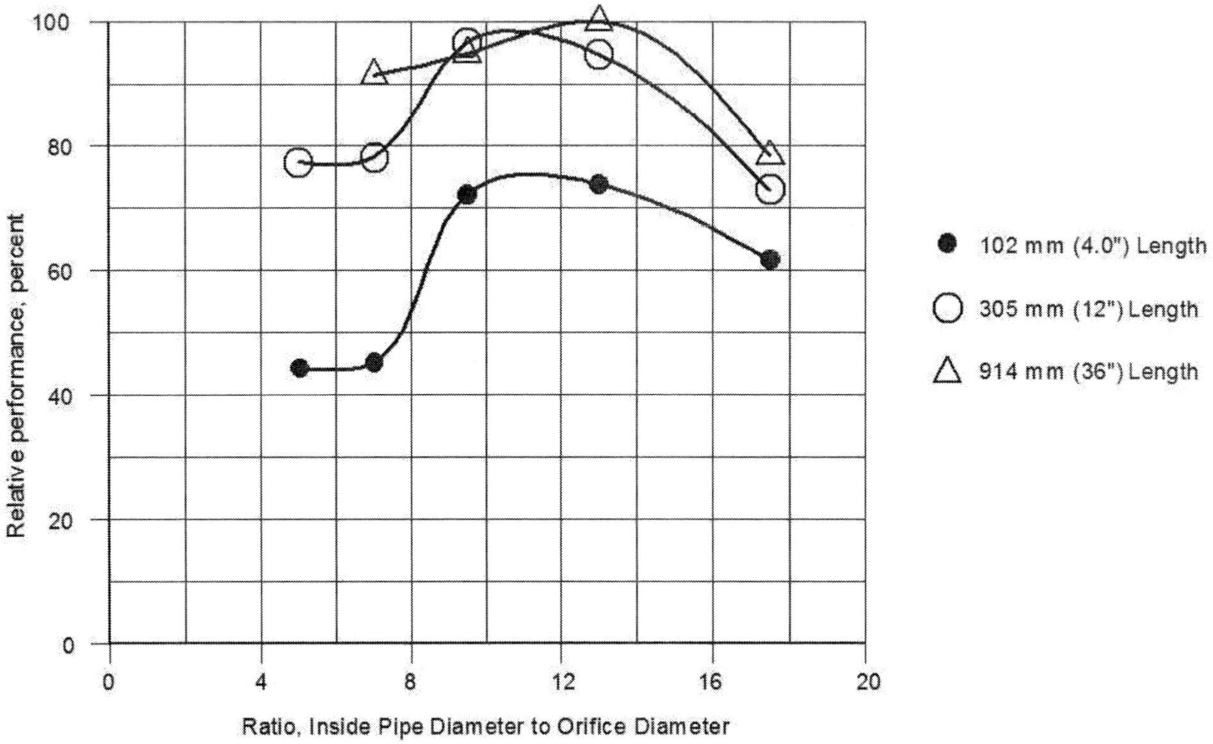

Relative Performance with Increasing Feeder Tube Inside Diameter at 1.6 mm (.063") Orifice Size
Figure 8.

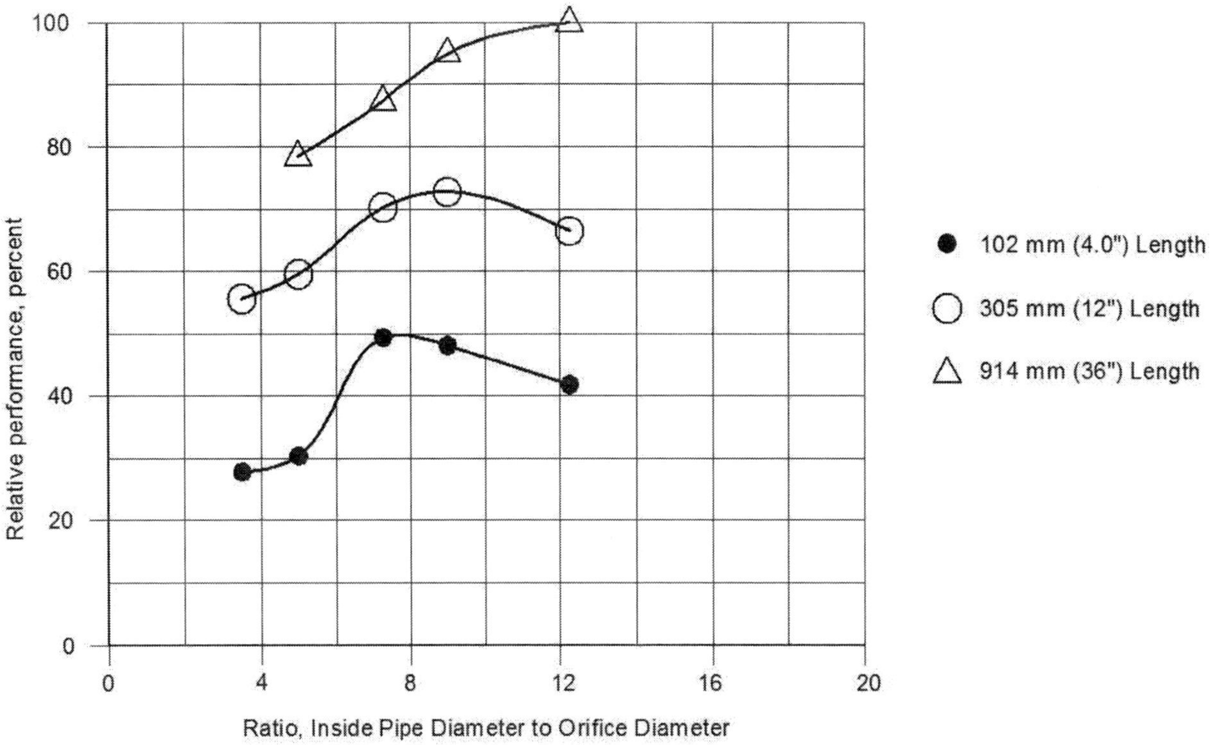

Relative Performance with Increasing Feeder Tube Inside Diameter at 2.3 mm (.090") Orifice Size
Figure 9.

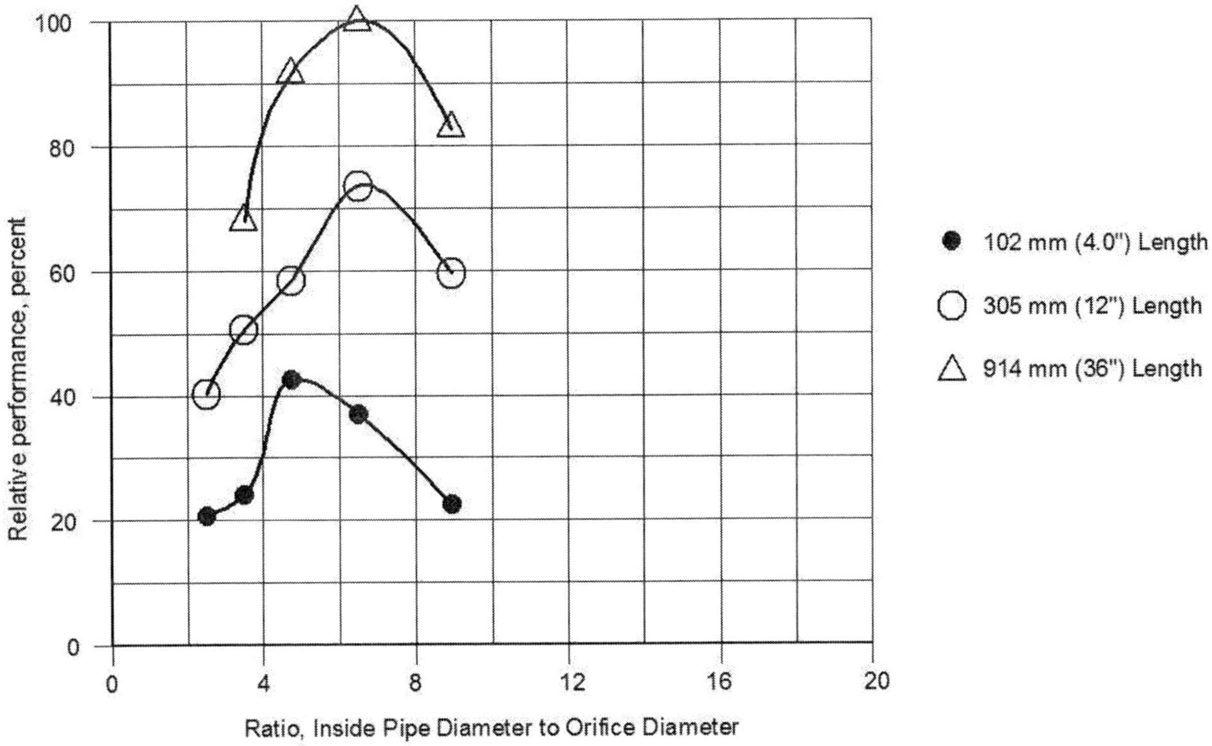

Relative Performance with Increasing Feeder Tube Inside Diameter at 3.2 mm (.125") Orifice Size
Figure 10.

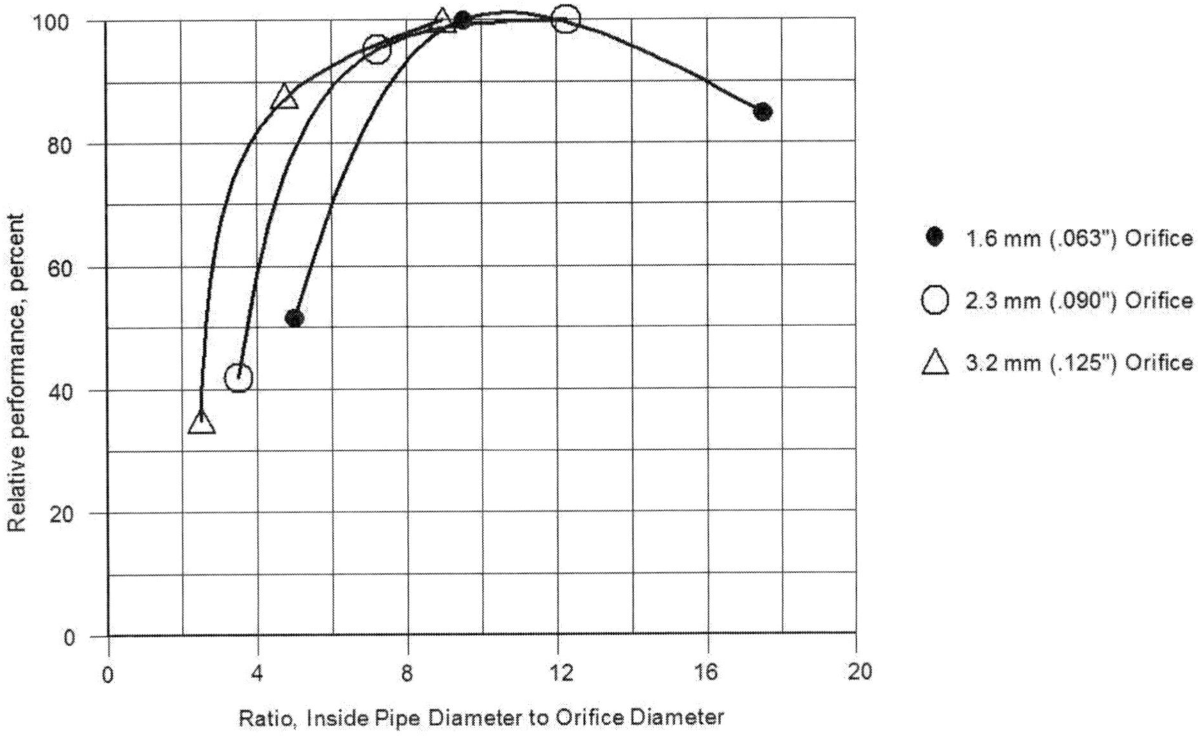

Relative Performance at a Feeder Tube Length to Inside Diameter Ratio of 12:1
Figure 11.

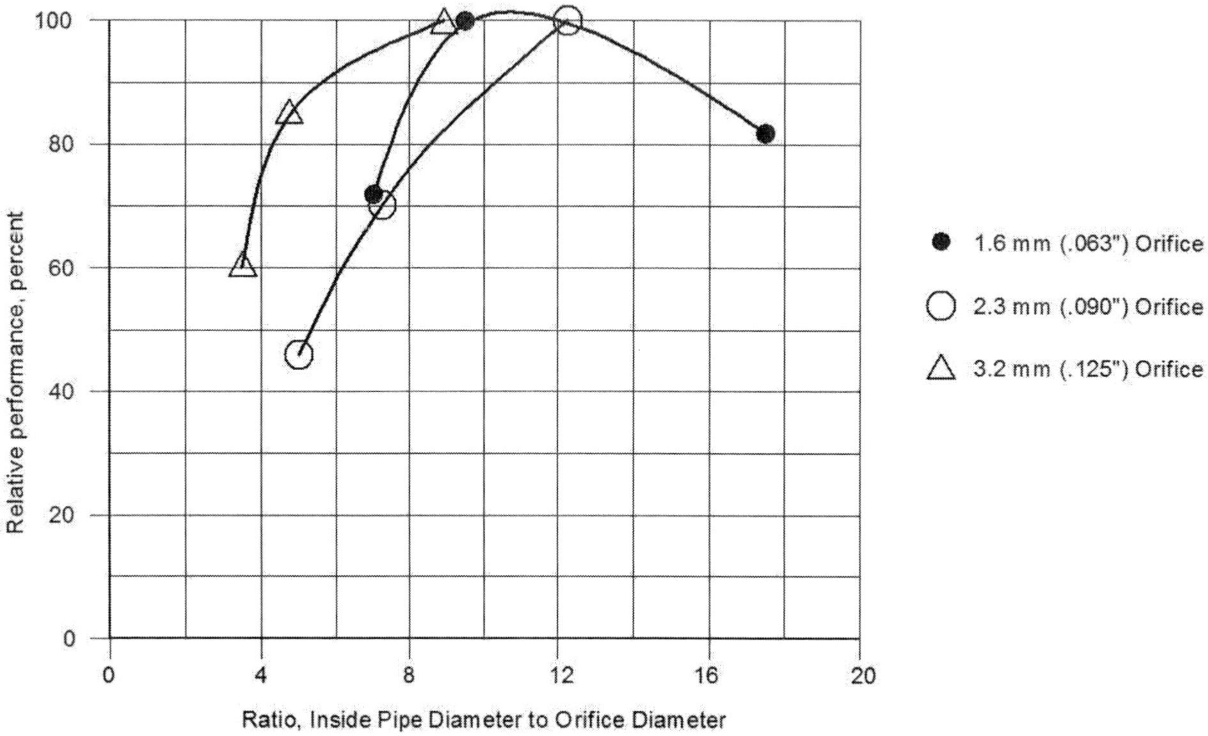

Relative Performance at a Feeder Tube Length to Inside Diameter Ratio of 20:1
Figure 12.

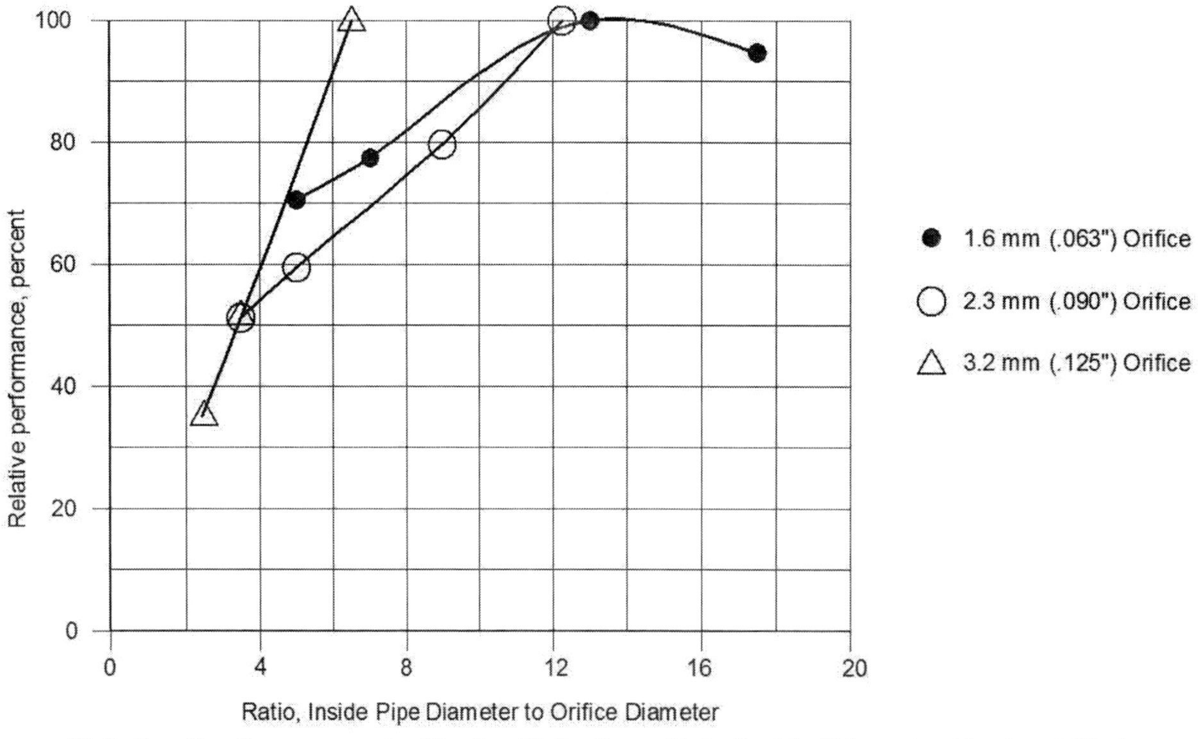

Relative Performance at a Feeder Tube Length to Inside Diameter Ratio of 28:1
Figure 13.

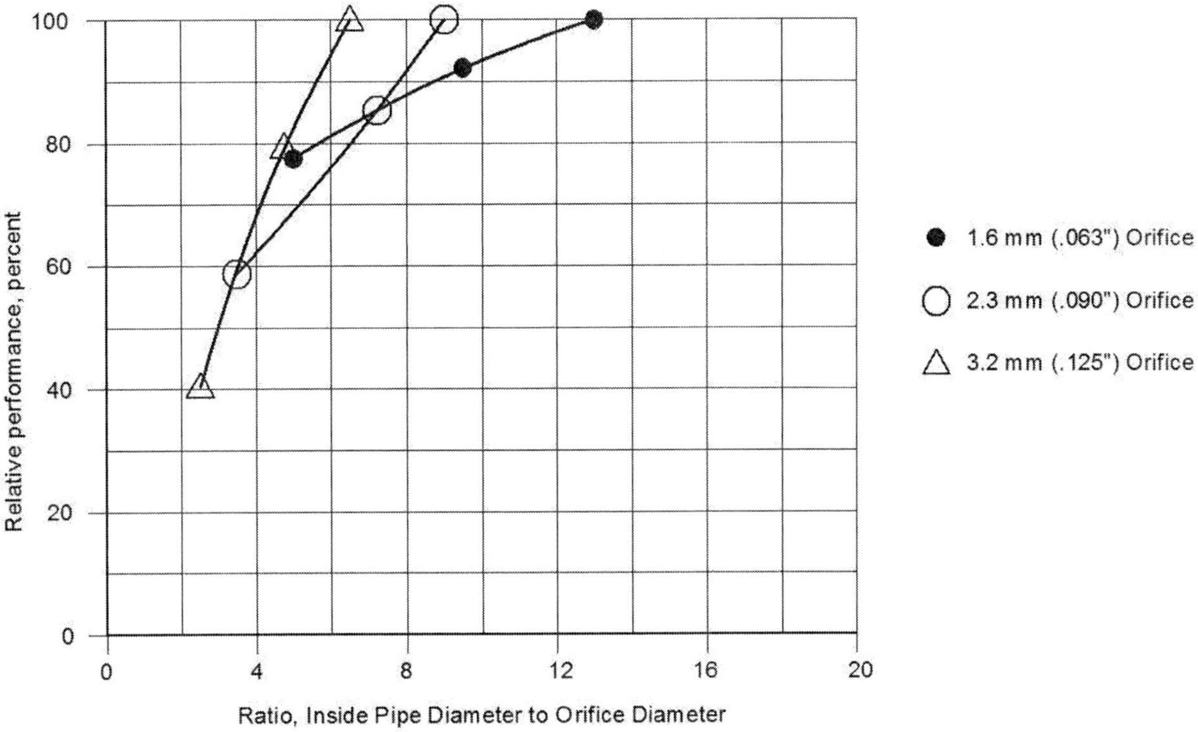

Relative Performance at a Feeder Tube Length to Inside Diameter Ratio of 40:1
Figure 14.

2015 WJTA-IMCA Conference and Expo
November 2-4 • New Orleans, Louisiana

Paper

DESIGN PROTOCOLS OF CAVITIES CREATED BY WATERJET BOREHOLE MINING OF THIN-SEAM DEPOSITS

Yasser Akbarzadeh and Hugh B. Miller
Colorado School of Mines
Mining Engineering Department
Golden, Colorado, USA

ABSTRACT

The economic extraction of thin-seam coal deposits are often problematic due to several significant limitations associated with conventional mining methods, operating practices, and equipment. It appears that the most prudent way to extract these resources in a more economical way is through the development of technology to remotely extract these resources from the surface, where waterjet borehole excavation represents a novel approach. There are a number of technical challenges that must be overcome to advance the concept of in-situ waterjet borehole mining of non-soluble resources to a commercially viable stage. Paramount among these include the continued technical advancement of drilling and excavation systems, the mechanisms used to crush, bail, and transport cuttings from the borehole, the required instrumentation to effectively control and monitor the mining process, and a technical understanding between cavity formation and stability for a given set of operating characteristics and geomechanical rock properties. Understanding the structural dynamics of these cavities is a key element in designing a mining system capable of sufficient resource recovery to economically justify the capital investment. The unintended collapse of these cavities could potentially result in the detrimental loss of mineral reserves, as well as surface subsidence.

Organized and Sponsored by WJTA®-IMCA®

1. INTRODUCTION

The economic extraction of thin-seam coal deposits with thicknesses of less than 1 meter are often problematic due to several significant limitations associated with conventional mining methods, operating practices, and equipment. Mines with low-seam heights are endemic of operations that possess low labor productivities, high operating costs, and relatively small production capacities. Furthermore, the ability to implement new equipment and automation in order to efficiently exploit these thin-seams is hampered by the limited cash-flow positions of most of these operations and the inability to amortize their high costs over a sufficiently large resource base. Consequently, these mines are usually small, labor intensive, and rely extensively on used and rebuilt equipment modified to operate in these challenging work environments.

Britton [2] has investigated the productivity of thin-seam mining, where he attributes the inherent low productivity of these operations to the small tonnage produced per unit length of linear advance and the challenges associated with mining in a confined work environment, particularly with regards to material handling, unit operations at the face, and logistics. Another pragmatic issue identified by Britton is ventilation, where obtaining the required airflow at specific work areas can be difficult due to the obstruction of a large percentage of openings by equipment.

Fotta et al [3] did research to identify the types of injuries common to operating mines that exploit thin-seam resources. Due to their research objectives, longwall operations and large mines that employing more than 50 workers were specifically excluded from this study. Fotta found a direct correlation between working height and worker safety principally attributed to restricted employee mobility, reduced vision and poor posture, and limitations of using protective canopies and other common engineering safety devices/controls as a consequence of space constraints. [3]

Building upon Fotta's research, Peters et al [4] studied significant potential hazards that are substantially dependent upon coal seam height. The study focused on accidents that occurred between 1990-1996, where 117 workers were fatally injured at small (non-longwall) underground bituminous coal mines. The three types of incidents that were responsible for a preponderance of these fatalities were associated with: 1) roof falls, 2) powered haulage, and 3) machinery. Based on the same research, the six most prominent types of incidents responsible for non-fatal day's lost (NFDL) accidents were: 1) handling materials, 2) machinery, 3) powered haulage, 4) slips and/or falls of person, 5) roof falls, and 6) non-power hand tools.

Figures 1 and 2 illustrate the fatal and non-fatal day lost incident rates for underground coal mines between 1998-2008. Based on the number of employees, U.S. underground coal mines have been divided into two distinct categories: small (employing less than 50 employees) and large (employing greater than 50 employees). Using data obtained from the Mine Safety & Health Administration (MSHA) Injury Experience Reports for Coal, Figure 1 shows, with the exception of 2007, the fatal incident rates for small underground mines are higher as compared to larger operations.

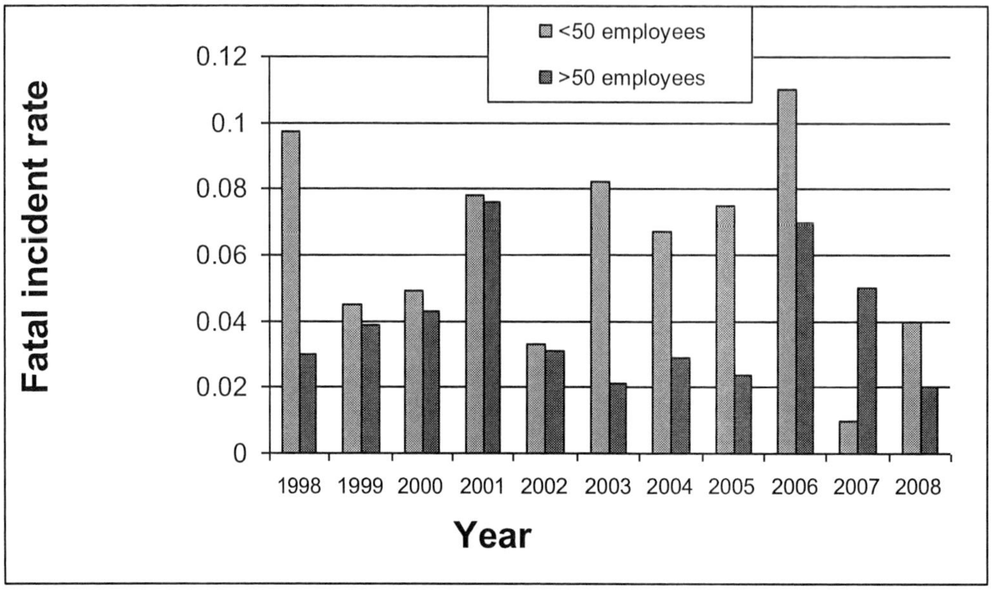

Figure 1. Fatal incident rate for small and medium/large underground coal mines for 1998-2008. [1]

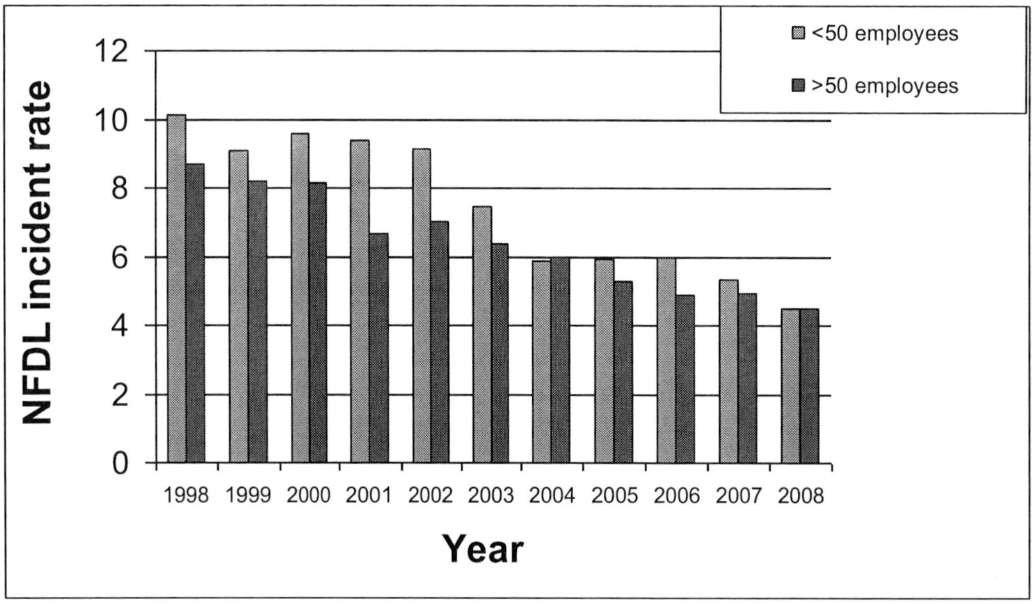

Figure 2. Non-fatal day lost incident rate for small and medium/large underground coal mines for 1998-2008. [1]

During 2008, an estimated 5.2 million tons of coal were produced in the U.S. from thin seams deposits with operating heights less than 1.06 m (42 inches). This represents approximately 14.5% of the total U.S. underground coal production. Of the 657 underground coal mines operating in 2008, nearly 70% of them have less than 50 employees. This is consistent with previous research that indicates 94% of mines operating in seams of 1.06 m or less employed fewer than 50 people [2].

Papas et al [5] conducted research that focused on injuries attributed to roof and rib falls that occurred during the period 1995-1998 in all the U.S. underground coal mines. As expected, this study concluded that mines operating in thin seams (< 1.09 m) tended to be smaller scale operations that solely used room-and-pillar extraction methods. The research showed that small mines in thin seams had a ground fall fatality rate that is 44% greater than the national average (Figure 3), while small mines operating in thick seams had a ground fall fatality rate that is 53% lower than the national average. The consequence of this data has led to the conclusion that small mines in themselves are not a contributing factor in the incidence of rock fall fatalities, whereas seam height appears to be a dominant and influential factor. The study also indicates that the potential root-causes of these fatalities largely stem from the lack of engineering controls specifically suited for these small, low seam environments, such as protective cabs and canopies on operating equipment. In these operating environments, workers are especially at risk should a massive roof or rib failure occur.

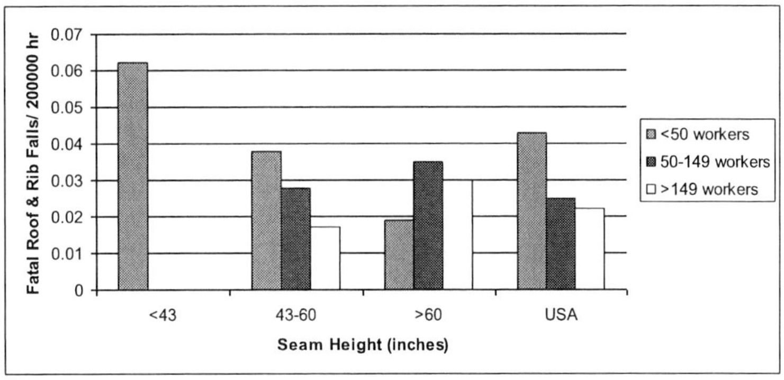

Figure 3. Roof and rib fall fatality rates by mine size and seam height for room-and-pillar mines, 1995-1998. [5]

Despite these challenges, a number of companies are focusing on ways to exploit these narrow, tabular deposits. In specific geographical locations whose economies are heavily dependent on coal mining, the depletion of thick seams have placed greater pressure to develop innovative technologies to economically recover coal resources from much thinner deposits. An examination of world coal resources also shows that a substantial percentage is contained in thin-seam deposits. Shiua [10] estimates that approximately 96% of China's total coal production is from underground mines, where more than half of these reserves are reside in thin seams less than 0.7m. As such, these thin-seams are of strategic importance and represent a significant source for supplying the country's future energy needs. However, advancements in safety and extraction methods are requisites for exploiting these resources. Coupled with competing demands for surface land-use, it appears that the most prudent way to achieve these objectives is through the development of technology to remotely extract these resources from the surface. Displacing workers from the underground work environment will eliminate the inherent hazards of mining these deposits and reduce the direct costs associated with ventilation, support, and equipment.

Over the last three decades, a number of researchers have been working on the development of in-situ applications for the extraction of coal and other economic minerals through boreholes. For example, Dr. George Savanick of the U.S. Bureau of Mines conducted experiments using in-situ borehole mining equipment on several different commodities including coal, gold, and industrial minerals, beginning in the mid-1980s [6]. His test results for the borehole mining of coal showed that it was technical feasibility but not economically viable at the time. Through his work, several advantages associated with this method were quantified, including improved worker safety and the mitigation of adverse environmental impacts [6]. Wiley et al [8] also identified several advantages associated with borehole mining including: safety, the ability to work in remote areas, a reduction in adverse environmental impacts, greater mobility, increased selectivity, lower capital cost, improved system simplicity, and the ability to operate in a variety of ground conditions. According to his work, borehole mining could economically compete with conventional methods in areas that possess tabular structures, low ore grades, complicated hydrogeological conditions, hazardous operating conditions, and inaccessible locations. Wang and Miller [7] also identified potential reductions in the cost of social and environmental impacts over conventional mining systems in numerous applications.

Despite these potential benefits, there are a number of technical challenges that must be overcome to advance the concept of waterjet in-situ borehole extraction of non-soluble resources to a commercially viable stage. Paramount among these includes the stability of the induced cavity during the mining process. In several empiric experiments and field trails, waterjets have shown the ability to excavate cavities that are inherently unstable for a given rock type because of geometry, cross-sectional dimensions, alteration/damage to the wall-rock, or applied stress as a consequence of the mining sequence. The unintended collapse of these cavities could potentially result in the detrimental loss of mineral reserves, as well as surface subsidence, the incidence of significant dilution, and the loss of equipment. In addition, adverse alterations in cavity geometry caused by failure in the surrounding host rock will significantly hamper the ability to recover and bail fragmented mineral from the borehole. Cavity stability, in turn, is the product of a complex set of multi-dimensional variables that include in-situ stresses, rock properties, cavity geometry, time, and the rate and manner of excavation. A number of additional confounding issues associated with the proposed excavation methods (e.g., fluid pressurization of the cavity) may also adversely influence the stability of these cavities, where their potential effects need to be quantified [11].

2. DEFINITION OF BOREHOLE MINING METHOD (BHM)

In general terms, in-situ mining is defined as the physical extraction of the valuable components of a mineral resource through a borehole. In the context of this paper, Borehole Mining (BHM) is defined as a remotely operated method of extracting (mining) mineral resources through one or more strategically placed boreholes by means of high pressure fluid jets. This process can be conducted through a variety of operating configurations and drilling platforms, including conventional surface locations, sites within existing mines (both open pit and underground), and from floating vessels/rigs. A borehole is drilled from the surface to a desired depth, where the actual mining process will take place. After the hole has been drilled, a casing column is then inserted into the hole. The purpose for using casing depends on the method in which the cuttings are bailed (removed) from the hole. In most conventional applications, the casing provides hole stability, minimizes material loss, and reduces the potential for dilution caused by wall erosion or borehole failure. Since a cavity is formed as part of the normal mining process, a casing shoe is strategically positioned within the seam. Similarly, an ejector is situated at the bottom of the projected cavity to collect the excavated material and an inflatable packer is generally placed above the seam to regulate borehole pressure and bailing velocity. The BHM excavation tool is mounted to the drill string and is oriented relative to a predetermined excavation strategy [6, 7 and 8]. Figure 4 illustrates a cross-section of a conventional borehole mining system [9].

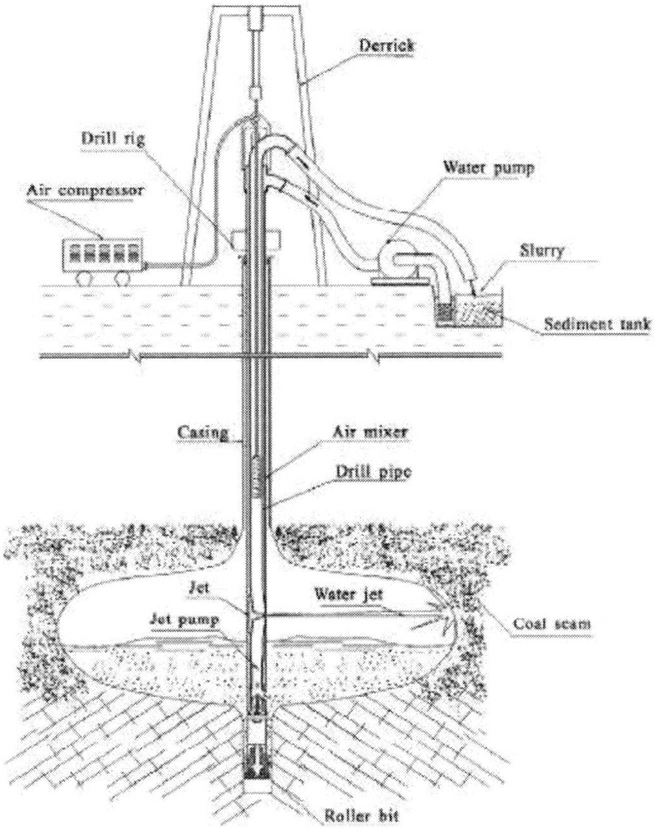

Figure 4. In-situ borehole mining procedure [9].

3. EFFECT OF CAVITY DIMENSION, INTERNAL PRESSURE, DEPTH OF COVER AND HORIZONTAL STRESS

The ultimate objective of this research was to establish a set of design protocols (guidelines) for estimating optimum cavity geometry and orientation for several critical excavation and geological factors. Based upon the positive results of the literature search and in an effort to address one of the prevailing technical challenges impeding the commercial applications of borehole mining in coal, a finite difference method (Flac2D) has been applied to do a parametric sensitivity study to investigate the effect of several parameters like internal pressure, cavity size, and in-situ stress, on the stability of a cavity developed through excavation of a mineral resource during borehole mining. In this stage, the research was focused on delineating the impact of critical factors associated with maintaining cavity stability during the mining process and how it pertains to the development of a design protocol.

This section describes the process of performing a stress analysis of a cavity excavated by BHM, and includes consideration of fluid pressurization of the cavity periphery. To accomplish this analysis, a finite difference method (Flac2D) was used. Figures 5 and 6 illustrate the geometry of the models. As Figure 5 shows, depth of cover, cavity length, and cavity height are represented as

H, L and D, respectively. To simplify the analysis, each side of the cavity is modeled independently and the application of internal fluid pressure is applied to only one side of the excavation. Figure 6 shows how internal pressure is applied inside the cavity.

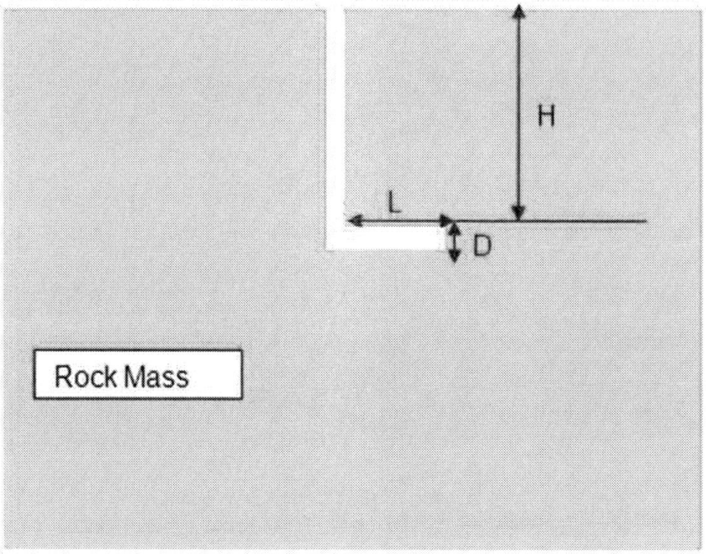

Figure 5. A view of model geometry [11].

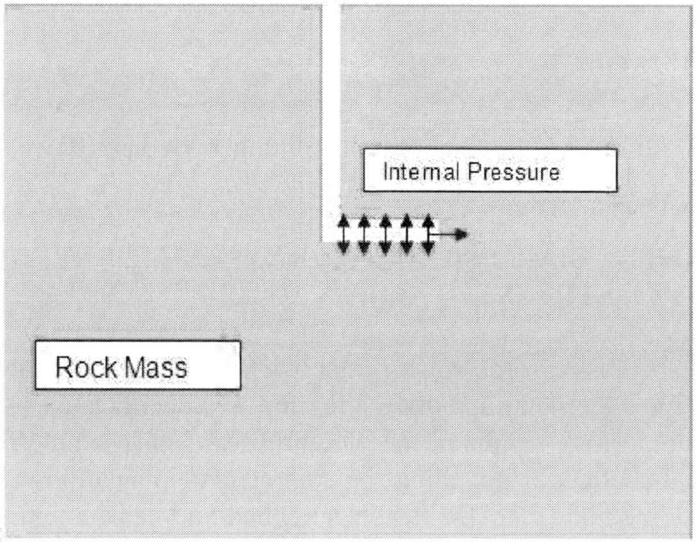

Figure 6. A view of under internal pressure (pressurization) [11].

Table 1 presents the mechanical rock properties that were applied in the models. For this study, the rock is assumed to be a homogeneous, isotropic material with no internal structure (e.g., bedding or jointing). The vertical in-situ stress is assumed to be gravitational, and equal to the unit weight multiplied by depth.

Table 1. Input mechanical properties [12].

Density	2700 kg/m^3
Elastic modulus (E)	11 Gpa
Poisson's ratio (ν)	0.3
friction angle (φ)	14.4°
cohesion (c)	38.4 MPa
tensile strength (σ_t)	14.4 MPa

While the impact of groundwater flow has been neglected in this analysis, the effects of pore pressure were considered. The CONFIG GW command and a groundwater bulk modulus of 1.9 GPa were used in this case. The density and tension limit of the water were assumed to be 1000 kg/m^3 and 2000 Pa. respectively, and the SET flow function was turned off in order to exclude groundwater flow. The INITIAL PP command sets the initial pore pressure distribution for this case. The geometry of the model was 60 × 120 m which was made of 30 × 60 grids. The modeling sequence consists of the following stages:

- Establish equilibrium conditions to initialize stresses.
- Excavate the borehole and initiating a cavity in horizontal direction.
- Add internal pressure to the cavity roof, floor and side-walls, and cycle to equilibrium.
- Alteration of different cavity sizes under different pressures and depth were modeled.

The bottom and sides of the mesh are fixed in both x and y directions. In addition, since the borehole will likely be cased in the actual application, the walls of the borehole were also fixed. The height of the cavity was considered as one grid equal to 2 m. Four different values were considered for cavity length ranging from 2 to 20 m (i.e., 2, 6, 10 and 20 m). The internal pressure values were selected as 1.5, 7.5, 22.5 and 75 MPa. The reason for choosing this range was to observe the sensitivity of different internal pressures on the maximum principal stress. In this analysis, two different depths (i.e., 30 and 40 m) and the ratio of horizontal to vertical effective stress (1.0) were also assumed. Figures 7 and 8 illustrate the maximum principal stress around the cavity roof for different internal pressures, cavity length, and depth of cover.

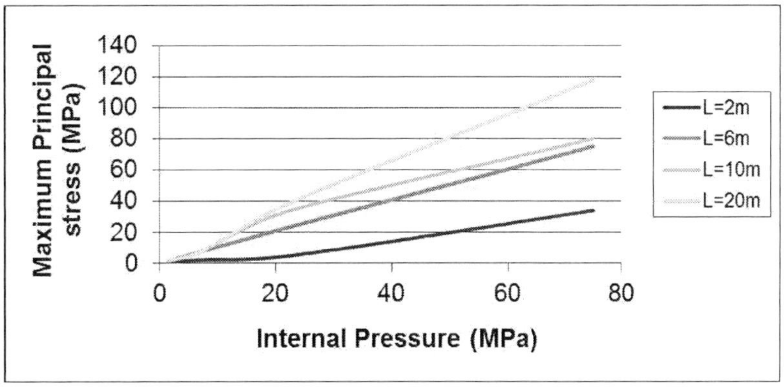

Figure 7. Maximum principal stresses versus internal pressure for different cavity length for depth of cover of 30 m [11].

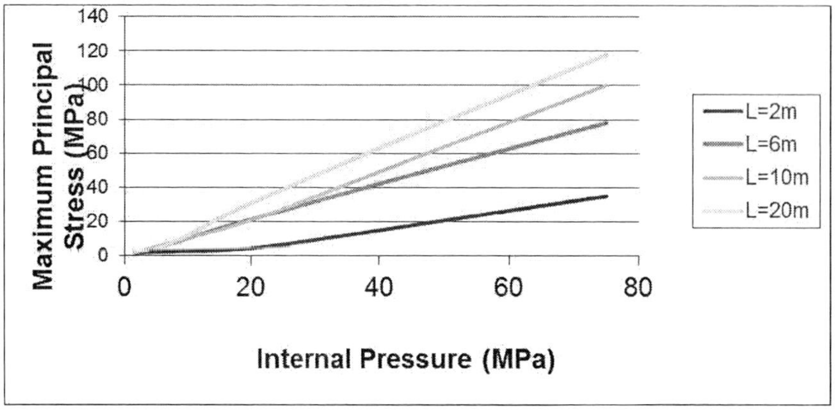

Figure 8. Maximum principal stresses versus internal pressure for different cavity lengths for depth of cover of 40 m [11].

As Figures 7 and 8 illustrate, increasing internal pressure and cavity length will significantly increase the induced maximum principal stress. Depth of cover plays an important role, but is not as significant as internal pressure and cavity size. As shown in the figures, situations involving low applied internal pressure (less than 5 MPa) will generate similar principal stresses for a variety of cavities size. As the figures illustrate, cavities with smaller dimensions are subject to fewer impacts derived from internal pressures as compared to larger cavities.

The influence of different values of K (horizontal stress/vertical stress) was also studied. While the range of different K values did not have significant effect on the maximum principal stress around the cavity; the difference between the maximum and minimum principal stresses changed significantly around the cavity periphery. Figure 9 illustrates the principal stress difference (σ_1- σ_3) for different values of K. Two different cases with internal pressure of 7.5 and 20 MPa were modeled, where cavity width was held constant as 10 m. As Figure 9 shows, increasing K will decrease (σ_1- σ_3) around the cavity roof corner. This implies that increasing the horizontal stress will result in a decrease of the induced shear stress at the cavity roof corners. According to Figure 9, increasing the internal pressure inside the cavity will translate to an increase in (σ_1- σ_3).

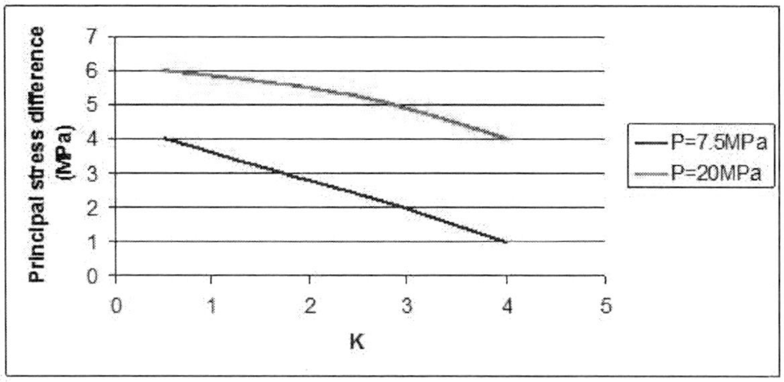

Figure 9. Principal stresses difference (σ_1- σ_3) versus different values of K [11].

Figures 10 to 14 show the effective principal stress before excavation (after stress is in equilibrium) and borehole excavation (before applying pressure) for a variety of cavity lengths (radii) (2 m, 10 m and 20 m). The internal pressure applied inside the cavity was 7.5 MPa. The rationale was that

7.5 MPa will give a more meaningful result as compared to 1.5, 22.5 or 75 MPa because an internal pressure of 1.5 MPa does not have a significant effect and that internal pressures of 22.5 and 75 MPa cause some failure. Depth of cover was held constant at 30 m. As seen in Figure 10 (before excavation), the major principal stress is vertical. As defined, these stresses linearly increase with depth. Figure 11 illustrates the effective principal stresses after the excavation was made and stresses have reached equilibrium, but prior to applying pressure around the cavity. It is apparent that the principal stress tensors rotate about the circumference of the excavation for a distance of approximately 1 cavity height.

Figures 12 to 14 illustrate the effective principal stress distribution for the cavity length (radius) of 2, 10 and 20 m under 7.5 MPa internal pressures. Comparing the unexcavated to excavated stresses, it can be seen that the maximum principal stress has increased around the cavity. Changes in principal stress tensor direction between the unexcavated and excavated states represent shear stresses due to excavation. As these figures indicate, the tensile-stress region expands as the cavity size is increased.

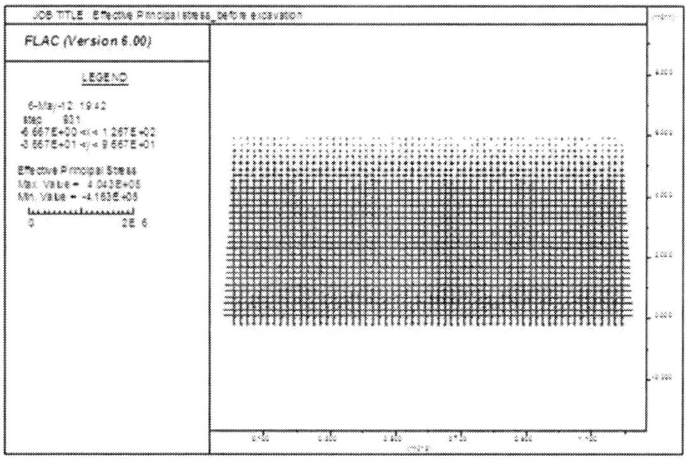

Figure 10. Effective principal stresses under gravitational loading.

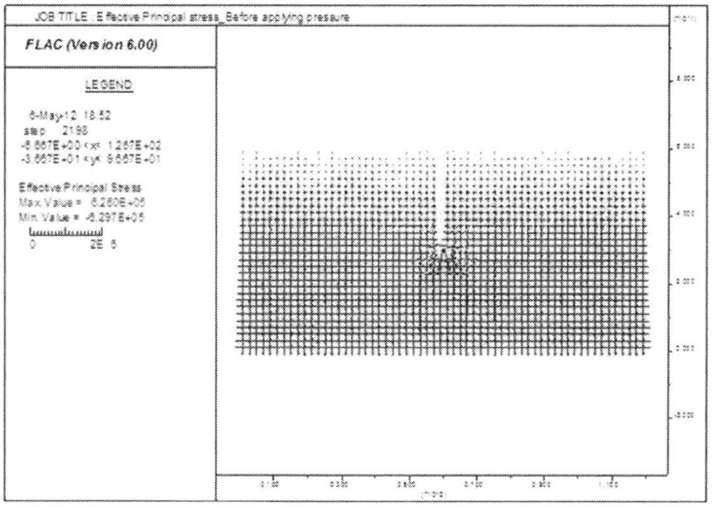

Figure 11. Effective principal stresses after borehole excavation and cavity initiation.

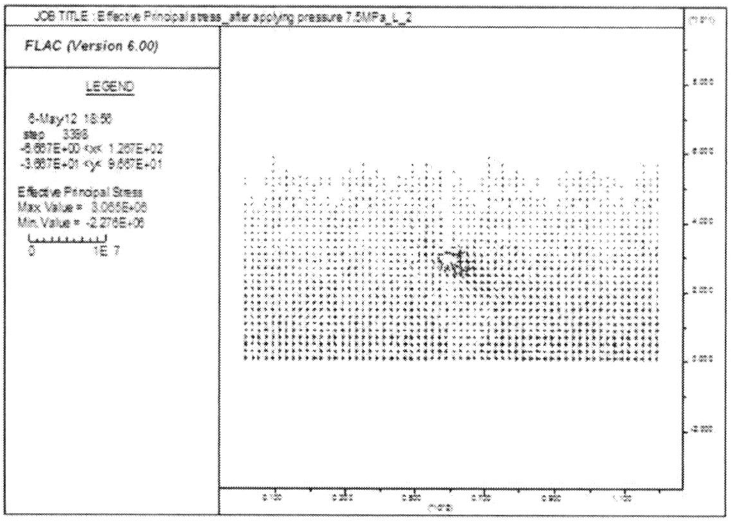

Figure 12. Effective principal stresses for cavity length of 2 m and internal pressure of 7.5 MPa.

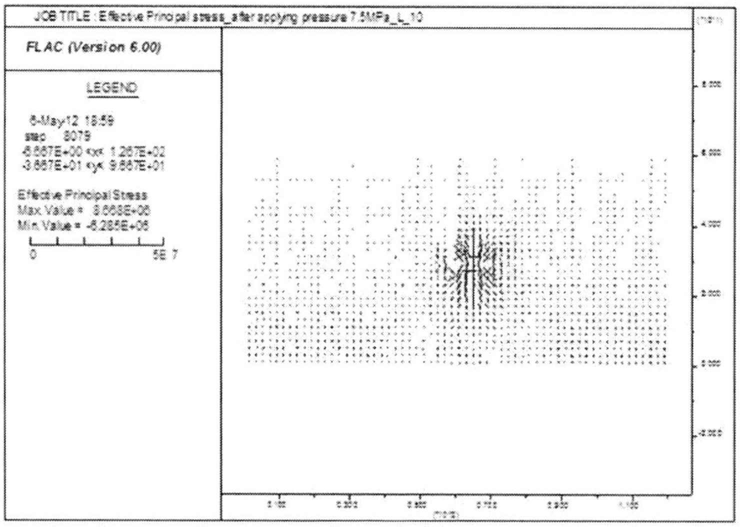

Figure 13. Effective principal stresses for cavity length of 10 m and internal pressure of 7.5 MPa.

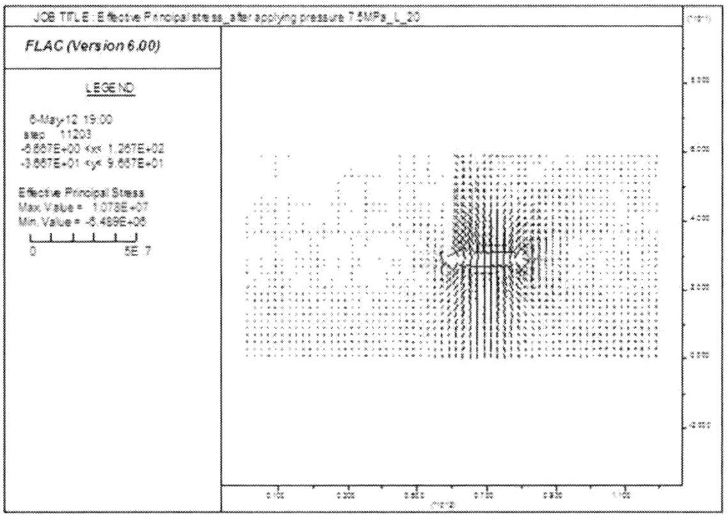

Figure 14. Effective principal stresses for cavity length of 20 m and internal pressure of 7.5 MPa.

4. SUGGESTED ALGORITHM TO PREDICT THE MAXIMUM CAVITY SIZE

Figure 15 presents a suggested algorithm for applying two dimensional finite difference method in order to predict the maximum cavity size that can be obtained without stability concerns.

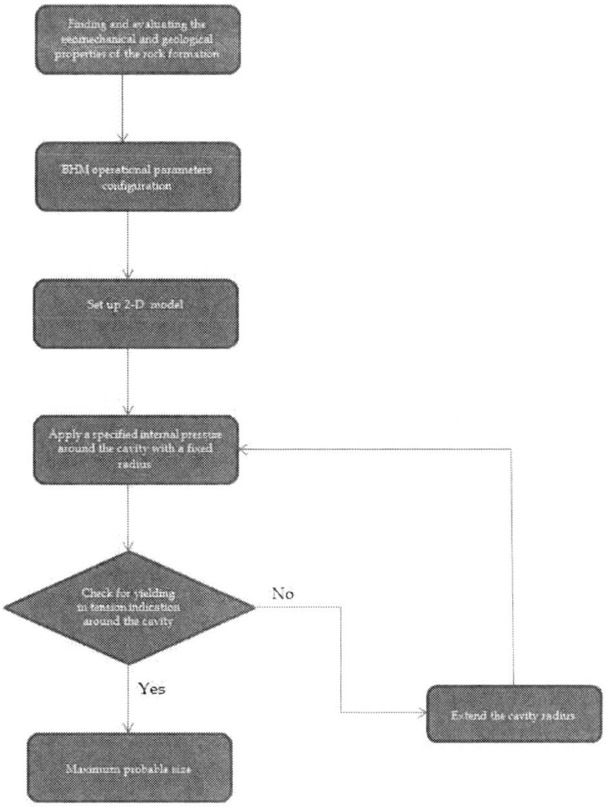

Figure 15. General design methodologies by applying finite difference method for prediction the maximum cavity radius [13].

5. OBSERVATIONS AND CONCLUSIONS

Several models were analyzed using Flac2D to perform a parametric study of factors that could potentially impact the borehole mining process, whereas information derived from the literature was useful in identifying several parameters that could possibly affect cavity design and the mining process. The results and observations of these studies (numerical modeling and literature search) led to the following proposed protocol design for the creation of stable cavities during waterjet borehole mining:

- While mitigating premature roof collapse is the primary motivation behind this research, it is also desirable for large cavities to collapse (fail) after the completion of mining so that adjacent cavities and pillars (segment between cavities) will be de-stressed. In some cases, coal seams lie in soft strata, such as claystone. These seams are usually characterized by geologically younger coking-coal deposits. Rapid extraction of the cavity may prevent development of accelerated deformation and collapse of cavity roof before coal has been extracted. The following are the general precautions that need to be considered regarding the immediate roof of a cavity as derived from literature search:

    a) Conglomerate: highest strength in the series of rock types that can be associated with a coal bearing strata,
    b) Sandstone: often resistant to caving after coal extraction,
    c) Siltstone: will usually cave after coal extraction,
    d) Shales and mudstones: susceptible to moisture, and may cave before coal extraction is completed,
    e) Clay rocks: often show low strength and higher moisture content with very low bearing capacity,
    f) Limestone: usually of moderate strength and good caving properties,
    g) Dolostone: similar properties to limestone,
    h) Shaly limestone: characteristically possess lower strength. Silty limestone is similar to shaly limestone and possesses lower strength.

- According to literature survey, if the immediate roof is thick and consists of strong sandy shale or sandstone, conglomerate and limestone, it can be left unsupported for extended period of time (up to 8 hours).

- Increasing internal pressure and cavity length will significantly increase the induced maximum principal stress around the cavity. Increasing internal pressure inside the cavity will also increase the induced shear stress at the cavity roof corner. In cavities with smaller dimensions, the impact of internal pressures is less than those with larger geometries. Depth of cover plays an important role, but in the case of shallow depths (below 40 m) it is not as significant as internal pressure and cavity size.

- During modelling, no plastic or shear yielding was observed for cavity radii of 2 m or less, even in environments with higher internal pressures. By keeping internal pressure up to approximately 10 MPa, no plastic and shear yielding were observed in all the modeled cavities (i.e. radii of 2, 6, 10, and 20 m).

- Applying internal pressure of 75 MPa caused plastic or shear yielding in all the cavities.

- Depending on the geomechanical properties of the rock formation (intact rock with the geomechanical properties similar to those outlined in Table 1 and a cavity height of 2 m), cavity radius of 10 m was determined to be the maximum size that could be reached without increasing stability issues, provided the internal pressure did not exceed a maximum of 30 MPa. It was observed that applying internal pressure of more than 30 MPa would induce tension failure around the cavity at a radius of 10 m. Applying internal pressure of 35 MPa will limit the cavity radius up to 6 m. In practice, the maximum borehole mining cavity radii has been empirically determined to be 26 ft (7.9 m) in sandstone.

- Designing a cavity with a radius of more than 10 m without having stability problems requires applying less internal pressure. In order to reach a cavity radius of 20 m, the internal pressure shall not exceed 10 MPa. Applying small internal pressure (extraction and pressurization) may have technical challenges associated with shorter standoff distance and material removal [13].

- The influence of changing pore pressure on the effective stress distribution, and the extent of the tensile stress region, were determined to be minor. However, variation in effective stress will impact permeability, and will increase the gas flow as a function of increasing permeability. The absolute permeability in coal can vary due to changes in the pressure of the formation. While borehole mining underwater has the advantage of providing additional roof support during the excavation process, the standoff distance that the waterjet can reach effectively will be reduced dramatically influencing the economics. As such, waterjet borehole mining is preferred in deposits above the water table in unsaturated environments.

- Cavity cyclic pressurization (bailing) will cause the stress concentration to move from cavity roof to the side-wall and corner. However, since negative pressure values are relatively minor compared to the original applied internal pressure, it should not impact cavity stability significantly.

- Increasing the horizontal stress will decrease the induced shear stress at the cavity roof corners.

- One of the major concerns in the design of borehole mining systems is the evaluation of cavity stability based on stress concentration and relaxation around the extraction area. The stress concentration generally depends on two parameters: the volume of removed coal and the load transfer to the cavity face as a consequence of mining. Cavity stability can be maintained by caving the mined-out area before the stress concentration on the face becomes critical. Designing suitable cavity shape and extraction orientation can be mitigating factors. Two extraction strategies under different k values, and different cavity shapes were studied.
- Reaching final optimum cavity size by using horizontal slices extraction has less stability concerns as compare to using vertical slices.

- In a borehole mining design, leaving the side walls in an arch shape (configuration), will increase the stability.

REFERENCES

1. MSHA injury experiences report.

2. Britton.S.G., improving productivity in thin-seam continous mining at Tanoma.Mining engineering 1991.

3. Fotta.B., Peters R., Mallet.L. 1999 Safety challenges at thin mines, HSA Bulletin, July/August 1999. pp 5-10

4. Peters.H.R., Fotta.B., Mallet.L. The influence of seam height on lost-time injury and fatality rate at small underground bituminous coal mines.1999.

5. Papas.D.M., Bauer.E.R. Mark.C. 1999. Roof and rib fall incident and statistics: a recent profile. 1999.

6. Savanick, G.A., 1987. Borehole (slurry) mining of coal, uraniferous sandstone, oil sands, and phosphate ore, 1987.

7. Wang.F.D., Miller.H.B. Waterjet Technology: Past achievements and future possibility.

8. Wiley, Marcus A., Abramov, Gregory. Borehole mining: getting more versatile. 2004, SME

9. Bairu, X., Z. Xiping, M. Zhixin. 2008. Research on one borehole hydraulic coal mining system, Earth Science Frontiers, 15(4):222-226

10. Shen Shihua., Coal mining technology in China, Energy Vol. 11, No. 1 l/12, pp. 1141-l 145, 1986, printed in Great Britain.

11. Akbarzadeh, Y., Miller, H.B. 2014. A Numerical Study of Critical Parameters That Affect Stability of a Cavity Induced by Borehole Mining, In Proceedings of the 48th US Rock Mechanics/Geomechanics Symposium, Minneapolis, 1-4 June 2014.

12. Itasca, Flac2D manual

13. Akbarzadeh, Y., Miller, H.B. 2015. A Parametric Study of Critical Factors That Affect Stability of a Cavity Induced by Borehole Mining in a Coal Deposit, In Proceedings of the 49th US Rock Mechanics/Geomechanics Symposium, San Fransisco, CA, USA, 28 June- 1 July 2015.

14. Miller, H.B. and Wang, F.D., IET Jet Technology, Inc. Personal Correspondence.

2015 WJTA-IMCA Conference and Expo
November 2-4 ● New Orleans, Louisiana

Paper

DEVELOPMENT AND TESTING OF ULTRA-HIGH PRESSURE DOWNHOLE PUMP DRIVEN BY VANE MOTOR

Shi Huaizhong, Li Gensheng, Li Hao, Zhang Yi
State Key Laboratory of Petroleum Resources and Prospecting, China University of Petroleum Beijing, Beijing 102249, China

ABSTRACT

Ultra-high pressure, jet-assist drilling techniques proved to be an efficient way to increase rock breaking efficiency in deep wells where rocks become harder and more difficult to drill. In this technique, the ultra-high pressure down-hole pump, which locates between the drilling collar and the drilling bit, is the key component. This paper presents the development and testing for a novel ultra-high pressure and jet-assist down-hole pump using hemispherical vane motor. The mechanical principle is that when drilling mud flows through the vane motor from the inlet to the outlet, it causes pressure loss and pushes the rotor and the crankshaft and connecting rod mechanism to rotate at a high speed along the axis. The crankshaft and connecting rod drive the plunger to reciprocate back and forth in the pump. The mechanism of energy transform is to convert the kinetic energy of mud into the rotating speed mechanical energy. First, the shell, the vane motor assembly, the crankshaft and connecting rod mechanism, the plunger, and cylinder liner etc. are redesigned to meet the purpose of increasing the rock breaking efficiency. Second, a numerical simulation is built to better understanding the flow behaviors of the novel pump. Third, the torque of motor, pressure loss, motor structure, and vane size and numbers are studied using CFD technique. At last, intensive laboratory experiments are designed to test the hydraulic parameters of the novel pump. The results show that: 1) the rotor torque increases with the increase of drilling mud flow rate. 2) The torque of the vane motor assembly is up to 400 N·m and the outlet pressure of the pump is about 14Mpa when the flow rate is 14L/s.

Organized and Sponsored by WJTA®-IMCA®

1 INTRODUCTION

In 1993, FlowDril and Gas Research Institute (GRI) developed downhole ultra-high pressure pump to carry out jet-assisted rock breaking research plan (Veenhuizen 4-6), in which booster increases the pressure of some fluid to help drilling. The experimental result shows that ROP can be improved 1.0~2.5 times.

At the end of 1994, the U.S. Department of Energy (DOE), FlowDril and GRI jointly developed the second generation of downhole pump prototype (Veenhuizen, Stang, and Kelley 1-4). Field test results shows that jet-assisted drilling with ultra-high pressure downhole pump can increase rate of penetration by 45%~100%, while the device's working life is short.

In 2001, Maurer Engineering Inc. conducted the research and development of high-pressure jet drilling system, indoor experiment field test (Cohen and Deskins 2-8). The system can send BHA into borehole through coiled tubing and also use improved conventional rotary drilling methods. The indoor experiment shows that jet kerf drilling technique can improve ROP by 1.0~2.0 times in different formations. Field tests in a number of different formations including sandstone, shale, arenaceous shale, limestone, etc. indicate that the mechanical drilling rate can be improved by 1.3~6.0 times.

In 2003, Li et al. put forward hydraulic pulsed self-oscillation cavitation jet and studied the jet characteristics experimentally (Li, Shen, Zhou, Zhang, and Yi 570-75). The results shows that the impact pressure peak value and pulse amplitude of self-oscillation cavitation jet is 37% and 24% higher than ordinary nozzle, respectively.

In 2005, Wang et at. completed the design of the first generation of fluidic downhole pressure booster (Wang, Xue, and Zou 17-19). Ground simulation test and whole machine pilot of the device shows that outlet pressure changes directly associated with the input displacement and throttling pressure drop, and the pressure fluctuation of stand pipe also reflects that the designed tool's pressure loss conforms with the actual experimental pressure loss.

Guan et al. designed a novel downhole pressure booster using drilling string's vibration as energy source (Wei 67-80). The device can take advantage of the axial vibration which drives plunger piston to reciprocate up and down. In this way, vibration during drilling can be turned into beneficial reciprocating motion of the plunger, and thus drilling mud is compressed and the pressure is increased. This device not only alleviates drilling string's vibration but also boosts the drilling mud's pressure.

Predecessors did a lot of work in this area (Kolle 61-69; Dou 1-7 151; and Fu 5-11).On the basis of previous studies, this paper puts forward a novel downhole pressure pump driven by vane motor which uses drilling fluid to drive the vane, and crankshaft turns torque into piston's reciprocating movement to provide local high pressure jet-assisted rock breaking capacity. The novel pressure pump can provide new ideas for downhole pressured jet assisted rock breaking method.

2 STRUCTURE AND WORKING PRINCIPLE

2.1 Structure

Hemispherical vane motor mainly includes the following five parts: shell, separator, vane motor assembly, crank-link mechanism and plunger pump assembly. Its upper end can connect drill collar or PDM, and the lower end can directly connect drill bit. In the process of drilling, drilling fluid coming from drill collar or PDM passes through the separator located at the upper part of the pressure pump, and then most of the fluid goes into the high pressure inlet of a pair of vane motors. Therefore, under the effect of differential fluid pressure, the vanes generate torque and drive the rotor. The rotor makes the crankshaft rotate, which makes the connecting rod and plunger piston reciprocate up and down. In this way, a small part of the high pressure drilling fluid coming from collar or PDM turns into ultra-high pressure fluid, and the ultra-high pressure jet comes out of the ultra-high pressure nozzle and assists downhole rock breaking process to improve the ROP.

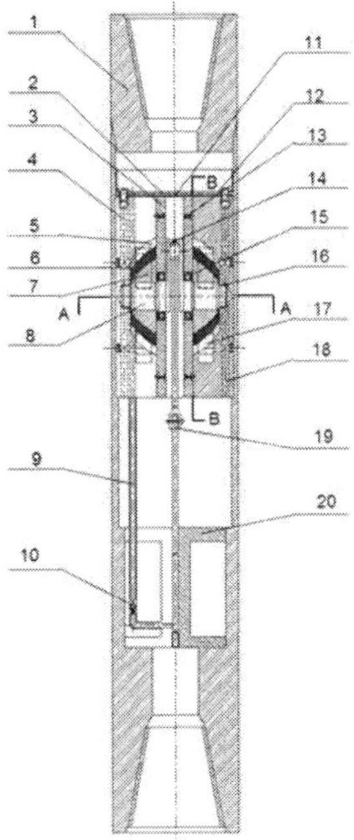

1. Body 2. Separator 3. Motor gland 4. Fluid inlet 5. Vanes 6.Spring 7. Rotor 8. Crank shaft 9. Channel 10. One-way valve 11. Spacer 12. Through-hole bolt 13. Gland bolt 14. Connecting rod and set bolt 15. Bearing and seal 16. Bearing bush 17. High pressure and low pressure chambers 18.Fluid outlet 19. Cardan shaft 20. Plunger pump

Figure 1. Structure of Downhole Pump Driven by Vane Motor

The Pump which drives the crank-link mechanism can continuously provide power for plunger pump and make ultra-high jet continuously act on the downhole. The structure is shown in Figure 1.

2.2 Working Principle

At the upper part of the downhole pump driven by vane motor there is a separator, which divides drilling fluid coming from drill collar or PDM into three parts, in which two parts of the fluid flows into the high pressure inlet section, and the third part goes into the plunger pump's inlet section through the channel opened at the motor stator.

A pair of symmetrical vane motors is mounted next to the separator under the limited downhole space and the stator's inside profile is semi-spherical. Two high pressure chambers and two low pressure chambers are installed at the motor stators. High pressure drilling fluid coming from the high pressure chambers flows into the low pressure chambers, so the pressure difference is produced driving rotor to rotate.

Spring is installed between vanes and rotor slot. When the vanes rotate to the position where the stator's inner diameter is the minimum, spring will be pressed thoroughly, while when the vanes reach the position where the inner diameter is the maximum, spring will push the vane's edge to the stator's inner surface. Sliding bearing, seal system and the crank-link mechanism are also designed according to the downhole space. The stators are installed on both ends of the crankshaft.

Plunger pump is located at the bottom and its main body sits on the shell. The drilling fluid at the inlet comes directly from the separator, and the outlet fluid flows through high pressure channel and moves out of bit nozzle to break rocks.

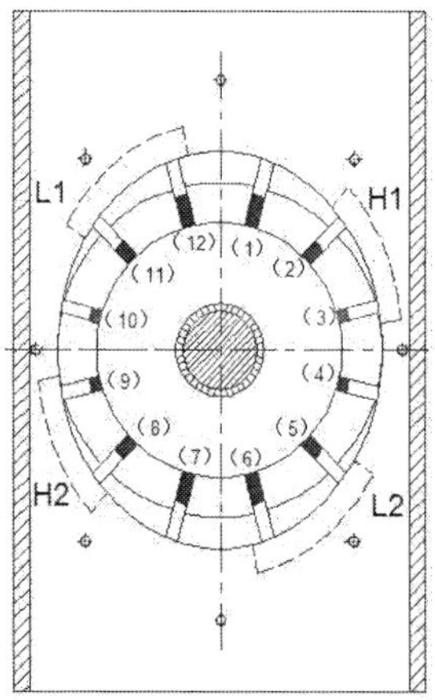

Figure 2. B-B Sectional View

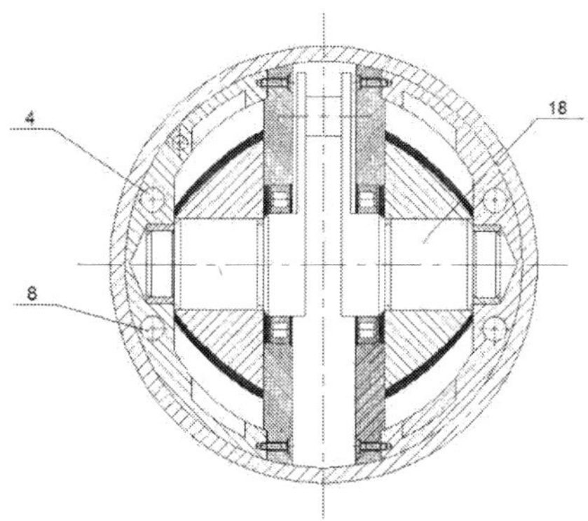

Figure 3. A-A Sectional View

The core section of the device is the vane motor, which can be more detailed observed from B-B and A-A sectional view. In the above figures, L1 and L2 indicate low pressure chambers; H1 and H2 indicate high pressure chambers.

3 NUMERICAL SIMULATION OF HYDRAULIC PARAMETER'S CHARACTERISTICS

Hemispherical vane motor is the key component of the whole device, and its structure is directly related to the energy transform efficiency of the pressure pump system, which is key to the design. This paper uses FLUENT software to study the hemispherical vane motor model. Similar numerical simulation and optimization for dwonhole boost compressor has been done (Wang 308-12; Xue 5-11; and Ai 92-4).

3.1 Geometric Model

Due to the symmetry of hemispherical vane motor, we only analyze the fluid flow of one vane. Each side of the motor has two inlets and two outlets, as shown in Figure 4. The blue section indicates velocity inlet and the red one indicates pressure outlet. The green part shows the shell and the grey one means the rotor. Since the top and bottom radii of actual motor are relatively small and contact seal is adopted between vane and shell, there is no space between top and bottom vanes during simulation and the space between vanes and shell on both sides is 1 mm. The rest of the parameters are of the same size as the actual test device.

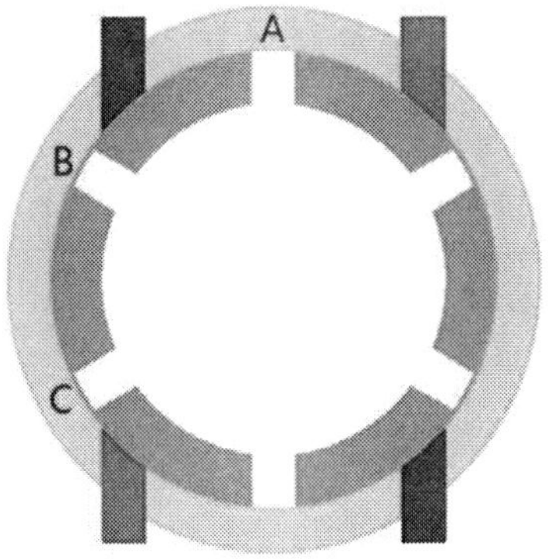

Figure 4. Two-dimensional Flow Model

RNG k-epsilon turbulence model is adopted due to the unsteady flow field. Unstructured mesh was used with the mesh size of 0.3, and the sliding mesh was adopted to simulate the rotation of the rotor in the stator.

3.2 Boundary Condition

Inlet boundary was set as velocity inlet; Outlet boundary was pressure outlet; Wall condition was set as stationary wall.

3.3 Sliding Mesh Settings

According to the measured data in practical experiment and the conditions in real operation, we assume the rotational speed of the rotor is 30 rpm (revolutions per minute). Between the two inlets and two outlets interfaces were set, and the space between vane and stator was 0 mm according to real design. Therefore, the rotation time that dynamic mesh can simulate is limited.

In the simulation, the velocity field that vane A rotates counterclockwise to the position 30° to the velocity inlet can be simulated. Due to the cyclic process of the whole vane's flow field, the flow field between this 30° can be used to speculate the pressure and torque during the whole rotation of vane motor.

3.4 Numerical Simulation Method

1. Flow rate

According to the flow rate of working condition, the inner flow fields of one side of the vane motor was simulated when the flow rate was 3L/s、5L/s、7L/s, respectively. The corresponding inlet velocities are shown in Table 1.

Table 1. Simulation Scheme of Different Velocities

Flow Rate(L/s)	3	5	7
Inlet Velocity(m/s)	4.3	7.1	10

2. Angle of rotation

When rotor rotates, the relative position among vane, inlet and outlet will change, which means the impact force will change accordingly. Analyzing the torque change at different angle of rotation can get the dependability of this tool.

3.5 Calculation and Analysis Results

3.5.1 Pressure Field Analysis

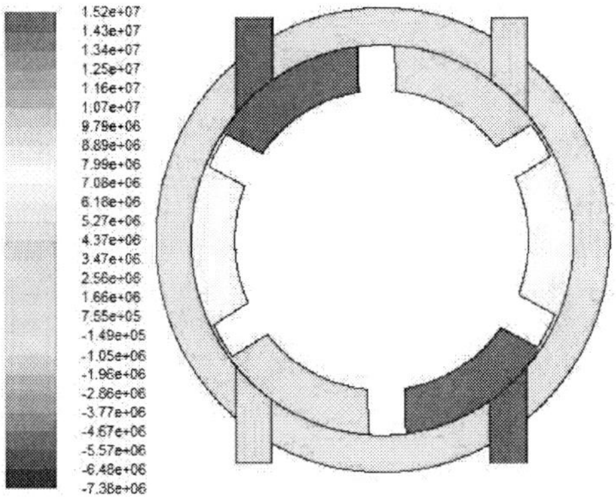

Figure 5. Pressure Field Distribution at the Beginning with Inlet Velocity of 10 m/s

Figure 5 is the pressure field of the vane motor with velocity of 10m/s at both inlets. In this figure, we can see that the pressure field of the vane motor is centrosymmetric and the pressure in channels on both sides attenuates from the inlet to outlet. Vane B and C provide torque for the whole vane motor and vane A in the real tool is compressed into the crank shaft of the vane motor, so when we calculate the torque, we don't take vane A into account. As shown in Figure 5, the pressure in space between vane A and B and vane B and C are respectively 15Mpa and 8Mpa, and the pressure at the outlet is 0.8Mpa.

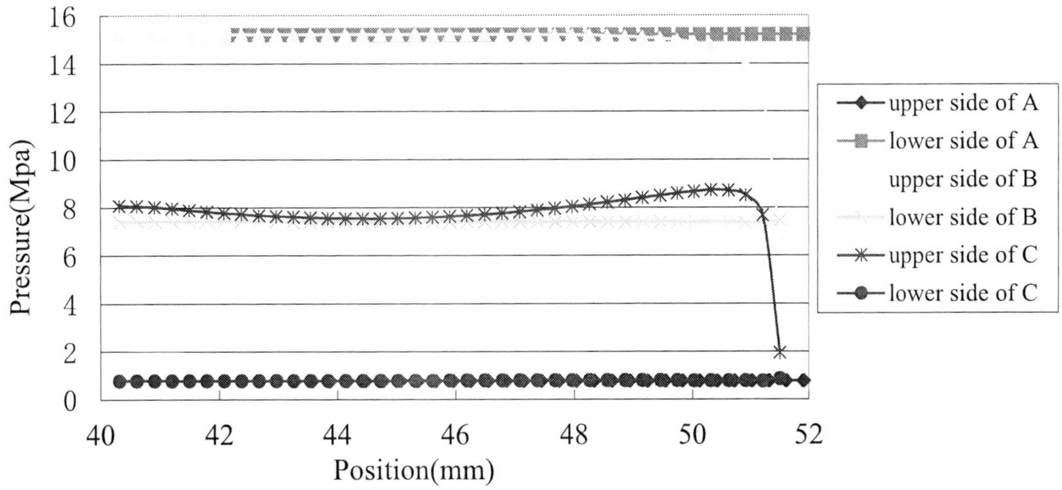

Figure 6. Pressure Distribution Curve for Different Vanes

Figure 6 shows the pressure curve of each vane on both upper and lower sides. As shown in Figure 6, the pressure on different sides of vane A shows great diversity, but in the real tool vane A does not generate torque. The pressure between two vanes have little difference on the upper side and the lower side. For example, the pressure on the lower side of vane B is approximately the same of that on the upper side of vane C. For vane B and C, the pressure shows a great decrease at the finger tip of the vane because the velocity at that point increases greatly.

3.5.2 Flow Field

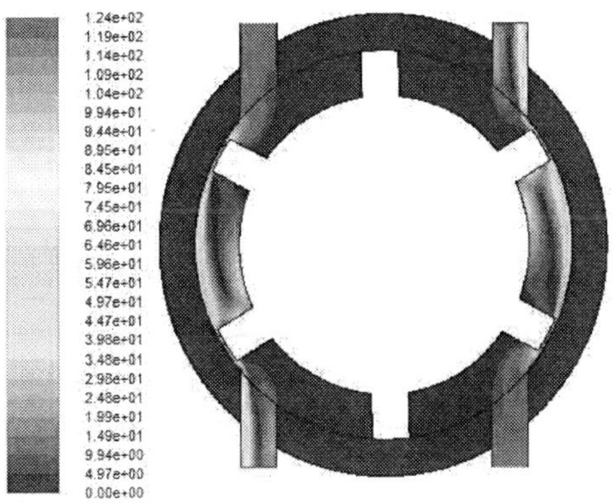

Figure 7. Flow Field with Inlet Velocity of 10 L/s and Gap of 1 mm

Figure 7 shows the flow field when fluid flows through the vane motor at velocity of 10m/s at initial moment. In this figure, we can see that since the gap between the vane and the stator is very small, velocity here is quite high—approximately 120m/s. What's more, the high-speed water flow impacts the vane of the motor and generates flow back at the upper side of each vane. At last, the fluid flows out of the outlet and the flow field is centrosymmetric.

3.5.3 Impact of the Inlet Velocity on Pressure Field and Torque

Keeping the rotational speed at 30rpm, the initial flow fields with inlet velocity of 4.3m/s, 7.1m/s and 10m/s were analyzed respectively. Since vane B and vane C generates most torque, taking vane as cantilever beam, the maximum torque can be calculated according to the differential relation between the load and the torque.

$$\frac{d^2M(x)}{dx^2} = \frac{dF_S(x)}{dx} - q(x) \tag{1}$$

Extracting data from both sides of vane B and vane C, torque map of the vane can be calculated by differential equation shown above. The torque map is shown in Figure 8.

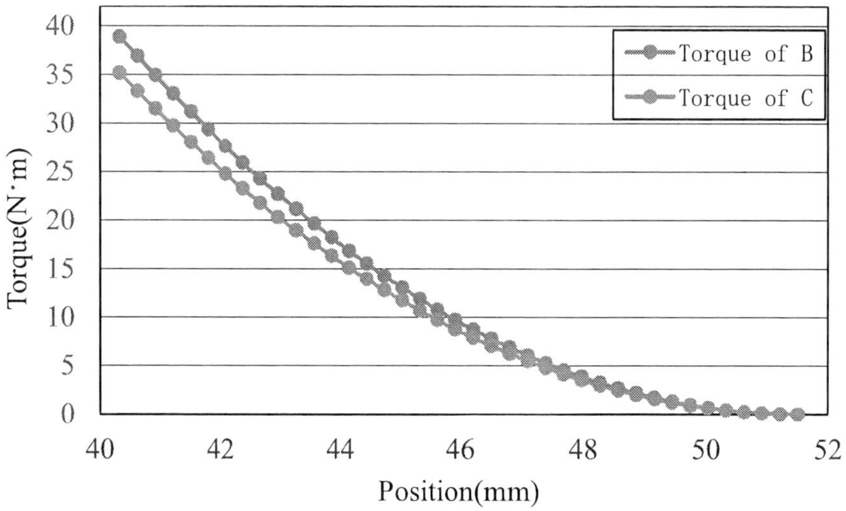

Figure 8. Torque Distribution Curve with Flow Rate of 10 m/s

As shown in Figure 8, the toque in both vanes increases from the finger tip to the root tip. The maximum torque provided by vane B and vane C are 40 N·m and 36 N·m, respectively. Since every side of the motor is centrosymmetric and one motor has a pair of vanes. The total torque is four times as much as the torque provided by vane A and vane B.

3.5.4 Torque at Different Rotation Angle

The rotation process of the motor is simulated to evaluate the motion state. As the motor rotates, the impact force will change. The torques of vane B and C at different position with inlet velocity of 10m/s is calculated respectively.

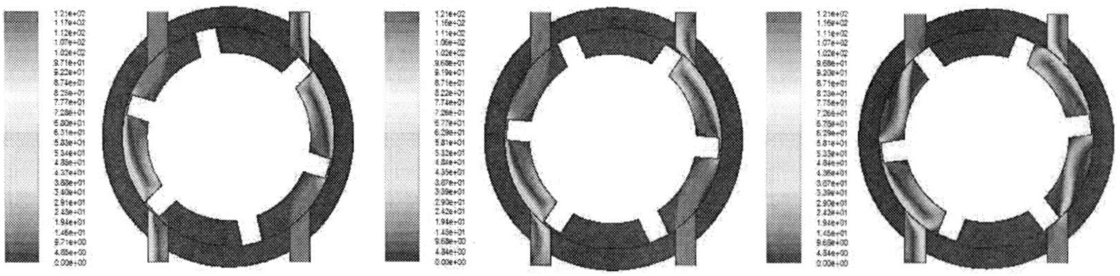

Figure 9. Fluid Field at 0.07s,0.14s and 0.21s

The figures above show the flow field at different rotation time. In these figures, we can see that the flow fields at different time have no obvious difference. At 0.21s, since vane A obstructs the inlet, a relatively high speed water flow forms at the inlet.

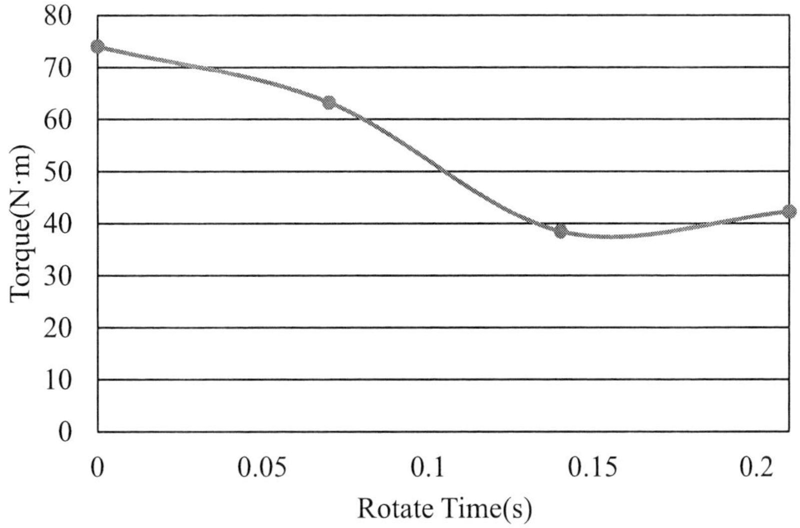

Figure 10. Torque Provided by Vane B and C at Different Rotate Time

The total torques of vane B and C at different rotation times are calculated and the results are shown in Figure 10. We can see that from the initial moment to 0.21s, the torque decreases from maximum 75N·m to 40N·m. This is mainly because vane C rotates across the outlet and the fluid impact on vane C decreases. The torque is provided mainly by vane B at 0.21s.

4 HYDRAULIC PARAMETER LABORATORY EXPERIMENT

4.1 Preparation

All components of the hemispherical vane motor are manufactured and assembled, and engine oil is used as the flow medium. Main components of the hemispherical vane motor are shown in Figure 11. The rotate speed and the torque generated by the motor are measured with the flow rate with one side of the motor changing from 3L/s-7L/s.

Figure 11. Hemispherical Vane Motor Components

The main components in Figure 11 are shell, stator, crank shaft. The stator, upper cover and lower cover are fitted together by pin bolt, creating a cavity with two inlets and two outlets. Crank shaft is placed inside the cavity with six retractable vanes fixed in it. These are the main components of the motor. The motor is fixed inside the shell by cover plate with its flow pass connected to the corresponding flow pass of the shell. Then one side of the testing prototype is assembled. The two sides are totally symmetrical with crank shafts of two sides connected by hexagonal joint and shell connected with bolt.

4.2 Experimental Scheme

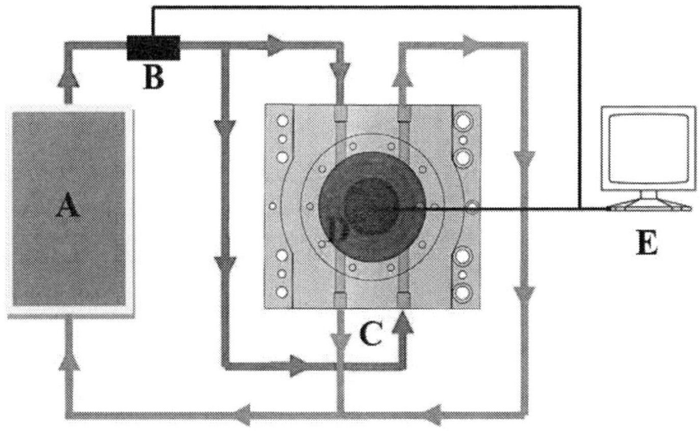

A Pump; B Pressure Transmitter; C Hemispherical Vane Motor Testing Prototype;
D Torque Measurement Device; E Data Acquisition System

Figure 12. Experimental Facilities Schematic Diagram

Figure 12 shows the schematic diagram of experimental facility. Pump A pumps engine oil out through the hose into the two inlets of the motor. Before the inlets a pressure sensor B is installed to measure the inlet pressure. Since a vane motor is composed of two identical sides, in this experiment only one side of the vane motor C is tested. The crank shaft is connected with torque measurement device D, which is connected to the computer E. Both pressure data and torque data can be gathered by computer. Real testing prototype is shown in Figure 13.

Figure 13. Hemispherical Vane Motor Testing Prototype Picture

4.3 Experiment Results

The inlet pressure, rotate rate and torque of the vane motor are measured and the results are shown in Figure 14.

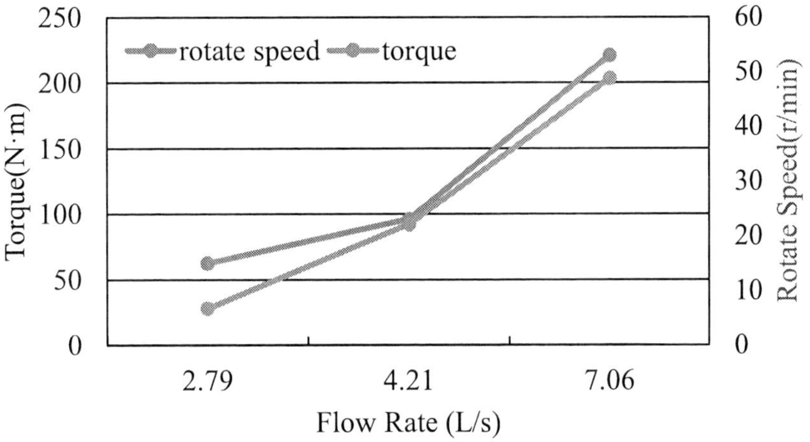

Figure 14. Test Result of One side of the Hemispherical Vane Motor

Test result and simulation result are basically unanimous and the value in test result is higher than simulation result. This is probably because in real test engine oil is used instead of water as the working fluid. What's more, the vane fixed in the crank shaft is retractable and can closely contact the inner wall of the motor.

Through the test, we find that this motor still need to be improved. By increasing the length of the vanes, more water energy can be transferred into kinetic energy of the motor. The parameter optimization needs further research.

5 CONCLUSIONS

In this paper, the structure and working principle of the novel pump driven by vane motor is illustrated. Through laboratory experiment and simulation, the inner flow field and working characteristics of this pump is studied. Through the above work, we come to the following conclusions.

1. A novel pump is designed and a testing prototype is manufactured. This novel pump prototype is proved feasible through laboratory experiment.

2. Through numerical simulation, the torque and the pressure loss provided by a single side of the motor are 150N·m and 14Mpa respectively. For vane B and C on a single side of the motor at the initial time, the torque provided by vane B is 40 N·m and that by vane C is 36 N·m. Moreover, when the motor rotates, the torque changes from 75N·m to 40N·m periodically.

3. In laboratory experiment, the torque and rotate speed at flow rate of 7L/s by a single side of the motor are 53rpm and 203 N·m. When flow rate is 4.21L/s and 2.79L/s, the test result shows that the rotate speed and the torque decrease as the flow rate decreases. The motor performs better in laboratory test than in numerical simulation.

4. The numerical simulation result and laboratory result show that downhole pump

driven by vane motor is feasible and this novel pump provides a novel method for ultra-high pressure jet-assist drilling technique.

6 ACKNOWLEDGEMENT

This work was part of the program on Investigation of Partial Underbalanced Drilling Hydraulic Characteristics modulated by Coiled Tubing funded under the national nature science foundation of China. The authors are grateful to all study participants.

REFERENCE

Ai, Chi, et al. "Calculating methods of principal structure parameters of diaphragm pressure converter." Journal of Daqing Petroleum Institute25.1 (2001) 92-94. Print.

Cohen, J H, and Deskins G. "High-pressure jet kerf drilling shows significant potential to increase ROP." SPE96557 (2005):2-8. Print.

Dou, Liangbin, et al. "Research progress on ultra—high pressure jet drilling by down—hole pressure boost." Special Oil & Gas Reservoirs3 (2012):1-7, 151. Print.

Fu, Jiasheng, et al. "Study on supercharging characteristic of downhole pulsed pressure intensified drilling device." China Petroleum Machinery1 (2015):5-11. Print.

Kolle, JJ, et al. "Development of a downhole separator and intensifier for coiled tubing jetting." Journal of Canadian Petroleum Technology47.12 (2008):61-69. Print.

Li, Gensheng, et al. "Mechanisms and tests for hydraulic pulsed cavitating jet assisted drilling." Petroleum Exploration and Development2 (2008):239-243. Print.

Li, Gensheng, et al. "An experimental study on impact pressure characteristics of self-resonant cavitating jets." Journal of Hydrodynamics (A)5 (2003):570-575. Print.

Li, Weicheng, et al. "Development and application of a new ultra-high pressure downhole supercharging system" Natural Gas Industry3 (2015):86-92. Print.

Sun, Wei, et al. "System design of the centrifugal downhole pressure device." China Petroleum Machinery34.3 (2006):36-38. Print.

Veenhuizen, S D. "Development and testing of a downhole pump for jet-assist drilling." Natural Gas RD&D Contractors Review Meeting (1995):4-6. Print.

Veenhuizen, S D, et al. "Development and testing of downhole pump for high-pressure jet-assist drilling." SPE38581 (1997):1-4. Print.

Wei, Wengzhong. "Study on vibration characteristic of bottom drilling string and design of drilling string shock absorption & downhole hydraulic pressurizing system." China University of Petroleum (Hua Dong) (2007):67-80. Print.

Wang, Zhiming, Xue Liang, and Zou Hejun. "Study of design theory of piston-type downhole booster." China Petroleum Machinery (2007):17-19. Print.

Wang, Zhiming, and Xue Liang. "Hydraulic parameter model for design of fluidics downhole boost compressor" Acta Petrolei Sinica29.2 (2008):308-312. Print.

Xue, Liang, Wang Zhiming, and Li Bangmin. "Numerical study of flow field for fluidic downhole boost compressor." Petroleum Drilling Technology1 (2015):5-11. Print.

Xu, Yi, et al. "Principle and design of screw downhole supercharger." Drilling and Production Technology34.3 (2011):71-73. Print.

2015 WJTA-IMCA Conference and Expo
November 2-4 • New Orleans, Louisiana

Paper

CNC WATER JET SET DESIGN FOR EDUCATIONAL PURPOSES AND ITS IMPLEMENTATION

F. Kartal
University of Kastamonu,
Kastamonu, Turkey

H. Gökkaya
University of Karabuk,
Kastamonu, Turkey

ABSTRACT

In today's manufacturing industry, any products are manufactured by using CNC (Computer Numerical Control) machine. CNC machine developed and sustained its improvement becoming an indispensable part of manufacturing sector. In order to educate operators that would use CNC machine which have important role for manufacturing industry; theoretical courses such as basic level machine utilization, machine programming are given at vocational schools, vocational high-schools and faculties. However; these information are not given practically in most of the educational institution. The reason for this is not being able to purchase expensive CNC machine by educational institution and maintenance services being ineffective. In this study, low-cost education set that would enable more active learning in which students who are taught CNC education can practice on their own was designed, its prototype was produced and application was done.

Keywords: Desktop CNC Machine, Educational CNC Machine Design

1. INTRODUCTION

Abrasive waterjet cutting process, as one of the non-traditional machining methods, is increasingly being preferred by the machine manufacturing industry and expanding its area of use continuously for its superiority when compared to other cutting technologies. Initially used in the mining industry, waterjet is now gained widespread use in other industries. Waterjet is categorized under two groups, namely, abrasive and pure waterjets. It is possible to cut brittle, ductile, composite and metal materials using abrasive waterjet. It is able to cut any engineering material no matter what the roughness level [1].

Abrasive waterjet cutters are of utmost importance for today's industry. Therefore, it finds its place in the curriculum of vocational high schools and faculties as a non-traditional manufacturing method. Abrasive waterjet cutting, as a non-traditional manufacturing method, has created an important employment opportunity and a new profession in the cutting industry. However, educational institutions cannot purchase costly abrasive waterjet cutters and are only able to provide theoretical information and illustrations about the operation of these machines leaving no choice for the student but to imagine an abrasive waterjet. It would be fair to say that there is almost no institution in Turkey which provide applied education with CNC waterjet. This study involves a CNC waterjet cutting machine which can be built by technical education institutions using their limited resources and its application in order to train students on basic subjects such as the workings of waterjet and cutting with waterjet as part of their non-traditional manufacturing classes. The literature scan performed in order to drive this study involved the following; Kim et al. addressed the lack of educational material in CNC education and manufactured a low-cost CNC milling machine [2]. Köbeloğlu designed an educational CNC milling machine for higher education studies and built a prototype [3]. The research of Alan lays emphasis on the CNC education in Turkey and the importance of CNC for vocational education and the industry and designed a CNC education set in order to be used in CNC classes [4]. Uyanık et al. worked on the integration of 3-axis surface milling machines with CNC [5]. Toroğlu et al. investigated the use of 3-D animation techniques to be utilized in technological education [6].

We have failed to find a study about an educational CNC waterjet suite and its prototype as a result of the literature scan. This study aims to be exemplary for institutions focusing on technical and engineering education with a CNC waterjet cutter design which is possible to build with resources readily available and it is expected to contribute to future research.

2. STRUCTURAL CHARACTERISTICS OF THE MACHINE

System was designed and put together in order to build an educational CNC waterjet. Initially, CNC vertical machining center was built using steel profiles in order to allow for computer controls on x, y, and z axes of the CNC waterjet cutter. The cutter was then welded as required using the measurements of the design.

Following the manufacturing of the frame of the cutter, the pressure unit was designed and manufactured. Industrial CNC waterjet cutters most commonly use a working pressure of 4000 bars. The pressure value was not increased as it would lead to increased costs, power consumption and safety issues when necessary measures are not taken for use with educational purposes. The operating pressure of the cutter was set to 110 bars which is sufficient for cutting process. A booster pump was used as the feed pump of the pressure unit. Water taken from still water container of 1 bar pressure feeds the high-pressure pump. Water from feeder arriving at the high-pressure pump with a pressure of 10 bars is then conveyed to the nozzle outlet with a pressure of 110 bars. A car wash machine was purchased as an alternative solution to obtain a pressure of 110 bars and it was dismantled in order to be used in the water pump of the waterjet mechanism. Figure 1 shows the pressure unit of the industrial waterjet while Figure 2 shows the pressure pump obtained from the pressure car wash machine.

Figure 1. Industrial water jet pressure unit view.

Figure 2. The pump unit is derived from pressure washer.

The pressure unit obtained from the car wash machine and the waterjet unit built using CNC vertical machining center were then assembled.

Figure 3 shows an industrial CNC waterjet cutter while Figure 4 gives an overview of the CNC waterjet cutter built for educational purposes.

Figure 3. The view of Industrial type CNC water jet machine

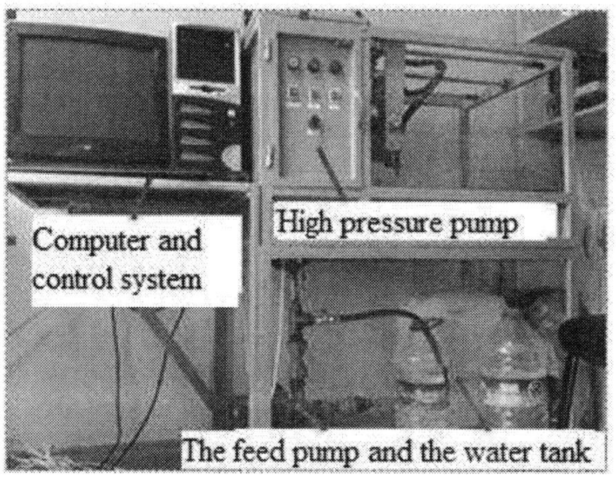

Figure 4. Educational purpose CNC water-jet bench overall image.

3. EVALUATION OF THE APPLICABILITY THE MACHINE

The prototype of educational CNC waterjet cutter was then subjected to tests and successful results were obtained. As the system was not fed with abrasive dust, it was only able to cut materials such as polystyrene. Thus, Stryrofoam insulation material was selected as the material to be cut. Drilling and cutting a workpiece made of insulation foam of 18mm thickness was successfully performed. Figure 5 shows the drilling process with waterjet while Figure 6 shows an image of the lateral view of the material being cut. Figure 7 shows the letters cut out of the styrofoam as a result of the waterjet cutting process.

Figure 5. Drilling operation of styrofoam with the water jet

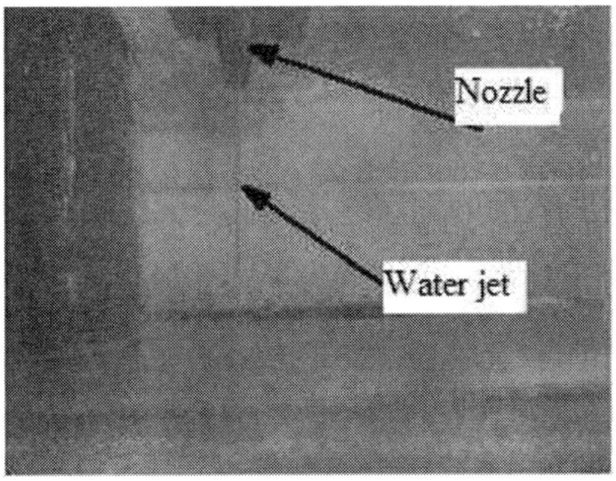

Figure 6. Side view of the styrofoam water jet drilling operation.

Figure 7. The image of the sample letters cut from styrofoam with the water jet.

4. CONCLUSION AND RECOMMENDATIONS

An educational CNC waterjet cutter was developed in this study combining a vertical machining center with a waterjet manufacturing system. The cutter involves a genuine waterjet nozzle. This design proved to be a low-cost educational set which can be manufactured by vocational high schools and faculties with readily available resources in order to be used in non-traditional manufacturing methods classes. As imported CNC waterjet cutters are expensive and it is hard to provide services for these cutters in Turkey, the educational set developed will eliminate these disadvantages. This study offers an opportunity for the students to work on a compact and low-cost experiment set individually as part of the non-traditional manufacturing methods classes. As a result of this study a more visual and realistic education for the waterjet subject once was only available in a theoretical manner will be possible with the educational set as shown in Figure 4.

The waterjet cutter developed does not include abrasive and abrasive feeder due to safety and monetary considerations. It is possible to improve the design with the addition of abrasive mixing tube and abrasive nozzle.

5. REFERENCES

[1] Wang, Jun. Abrasive waterjet machining of engineering materials. Trans Tech Publications Ltd, 2003.

[2] Kim,J., Kim,M., "A Study On The Design Of CNC Lathe For Education And Application" International Journal of Production Economics 25 169-180, 1991

[3] Köbeloğlu A. "Eğitim Amaçlı Masaüstü CNC Torna Tezgahı Tasarımı ve Prototipi" Karabük Üniversitesi Yüksek Lisans Tezi" 2010 Karabük

[4] Alan, S. (2006), Selçuk Üniversitesi, "CNC Eğitim Seti Tasarımı" Yüksek Lisans Tezi

[5] Uyanık. A, Şimşek. İ, Aytan. İ, Onat. M, Erdal. H (2009), "3 Eksenli Yüzey İşleme Tezgâhının Bilgisayar İle Kontrolü", , IATS'09, 13-15 Mayıs 2009, Karabük

[6] Toroğlu, A. , İçingür, Y. (2007), "Üç Boyutlu Bir Animasyon Sisteminin Tasarımı ve Teknoloji Eğitiminde Kullanılması", Politeknik Dergisi Cilt:10 Sayı: 3 s.247-252

2015 WJTA-IMCA Conference and Expo
November 2-4 ● New Orleans, Louisiana
Paper

A STUDY ABOUT THE EFFECT OF CONFINING PRESSURE ON HIGH PRESSURE JET

Li Jingbin, Li Gensheng, Huang Zhongwei, Song Xianzhi
State Key Laboratory of Petroleum Resources and Prospecting, China U. of Petroleum
Beijing, Beijing 102249, China

ABSTRACT

During the drilling process for oil and gas resources, the high-pressure water jet is a key factor which can improve the rate of penetration (ROP) via breaking rock, cleaning well bore, carrying cuttings and so on. But it is always high confining pressure at the bottom of wells, which may significantly weaken the effect of high pressure water jet, especially for the deep (≥4500m) and ultra-deep (≥6000m) wells. It is crucially important to study the effect of confining pressure on high pressure water jet. A high pressure water jet simulator that could generate low confining pressure (≤10MPa) by controlling the opening of outlet valve was designed. The jet pressure and impact pressure of a classical cone nozzle with different flow rate, confining pressure, and standoff distance were measured via experiment. Results show that jet pressure hardly changes until confining pressure exceeds its 0.51 times before it linearly increases with confining pressure; numerical fitting analysis suggests that axial hydrostatic pressure is proportional to the 3.3 power of confining pressure, and increases linearly with standoff distance; axial impact pressure is inversely proportional to the 0.15 power of confining pressure, and decreases linearly with standoff distance; but when dimensionless confining pressure exceeds the threshold value which is between 0.6 and 0.7 in this research, hydrostatic pressure is in accord with confining pressure, and impact pressure won't change which means the jet is already stable. This study provides the optimization of hydraulic parameters for drilling and sand-flushing operation and the design of highly effective nozzle with helpful instructions.

Organized and Sponsored by WJTA®-IMCA®

1 INTRODUCTION

During the drilling process for oil and gas resources, the high-pressure water jet is a key factor which can improve the rate of penetration (ROP) via breaking rock, cleaning well bore, carrying cuttings and so on. A detailed survey funded by American National Science Foundation of over 25 novel drilling techniques revealed that hydraulic jet had the potential for drilling oil wells economically in the late 1960 s (Maurer et al. 1969).

To improve the drilling rate, the mechanism of high pressure water jet assisted rock-breaking, assembly of nozzle, optimization of fluid, structure design of nozzle, and etc. had been researched (Feenstra et al. 1964; Feenstra et al. 1973; Kolle et al. 1991; Khorshidian et al. 2014). Meanwhile, studies on velocity and pressure distribution of high pressure water jet are underway (Albertson et al. 1950; Mclean et al. 1964; Shen. 1988). But it is always high confining pressure at the bottom of wells which may tremendously weakens the effect of high pressure water jet especially for the deep (≥4500m) and ultra-deep (≥6000m) wells. Robinson (1958) researched effects of confining pressures on failure characteristics of sedimentary rocks. Brighenti (1989) reported effect of confining pressure on gas permeability of tight sandstones. Alberts (1966) evaluated the performance of two types of abrasive cutting jets in chambers that simulated ocean depths to 6100m. Surjaatmadja (2010) conducted a series of tests to define new best practices for hydrajet perforating of rock under high confining pressure. But few studies report the impact pressure attenuation of water jet under confining pressure. With the development of exploration and drilling technology, deep and ultra-deep drilling, and deepwater drilling have become the focus of drilling. It is crucially important to study the effect of confining pressure on high pressure water jet.

The impact pressure of high pressure water jet with different confining pressure is measured by experiment. The exact expression of hydrostatic pressure and impact pressure with confining pressure and standoff distance are derived by numerical fitting analysis. This study provides the optimization of hydraulic parameters for drilling and sand-flushing operation and the design of highly effective nozzle with helpful instructions.

2 FACILITIES

2.1 High Pressure Water Jet Simulator

To study the effect of confining pressure on high pressure water jet, a high pressure water jet simulator which can generate low confining pressure (≤10MPa) is designed independently. The sketch and the photo of the simulator are shown in Figure 1 and Figure 2 respectively. It mainly consists of confining pressure cylinder, nozzle, pressure tap, standoff distance adjustment lever, high pressure outlet valve. Pressure gauge and pressure sensor are simultaneously installed at two holes at the inlet and outlet for observation and accurate measurement respectively. As the main part confining pressure cylinder resists the pressure. The standoff distance adjustment lever can change distance from pressure tap to the nozzle with no interruption to experiment. The high pressure outlet valve can generate confining pressure by controlling its open. A transparent pipe is used to recycle the water. The pressure tap connected with pressure sensor is a plane with a 1mm diameter hole in the center. This method to generate confining pressure is generally adopted in studies on confining pressure (Feenstra et al. 1973; Alberts et al; 1996; Surjaatmadja et al. 2010).

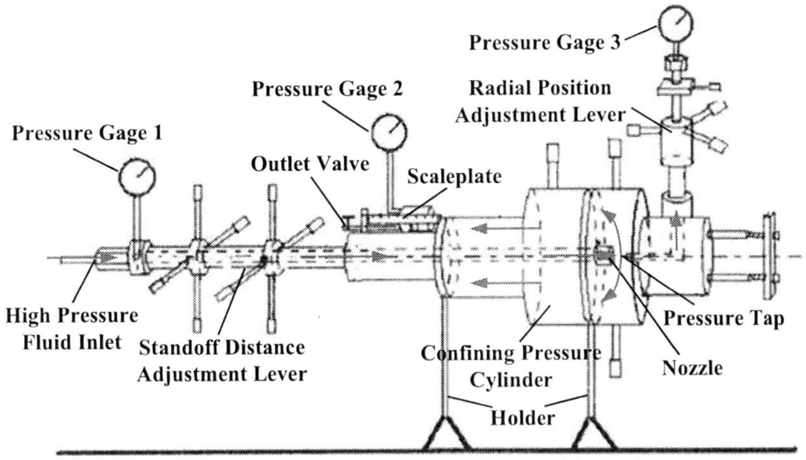

Figure 1. Sketch of High Pressure Water Jet Simulator

Figure 2. Photo of High Pressure Water Jet Simulator

2.2 Measurement Principle

As is shown in Figure 1, high pressure water flows through the inlet into the simulator and issues from the nozzle. The pressure gauge 1 and pressure sensor 1 which are installed close to the nozzle show and record the jet pressure labeled as P_1. After that, high velocity water impacts on the pressure tap, and total pressure labeled as P_3 is obtained by pressure sensor 3. At the end, water flows through the high pressure outlet valve which generates confining pressure. Confining pressure labeled as P_c is displayed and recorded by pressure gauge 2 and pressure sensor 2. According to the theorem of momentum (Yuan. 1986), impact pressure on a plane can be calculated by:

$$P_{impact} = \rho v^2 \tag{1}$$

Then the total pressure obtained by pressure sensor 3 can be expressed as:

$$P_3 = P_{static} + P_{impact} = P_{static} + \rho v^2 \quad (2)$$

2.3 Apparatus

A high-pressure plunger pump is used as the power source with a rated pressure of 60 MPa and a certified capacity of 100 L/min. A hydro-pressure sensor with a measuring range of 30 MPa, an output current of 4~20 mA, and an accuracy of 0.1% F*S is used to measure the pressure. A 10 MPa pressure gauge and a 30MPa pressure gauge are used to read the pressure. A high-pressure valve is used to control the confining pressure. A data acquisition system in which the National Instruments multi-channel data acquisition card is installed is used to collect and store the pressure data (Figure 3). Because the purpose of experiment is to investigate the effect of confining pressure on impact pressure, the general cone nozzle is the most suitable. The nozzle has an equivalent diameter of 3mm, contraction angle of 120 °, and outlet cylindrical section of 5mm. To ensure concentricity of the nozzle and the pressure taps, the inlet of cone nozzle connected with the inflow pipe with a compress cap and a O-ring seal is designed as a conical surface, as shown in Figure 4.

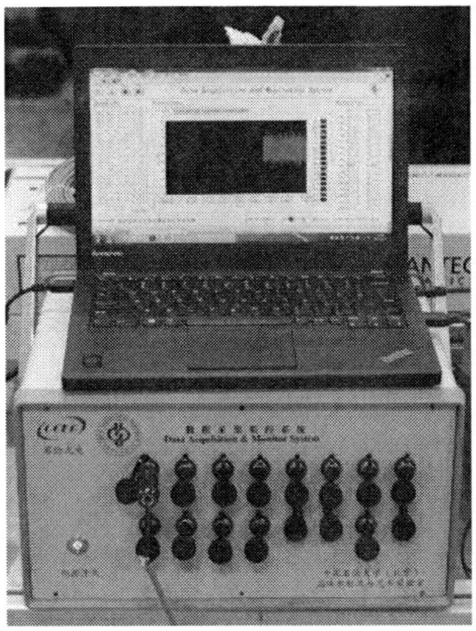

Figure 3. Multi-channel Data Acquisition System

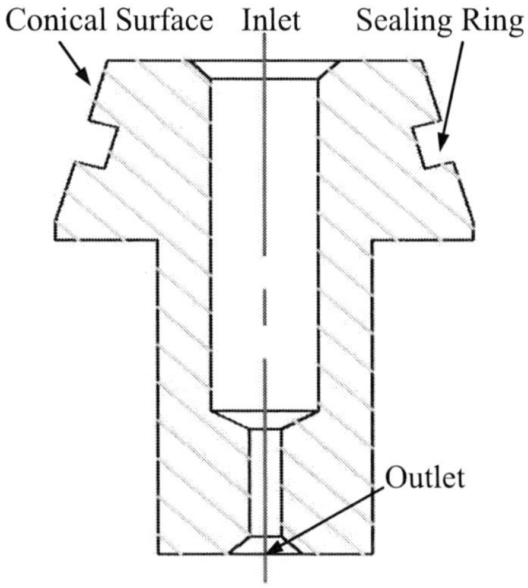

Figure 4. Sketch of Cone Nozzle

2.4 Projects

To study the impact pressure attenuation of high pressure water jet, flow rates are set as 0.6217, 0.6717, 0.7717L/s. Standoff distances are set as 1 to 7 times the nozzle diameter and are marked as 1d, 2d, 3d, 4d, 5d, 6d, 7d. Confining pressures are set as 0, 1, 2, 3, 4, 5, 6, 7MPa. All pressure data at different flow rates with different confining pressure and standoff distance are stored in the data acquisition system.

3 ANALYSES OF EXPERIMENT RESULTS

3.1 Validation

To validate the experiment method, the total pressure P_3 obtained without confining pressure is analyzed. According to Eq. 1, velocities at different standoff distance with different flow rates are obtained. The dimensionless axial velocity-dimensionless standoff distance plot is given in Figure 5. As is shown, the ordinate axis is the ratio of the measured axial velocity and the maximum axial velocity. Horizontal axis is logarithmic ratio of standoff distance to the nozzle diameter. The attenuation trends of the axial velocity with different flow rates are in good consistency with each other. According to Albertson's (1950) study, the intersection of two lines indicates the end of the potential core. The length of the potential core is about 4.6 times the nozzle diameter, which agrees with the fact very well. In conclusion, the experiment method is reliable to measure the impact pressure.

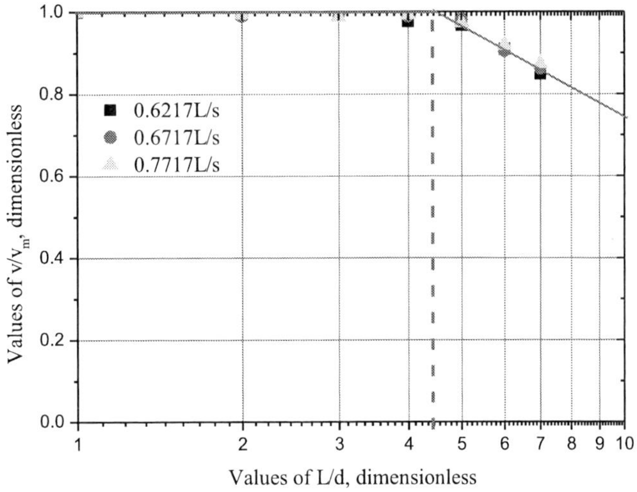

Figure 5. Axial Velocity of High Pressure Water Jet without Confining Pressure

3.2 Effect of Confining Pressure on Jet Pressure

According to the fluid dynamic energy equation, pressure drop of the nozzle (Chen et al. 2006) can be expressed as

$$\Delta P = \frac{0.05\rho Q^2}{C^2 A^2} \quad (3)$$

According to Eq. 3, pressure drop of a special nozzle is constant for a certain flow. During the experiment, the pressure drop labeled as P_0 is equal to the jet pressure without confining pressure for different flow rates. Relationship between jet pressure and confining pressure is given in Figure 6. The ordinate axis is the ratio of measured pressure (P_3) to pressure drop (P_0). The horizontal axis is the ratio of confining pressure (P_c) to pressure drop (P_0). According to submerge theory of high pressure water jet without confining pressure, the hydrostatic pressure in the flow field equals to environmental pressure. It can be inferred that at any cross-section of water jet the hydrostatic pressure is the same, as well as the axial direction (Shen. 1998). According to the law of conservation of energy, the jet pressure should linearly increase with the confining pressure. In fact, as is shown in Figure 6, jet pressure hardly changes until confining pressure exceeds its 0.51 times before it linearly increases with confining pressure. The effect of confining pressure on high pressure jet is complicated and needed to be further studied.

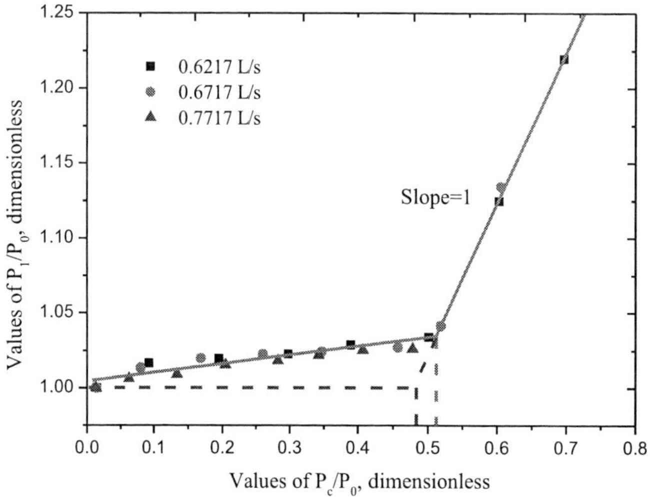

Figure 6. Effect of Confining Pressure on Jetting Pressure

3.3 Effect of Confining Pressure on Impact Pressure

To figure out the effect of confining pressure on high pressure jet, measured pressure P_3 and confining pressure P_c curves with different flow rates are given in Figure 7, Figure 8, and Figure 9 respectively. According to Eq. 2, measured pressure P_3 equals to the sum of hydrostatic pressure and impact pressure. It is easy to understand that hydrostatic pressure will increase with confining pressure which accelerates the attenuation of impact pressure. The variation of measured pressure P_3 is complex, but the slope of curves won't exceed 1. As shown in figures, dimensionless measured pressure P_3 at 1d has the same change law with jet pressure. It can be inferred that the confining pressure has no influence on high pressure jet within the standoff distance equals to 1 times nozzle diameter; as the emergence of confining pressure, measured pressure P_3 decreases quickly; with the increase of confining pressure, gradually the measure pressure P_3 stops decreasing but increases slowly instead; when the dimensionless confining pressure exceeds 0.51, measured pressure P_3 increases rapidly; as confining pressure continues to increase, the change of measured pressure tends to be linear with a 1:1 slope, which means flow field of water jet is already stable; there is a dotted line with a 1:1 slope for reference in each figure; the slope of measured pressure P_3 curve is obviously larger than 1 at the terminal section that is different with the previous analysis. We can tell that hydrostatic pressure is not evenly distributed; after dimensionless confining pressure exceeds the threshold which is in the range of 0.6~0.7 in our study, water jetting is already stable, and the measured pressure increases linearly with confining pressure.

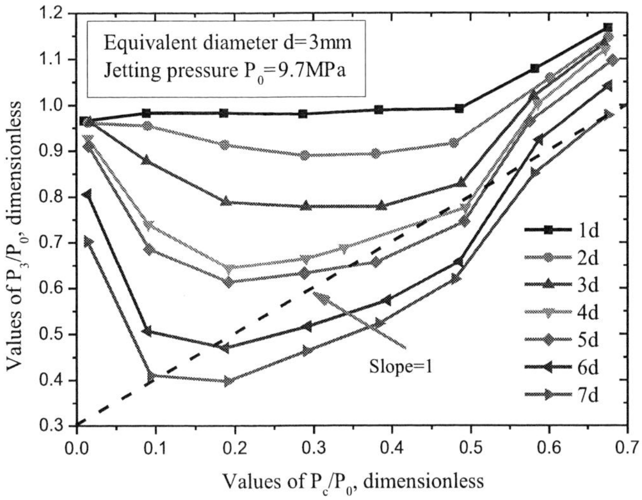

Figure 7. Effect of Confining Pressure on Measured Pressure P_3 When Flow Rate is 0.6217L/s

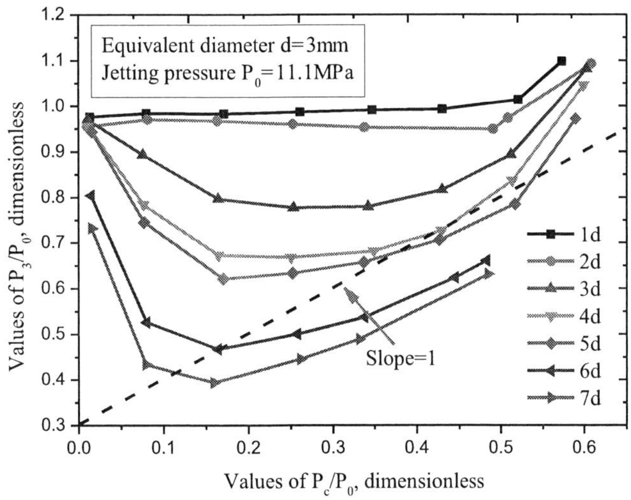

Figure 8. Effect of Confining Pressure on Measured Pressure P_3 When Flow Rate is 0.6717L/s

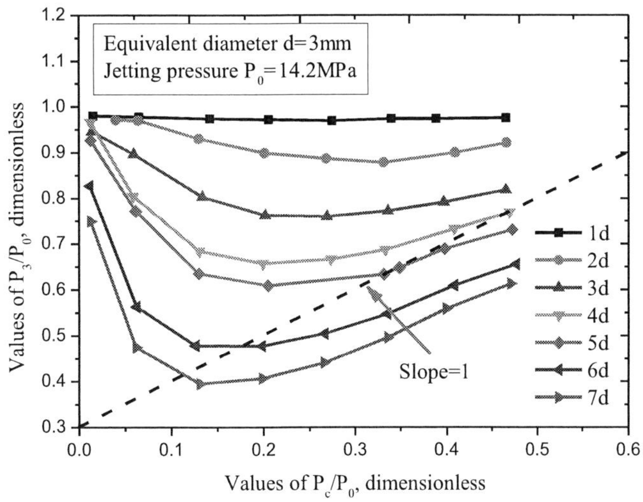

Figure 9. Effect of Confining Pressure on Measured Pressure P_3 When Flow Rate is 0.7717L/s

3.4 Numerical Fitting Analysis

To get the explicit effect of confining pressure on high pressure water jet, the measured pressure P_3 with different flow rates is summarized to Figure 10. As is shown in Figure 8, the measured pressure at 2d standoff distance differs from that in Figure 7 and Figure 9, so it is not included in Figure 10. On the basic of the analysis above, effect of confining pressure at 1d standoff distance is neglected in Figure 10, too. As shown, measured pressures with different flow rates agree with each other very well. Curves are fairly smooth which can be fitted by a single function. But when dimensionless confining pressure is larger than the threshold value, the smoothness of curves are destructed. When the dimensionless confining pressure equals to 0.7, high pressure water jet is already stable, so this part measured pressure is not considered in following parts. It is easy to understand that hydrostatic pressure is proportional to confining pressure and inversely proportional to standoff distance, while impact pressure has the opposite law. Based on above analysis, following function is used to fit the data; results are shown in Figure 11.

$$y = a * x^{\wedge} c + b / x^{\wedge} d \tag{4}$$

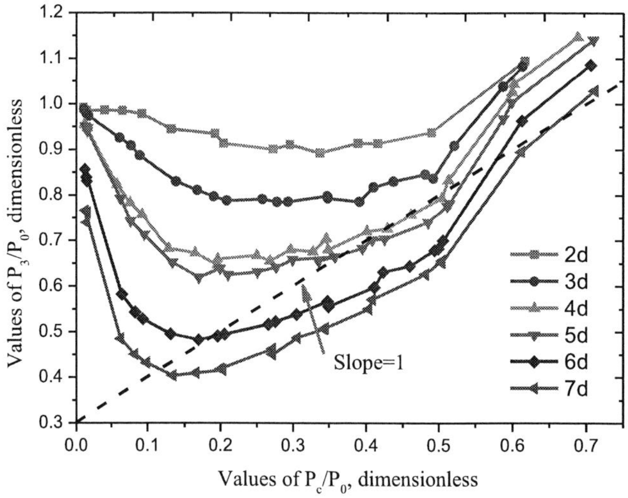

Figure 10 Effect of Confining Pressure on Measured Pressure P_3

As shown in Figure 11, the fitted curves smoothly across most of data. It proves the validity of the function which can describe the effect of confining pressure. At the same time, indexes of the function are proved to hardly change except for the 2d standoff distance. For convenience, the average of indexes is used, and Eq. 4 can be written as

$$P = a * \left(\frac{P_a}{P_0} \right)^{3.3} + b / \left(\frac{P_a}{P_0} \right)^{0.15} \tag{5}$$

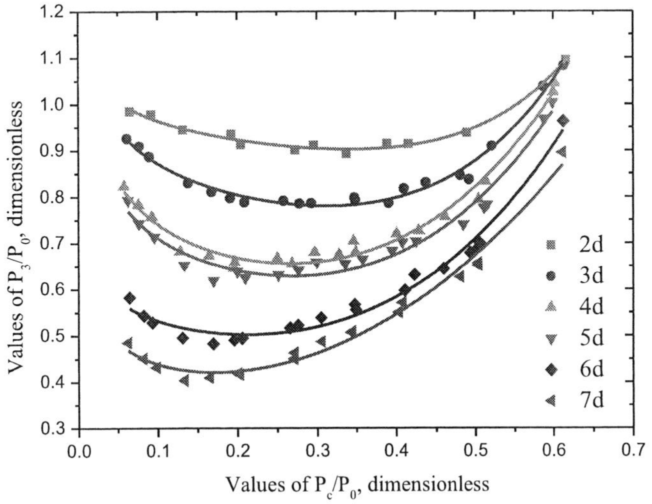

Figure 11. Results of y Function Fitting

Eq. 5 is used to fit the data again. Results are shown in Figure 12. As is shown, the fitting is preferable. The parameters of Eq. 5 turn out to be a function of dimensionless standoff distance. The fitting results are given in Figure 13. Expressions of parameters are given as follow:

$$a = 0.2244 * \frac{L}{d} + 1.371, \quad b = -0.0743 * \frac{L}{d} + 0.8413 \tag{6}$$

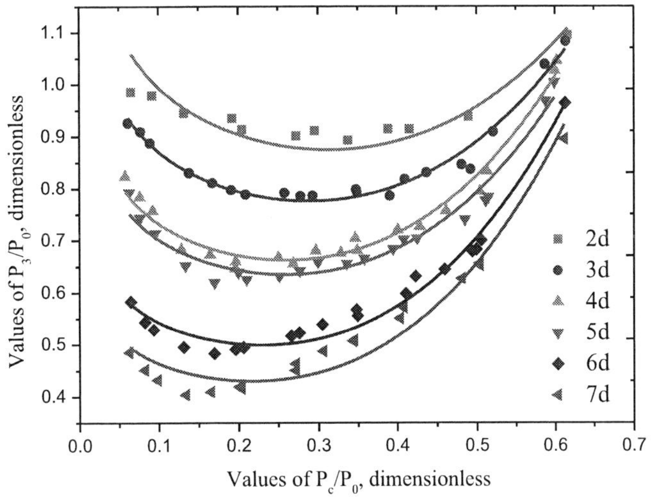

Figure 12. Results of P Function Fitting

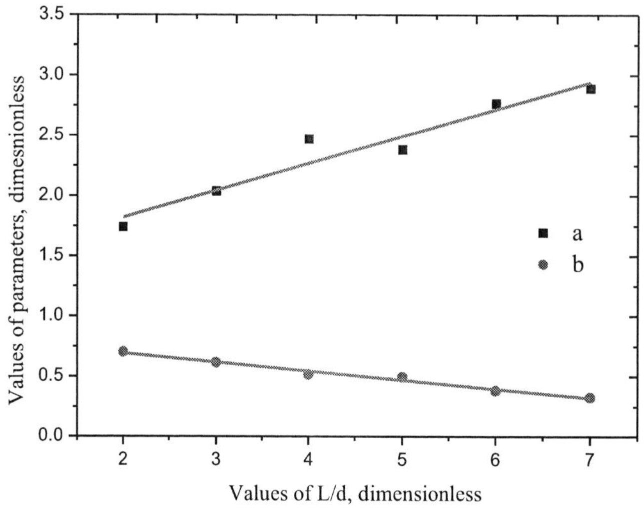

Figure 13. Results of Parameter a and b Fitting

Thus, measured pressure P_3 with different confining pressure and standoff distance can be expressed as:

$$\frac{P_3}{P_0} = \left(0.2244 * \frac{L}{d} + 1.371\right) * \left(\frac{P_c}{P_0}\right)^{3.3} + \left(-0.0743 * \frac{L}{d} + 0.8413\right) * \left(\frac{P_c}{P_0}\right)^{0.15} \quad (7)$$

According to Eq. 1 and Eq. 7, the distribution of hydrostatic pressure and velocity with different confining pressure and standoff distance are obtained (see Figure 14 and Figure 15).

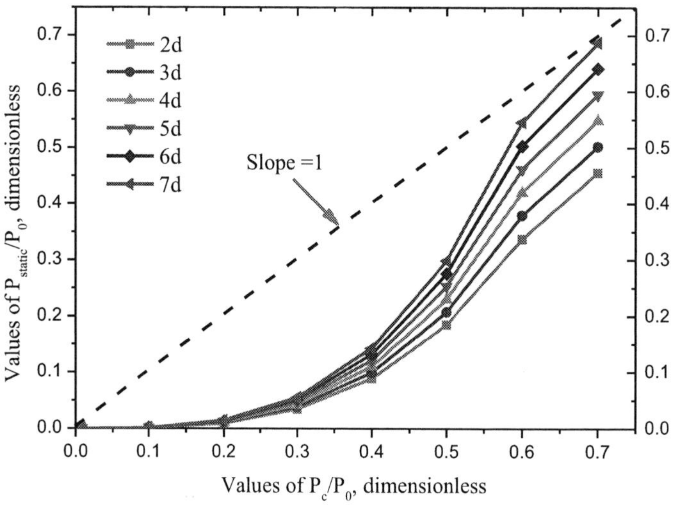

Figure 14. Changing Law of Hydraulic Static Pressure with Confining Pressure

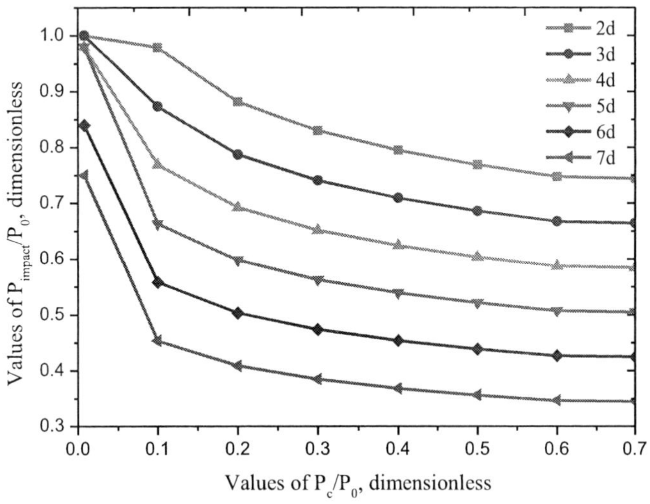

Figure 15. Changing Law of velocity with confining pressure

As shown in Figure 14, when P_c/P_0 is less than 0.2, hydrostatic pressure is tiny which is hardly affected by confining pressure; with the increase of confining pressure, hydrostatic pressure grows increasingly; when dimensionless confining pressure exceeds the threshold which is between 0.6 and 0.7 in this research, it tends to be linear with a 1:1 slope. On the contrary, impact pressure significantly decreases with confining pressure at the beginning; with the increase of confining pressure, the impact pressure attenuation gradually slows down; when dimensionless confining pressure exceeds the threshold, impact pressure basically remains unchanged, which means water jet is already stable. Figure 14 and Figure 15 explain the change law of Figure 10, which means that at the beginning of curves, the axial velocity dominates the change of measured pressure; at the end of curves hydrostatic pressure dominates the change of measured pressure.

4 CONCLUSIONS

The effect of confining pressure generated by controlling outlet valve's opening on high pressure is researched. The change law of high pressure water jet is obtained by experiment and theory analysis. The following conclusions are achieved:

(1) Jet pressure hardly changes until confining pressure exceeds its 0.51 times before it linearly increases with confining pressure.

(2) Hydrostatic pressure is not evenly distributed in the zone of high pressure jet. It is proportional to the 3.3 power of confining pressure, and increases linearly with standoff distance.

(3) Axial impact pressure is inversely proportional to the 0.15 power of confining pressure, and decreases linearly with standoff distance.

(4) There is a threshold which is between 0.6 and 0.7 in this research for dimensionless confining pressure. After that jet pressure will linear increase with confining pressure, hydrostatic pressure is in accord with confining pressure, and impact pressure won't change which means the jet is already stable.

ACKNOWLEDGMENTS

This study was funded by the National Science and Technology Major Project (No. 2011ZX05022002), the National Science and Technology Major Project (No. 2011ZX05036002) and Science Foundation of China University of Petroleum, Beijing (No. 2462013BJRC002). This support is gratefully acknowledged by the authors, who are also grateful to the reviewers of this paper for their detailed comments.

REFERENCES

Chen Tinggen and Guan Zhichuan. Drilling Engineering Theory and Technology. Dongying : China U. of Petroleum Press, 2006. 145-146. Print.

D.G. Alberts and M. Hashish. Evaluation of Submerged High-Pressure Waterjets For Deep Ocean Applications.The Sixth International Offshore and Polar Engineering Conference, 26-31 May, Los Angeles, California, USA. International Society of Offshore and Polar Engineers, 1 January 1996. Print.

G. Brighenti. Effect of Confining Pressure On Gas Permeability of Tight Sandstones. ISRM International Symposium, 30 August-2 September, Pau, France. International Society of Rock Mechanics, 1 January 1989. Print.

H. Khorshidian, S.D. Butt and F. Arvani. Influence of High Velocity Jet on Drilling Performance of PDC Bit under Pressurized Condition. The 48th U.S. Rock Mechanics/Geomechanics Symposium, 1-4 June, Minneapolis, Minnesota. Alexandria: American Rock Mechanics Association, 8 August 2014. Print.

Jim B. Surjaatmadja, Andrew J. Bailey and Silverio A. Sierra. Hydrajet Testing Under Deep-Well Conditions Points to New Requirements for Hard-Rock Perforating. SPE Drilling & Completion 1 September 2010: 372-379. Print.

Kolle, J.J., R. Otta and D.L. Stang. Laboratory and Field Testing of an Ultra-High-Pressure, Jet-Assisted Drilling System. Society of Petroleum Engineers, 1 January 1991: 847-857. Print.

L.H. Robinson. Effects of Pore and Confining Pressures on Failure Characteristics of Sedimentary Rocks. The 3rd U.S. Symposium on Rock Mechanics (USRMS), 20-22 April, Golden, Colorado. Alexandria: American Rock Mechanics Association, 1 January 1959. Print.

M. L. Albertson, Y.B. Dai, R. A. Jensen, and Hunter Rouse. Diffusion of Submerged Jets. Transactions of the American Society of Civil Engineers 1 January 1950: 639-664. Print.

R. Feenstra and J.J.M. Van Leeuwen. Full-Scale Experiments on Jets in Impermeable Rock Drilling. Society of Petroleum Engineers, 1 March 1964: 329-336. Print.

R. Feenstra and J. van Steveninck. Rock Cutting By Jets: A Promising Method Of Oil Well Drilling. Society of Petroleum Engineers, 1 January 1974:1-15. Print.

R.H. McLean. Crossflow and Impact Under Jet Bits. Society of Petroleum Engineers 1 November 1964: 1299-1306. Print.

Shen, Zhonghou and Sun Qingxiao. Study of Pressure Attenuation of a Submerged, Nonfree Jet and a Method of Calculation for Bottomhole Hydraulic Parameters. 1988. Society of Petroleum Engineers 1 March 1988: 69-76. Print.

Shen Zhonghou. 1998. Theory and Technology of Water Jet. Dongying: China U. of Petroleum Press, 1998. 269-270. Print.

William C. Maurer and Joe K. Heilhecker. Hydraulic Jet Drilling. Society of Petroleum Engineers, 1

January 1969: 213-217. Print.
Yuan Enxi. Engineering Fluid Mechanics. Beijing : China Petroleum Industry Press, 1986. 77-78. Print.

NOMENCLATURE

C	Nozzle discharge coefficient, dimensionless
P_{impact}	Impact pressure on the pressure tap, MPa
P_{static}	Hydrostatic pressure at the measured point, MPa
P_0	Jet pressure without confining pressure, MPa
P_1	Jet pressure measured by pressure sensor 1, MPa
P_c	Confining pressure measured by pressure sensor 2, MPa
P_3	Total pressure equals the sum of impact pressure and hydrostatic pressure, MPa
Q_i	Flow rate (i=0, 1, 2, 3), L/s
v	Average velocity of the outlet, m/s.
ΔP	Pressure drop of the nozzle, MPa
ρ	Density of water, Kg/m^3

2015 WJTA-IMCA Conference and Expo
November 2-4 • New Orleans, Louisiana

Paper

CUTTING AND DRILLING WITH SUPERCRITICAL FLUID SLURRY JETTING

K. Oglesby
Impact Technologies LLC,
A. Padsalgikar, Ph.D. and R. Mohan, Ph.D.
University of Tulsa
Tulsa, Oklahoma, U.S.A.

ABSTRACT

Use of solids-entrained-in-polymer water slurries (ASJ) pressurized through a 'pump' have been limited to offshore surface casing and pylon cutting and for downhole cutting of casing, tubing and other 'junk' in the borehole. Such high pressure (69-104 MPa) slurries are currently generated in batch mixed tubes or in large bore hydraulic or triplex pumps, accepting the inherent abrasive wear. New slurry pumps equipped with low wear patented HPSP technology have been utilized by Impact. New cutting and drilling techniques using the patented FLASH ASJ™ system, that uses supercritical gases, have been demonstrated for cutting steel, granite and sandstone and for drilling through sandstone. New patented nozzles have been designed for optimum expansion of the supercritical fluid slurries to obtain the desired cutting dimensions (width and depth). Low specific energies were found required to cut even the hardest material using CO_2, even with pressures below 5000 psig. Other supercritical gases (including nitrogen, air, propane, methane and even compressed air) are to be tested in the future. Targeted applications include mining, tunneling, trenching, excavation, fabrication material cutting and wellbore drilling in hard rocks.

1. INTRODUCTION

1.1 Water Jet Cutting Development

The ability of water to erode materials has been known since ancient times by observing running water cutting wider flow channels in soils and rocks. In current times we have enhanced that natural capability of water by pressurizing and controlling fluid systems at increasing pressures and rates. Later it evolved into utilizing additives to enhance the eroding power of the jet. The methods now available for a broad range of applications are: High Pressure Water Jetting (WJ), Abrasive Water Jetting (AWJ) and Abrasive Slurry Jetting (ASJ).

1.1.1 High Pressure Water Jetting

This method utilizes only pressurized water through a single nozzle focused onto a target to cut holes, slots and other penetrations into that target material. WJs can cut materials without heat treating or heat affecting the material, which is not possible with lasers or EDM [1].

Polymers are added to the water to keep the jet coherent for longer distances. Pressures and rates of such systems have increased as pump capability has allowed. Pressures in excess of 4400 bars (65,000 psi) are now routinely used in machine and job shops internationally. Nozzle materials and smoothness have improved for longer useful and economic life. The limitation of WJ is that water only jets cannot efficiently cut harder materials. WJ systems typically make very narrow cuts, which is not adaptable for drilling systems, except for kerfing to aid mechanical rotary bits [2].

1.1.2 Abrasive Water Jetting (AWJ)

In 1980 a method for injecting abrasive into the fluid jet stream was developed that made it possible to cut a much greater range of material at lower, though still considerable pressures. That development has led to abrasive waterjet cutting for precision cutting in a wide range of materials that includes metals, glass and rocks. AWJ technology adds the abrasives via a dry feed to the water stream at an induction nozzle, using air as the carrier fluid of the abrasive. The pressurized water, abrasives and air then passes through a mixing chamber / second focusing nozzle to the target material. As with WJ, considerable pressure is required for both the efficient induction of the abrasives and for cutting. This method has some performance issues since the air used as the solids carrier fluid must also be accelerated in passing through the mixing chamber. This required configuration also imposes some design problems in developing an AWJ drill capable of drilling in the Earth. AWJ systems also typically make very narrow cuts, which is not adaptable for earth drilling applications. AWJ has limitations of slow cutting rates (compared to lasers or EDM, but that gap is narrowing with the higher pressure pumps now available), thin kerf and small fillet radii (below 025mm), and not efficient for cutting carbides, ceramics and other hard materials [1].

1.1.3 Abrasive Slurry/ Suspension Jetting (ASJ)

To overcome the efficiency problems of AWJ and to provide a simpler nozzle system the direct use of abrasive slurries was developed, ASJ. Although the initial design for the abrasive slurry jet has been assigned to Fairhurst [3, 4] in 1982, there was significant reported research available prior to that time [1]. In 1981, Fair may have made the first large scale attempt to directly use pumped abrasive slurries for oil well drilling [1]. Hashish [5] reviewed the possible ways of making abrasive suspension jets and described a concept to directly pump abrasive slurries. Leach and Walker [6] worked on nozzle design including the need for high levels of surface finish and smoothness of flow in the nozzle construction. Selberg and Barker [7, 8] confirmed these conclusions and added consideration of the entrance flow path. Hollinger et.al. [10, 11], Hashish [12] and Resnick [13] further developed ASJ systems. Their studies showed that the cutting efficiency of short nozzles is higher than that of long nozzles and that increasing the length of the cylindrical part of the nozzle increases the jet coherence and cutting distance. However, it was later found that the use of long nozzle designs had little benefit when the nozzle was very close to the target.

Summers, et. al. [14] showed that a properly designed ASJ system at 700 bar will give as much energy to the abrasives as a 2,800 bar AWJ system, or a 4x improvement of ASJ over AWJ. The removal of the air carrier fluid from the system, and the concurrent acceleration of the water and abrasive through the cutting nozzle, and only one nozzle, gave the system much more efficient energy transfer between the water and the abrasive. The result was a jet that could cut through the hardest rock, with a driving pressures of only 5,000 psi and up to 12 inches of depth. It also has advantages in submerged water applications.

The limitations of ASJ are the narrow jet (typically less than 0.1-inch in diameter), which is too small for the nozzle to advance into the cut hole to advance the cutting depth, such as required in drilling. One disadvantage in both AWJ and ASJ is the amount of abrasive required per volume or mass of material cut, although used abrasive recycling systems are available. The next limitation of both methods comes from the difficulty in creating the pressurized slurry, specifically the wear on valve and seats in pumps. Lastly is the concern for continuous operation and smaller equipment for mobilization.

1.2 ASJ Slurry Pumps

In 1981 Fair probably demonstrated the first large scale attempt to directly pump ASJ slurries for oil well drilling. Hashish [5] reviewed the possible ways of making abrasive suspension jets and described a concept to directly pump abrasive slurries. The first development of the ASJ system occurred in 1984, when Fairhurst [4] enclosed the abrasive within a pressure vessel, and through control of feed water flowing into that chamber, was able to mix abrasive into the feed line between the high pressure water pump and the primary nozzle. From that he developed a low pressure (69 MPa) batch pumping system based on a fluidized bed concept.

Oilfield service companies have utilized low pressure, high rate abrasive slurry systems for downhole cutting of casings and tubings for many decades. In these applications they often utilize large triplex pumps to pressurize the low concentration slurries. Others have utilized modified versions of the Fairhurst system. Surface cutting of casings and wellheads for well control have also utilized low pressure abrasive slurry systems since there is inherent spark control.

1.3 Erosion Mechanics for ASJ Systems

Erosion mechanisms are governed by many different parameters which need to be studied individually. Meng and Ludema [15] undertook a comprehensive study of many previous researches done in the area of erosion modeling up to that time and came to the conclusion that so far no unifying erosion model had been proposed that could predict erosion for all materials and flow conditions. Due to this, there are many publications in literature which look at different parameters that affect erosion. Various people have looked at mechanisms of erosion and the effects of particle size, velocity, material hardness, and fluid viscosities. Finnie [16] concluded that ductile and brittle materials exhibit different mechanisms of erosion. Ductile erosion is described as resulting from plastic deformation of material. Finnie [17] reported that the plastic deformation is due to multiplication of a surface dislocation caused by impacts. According to Levy [18] erosion takes place due to the shear deformation that occurs at the point of impact of abrasive particles and metals. Due to repeated impacts the material is heated close to its annealing temperature. Small distressed platelets are formed in this softened layer. The metal layer under this softened layer remains hard and gets cold worked by plastic deformation. The erosion efficiency actually is increased due to this hard layer. Consecutive impacts remove the platelets formed by the initial impacts.

One of the major factors affecting erosion is particle impact velocity. Equation 1 describes the empirical power law relation between the erosion ratio and the velocity of the impacting particle.

$$ER \propto V^n \qquad \text{(Equation 1)}$$

where ER is the erosion rate, V is the velocity of the particles and n is an empirical constant. Finnie [19] explained that the exponent 'n' discussed should increase with impact angle for a range of velocities and ranges from 2.05 to 2.44. Lindsley and Marder [20] studied the empirical power law relation described in Equation 1 and showed that the value of 'n' was 2.9 for two different erosion mechanisms namely brittle cracking and plastic deformation mechanisms.

Another factor affecting erosion rate is the impact angle of the particles. Finnie [21] derived an angle function for rigid particles impacting ductile materials. The model shows that maximum erosion for aluminum alloys is at 13 degrees and becomes nonexistent at zero and ninety degrees. The discrepancy between measurement and model is explained by the existence of different mechanisms for erosion at high and low angles. At low angles the erosion is due to cutting mechanisms while at high angles it is due to surface roughening and low cycle fatigue fracture. Winter and Hutchings [22] studied another factor that affects erosion rate, the particle orientation before impact. They studied the rake angle of the particles when they impacted the surface. They found that when the impacting particles had positive or small negative rake angles the erosion occurred via a cutting mechanism. However, at large negative rake angle values, the erosion was due to a ploughing mechanism. Palasamudram and Bahadur [24] furthered this work and proposed another parameter called the angularity parameter 'An'. This parameter is calculated using geometrical parameters of abrasives such as shape, size, sharpness, etc. Erosion rates of 1020 Steel were measured and it was found that the erosion rate increased with an increase in n when particle size was maintained at a constant value. It was also observed that the erosion increased with particle size if n was kept constant.

An Erosion Model was developed at the Erosion/Corrosion Research Center of The University of Tulsa [25]. Since erosion is dependent, among other things, on the target material, the equations developed are different for different materials. Even though this equation was developed for a specific material, it can be used for other materials if the constants C, n, and the angle function are determined by erosion testing of that material. In general, the erosion equation can be expressed as follows:

$$ER\ (kg/kg) = F_S \cdot C \cdot V^n \cdot F(\theta) \qquad \text{(Equation 2)}$$

where ER is the erosion ratio (material mass loss/mass of impacting particles), C and n are empirical constants, V is the particle impact velocity in m/s, F_S is particle shape factor (0.2 for fully rounded, 1.0 for highly angular), and $F(\theta)$ is a function of the impingement angle where the angle, θ is measured in degrees. This full model development will be reported separately [25].

1.4 Drilling Boreholes with Abrasive Fluids

Gulf Oil Company and others successfully tested abrasive cutting combined with rotary drilling in the field to depths of 15,000 ft. High rates-of-penetration (ROP) were achieved, but problems with high wear on the pumps, pipes and connections leading to higher costs outweighed the benefits of the higher ROP [2]. There is a push in the oilfield and geothermal industries to go to smaller bores, which pushes the limits to what can be done only mechanically. Typical oilfield industry wells use standard 6-1/4" to 24" OD drilling bits. Geothermal wells target bore sizes ending at depth with 8-1/2" and up to 36" at the surface. Those sizes require a lot of rock to be removed taking a lot of energy and very large equipment. The next step down in borehole size is called 'Slimhole Drilling' with bit sizes from 4" to 6-1/4", with 4-3/4" as standard. Slimholes have found acceptance in selected industry applications, primarily in oil and gas directional and horizontal drilling using conventional rotary bit drilling systems [27].

Microbores are bores of less than 4" in diameter. This is a new technology frontier where the conventional rotary bit drilling systems that require torque and weight on bit cannot perform satisfactorily. The potential benefits of microbores are:

1. Less rock removed, thus faster drilling;
2. Less fluids, muds and chemicals required and less volume for disposal;
3. Smaller tubing and casing required for reduced cost;
4. Smaller operational environmental footprint;
5. Smaller (hence less expensive) tubulars and rigs, even allowing for coiled tubing systems.

But microbores are hard to achieve with conventional drilling due to the limited weight-on-bit and torque that is possible with small diameter pipe. This is where ASJ can provide significant benefits due to its low reactive force, no torque and small tool sizes. However full bore cutting or rotation of the nozzle is required, but not possible with the narrow cut of conventional ASJ.

2. DISCUSSION

2.1 Impact HPSP Slurry Pumps

The wear on the ball and seat of all type pumps has been the concern for ASJ systems in reaching the desired pressures. A valve closing on hard, dense, angular abrasive particles can cause rapid wear of the valve and seat faces, as well as on any non-metallic seals. High pressure slurry pump (HPSP) systems were developed to flush abrasive from the valves just prior to closing to minimize wear. HPSP pumps are patented under US11/033,615 for piston and US11/705,222 for plunger versions, as well as internationally [28, 29, 30]. Some examples of HPSPs with clean fluid valving are given in Figures 1, 2 and 5.

Figure 1. HPSP1 Piston Pump, hydraulically driven

Figure 2. HPSP2 Plunger Triplex Pump, hydraulically driven

Such pumps have been used for a variety of applications, including for abrasive cutting up to 15,000 psi, 5000 psi FLASH ASJ^{TM} abrasive cutting using supercritical fluids, and abrasive bio-pulp processing. The limitations of HPSPs are that a relatively long stroke, meaning a larger pump size, is required. In addition, a viscous clean flush fluid and more viscous carrier fluid are required for high concentration abrasive systems to carry and control abrasive particles when required. This requires a pressurized suction system.

2.2 Advancements beyond ASJ

The need for a wide cut from the jet nozzle has not been a big concern or need in job shop machining or demolition. However, it is needed for full bore microhole drilling systems or for trenching to install utility cables or small pipes. For drilling, a bore that is wide enough for the nozzle to advance and to allow unimpeded return flow to the surface is required. Lower pressure systems would also be advantageous to reduce the cost of tubulars and pumps. To that end FLASH ASJ^{TM} technology systems were developed and tested by Missouri University of Science and Technology, Impact Technologies LLC and Oak Resources Inc.[27]. Such systems utilize a slurry jet created with supercritical fluids, including carbon dioxide (CO_2), nitrogen (N_2), propane, methane and steam. Only one simple nozzle is required. The process relies on the pressure reduction to drive jet, plus expansion of the supercritical fluid through the nozzle. In addition to the higher particle velocities delivered to the target, a wider pattern is developed, the standoff distance (nozzle to target) now has a gas phase for improved particle-target interactions, and a lighter gas phase exists for aiding and clearing the wellbore of solids. FLASH ASJ systems are trademarked and patented under US12/400,507 for method and fluids and US13/589,626 for nozzle and methods [32, 33]. The new process has been tested at MS&T in 2002-2008 [30] and at Impact Technologies LLC in 2005-2013, both reported now.

Potential applications of this new technology are: oilfield enhanced oil recovery and tight unconventional resources that require small drainholes; shallow (60 meter depth) geothermal heat pump heat exchange bores; shallow instrumented wells; trenching in hard rocks (granites, basalts, limestones, etc.) for utilities, cables and pipes; and tunneling, especially the cut along the finish wall. Related technology developments are- Inverted Motors for downhole rotation of the nozzle using pressurized abrasive fluids, Inverted Drainholes for low pressure recovery of oil or environmental pollution, and TOP Plate for multiple bores from one surface site.

Impact and Missouri University of Science and Technology (MS&T) initially developed the supercritical gas jet system under a Department of Energy (DOE) project, DE-FC26-04NT15476 [27], for very fast full-bore drilling without rotation, no torque, no weight on bit, and low (less than 5000) standpipe pressures. The system consists of abrasives, supercritical fluids, chemical additives, a specialized and patented nozzle [33] and modified patented HPSP pumps [29, 30]. Examples of the desired supercritical fluids include water as steam, carbon dioxide, methane, propane, butane and nitrogen. The supercritical fluid must be in its liquid or near liquid state inside the drill string and into the nozzle to suspend carry the abrasive particles. It then transitions to a low-density gas or fluid across the nozzle, expanding in volume by 5 to 12 times and further accelerating the abrasive particles and carrier fluid, in addition to the normal flow area restriction of the nozzle. With proper nozzle design, expansion of the gas phase propels the abrasive into a wider diameter pattern creating a wider cut bore.

2.3 Bench Testing at Missouri University of Science and Technology

Using a modified DiaJet pump system to create the supercritical CO_2 abrasive fluids, MS&T tested the new abrasive cutting system on a variety of rocks and materials, including limestone, sandstone, basalt, cement and steel. This work was performed in 2004 to 2008. The results of those tests can be seen in Figures 3-4 and Table 1 [27].

Figure 3. Drilling sandstone using 4,000 psi CO_2 at MS&T

Figure 4. MS&T Hole through steel, concrete and rock using CO_2 abrasive system

The rate in these tests was highly variable due to the dry ice CO_2 source, but the success in drilling full bores in a variety of rocks, including basalt was considered significant. In addition, extremely low cutting specific energies (SE) were found.

Table 1. Cutting Performance of a CO_2 Abrasive System at MS&T.

Rock	Jet Pressure (MPa)	Nozzle Dia. (cm)	ROP Max (m/min)	ROP Min (m/min)	Specific Min (J/mm³)	Energy Max (J/mm³)	Hole Dia. (cm)
Roubideaux	24.15	0.111	4.85	1.07	0.133	0.560	2.54
Roubideaux	24.15	0.111	3.41	0.64	0.150	0.810	2.22
Joachim lls	27.8	0.099	4.21	1.19	0.410	1.443	1.52
Joachim lls	27.8	0.099	2.71	2.71	0.360	0.360	2.03
Joachim lls	27.8	0.099	3.60	0.76	0.334	1.560	1.80
Joachim lls	27.8	0.099	4.85	0.61	0.560	4.510	1.25
Indiana lls	27.8	0.099	3.0	1.19	0.207	0.519	2.54
Missouri do	27.8	0.099	4.51	0.64	0.216	1.488	2.03
Missouri do	27.8	0.099	4.51	0.98	0.216	0.992	2.03
Missouri do	27.8	0.099	2.71	0.61	0.736	2.210	1.40
Basalt	**27.8**	**0.099**	**0.91**	-	**3.00**	-	**1.25**

2.4 Full Scale FLASH ASJ Testing at Impact

Impact Technologies' facility was designed to provide material removal information for various rock targets at a variety of rates and slurry/gas ratios. Figure 5 shows the experimental facility which consisted of two separate pumping systems- a high pressure gas pump to raise the CO_2 pressure and a high pressure slurry pump with HPSP clean fluid valving to deliver the high concentration abrasive slurry to the desired operating pressure. The slurry pump was driven by a high pressure water triplex pump. The water – polymer – abrasive slurry mixture was created in a storage tank prior to the test to control quality. The gas was pumped through a 2-in insulated stainless steel pipe to the mixer. A concurrent flow water jacket was placed around the pressurized gas pipe to raise the temperature of the gas to ensure that the gas stays in the super critical region. The super-critical gas and slurry mixture was delivered to the nozzle from the mixer via a 2-in stainless steel pipe. All water and gas rates were metered

The target rock was positioned on a moveable table at a fixed distance from the nozzle. All tests had a 2" standoff, except for tests 39 and 40 which had 6" standoffs. The table allowed the rock to be positioned at various standoff distances to the nozzle and also to allow transverse or movement of the target toward the nozzle (simulating drilling). The table transverse speed was adjustable, but was fixed during all tests at 0.1 meters/ minute. For transverse tests, the target rock was cycled across the nozzle on a continuous loop and on the same line until the test ended.

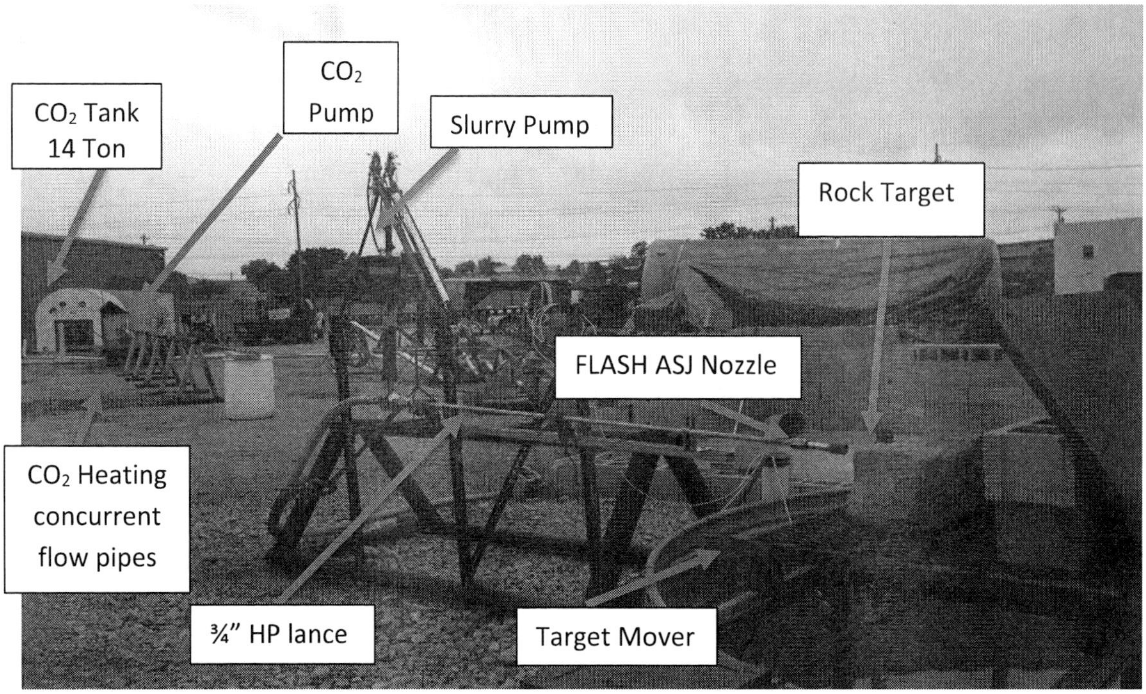

Figure 5. Impact Full Scale Test Facility

Figure 6, a, b, c. Rock Target Test Results

Both red and gray granites were tested with approximate properties of:

Density -	2700	kg/m^3
Young's Modulus (E)-	10-70	GPa
Modulus of Rigidity-	24	GPa
Unconfined Compressive Strength-	100-250	MPa
Shear strength-	14-50	MPa
Tensile strength-	7-25	MPa

Also, a local Oklahoma medium strength Red Sandstone was used in a few tests. Garnet abrasive sizes used were 40& 60 mesh, with the 60 mesh the most commonly used. Except for the tri-nozzle, all nozzle holders had outer diameters of about 3.2 cm (1.25 inches). A variety of nozzle designs were used, which contributed to the spread in the test data. A summary of these tests are given in Table 2 and Figures 7-10 in the evaluation section. A "B" on test number indicates it was a bore/ drilling test. Prior to CO_2 slurry cutting tests, steam tests were conducted with pressurized water heated to about 204°C (400°F), but these steam tests are not included in this report as different nozzle designs were utilized.

2.5 Evaluation of Bench Test Results

Figures 7-10 and Tables 1 and 2 were utilized to evaluate the bench test data. The first goal of these tests was to obtain operating proficiency of the combined supercritical abrasive system. The next goal was to obtain as wide a cut as possible with the small fixed nozzle. In this regard 20-107 mm widths were obtained, mostly on the first pass of the nozzle. Secondary passes only increased depth of cut. This is 6 to 33 times the diameter of the nozzle holder and would allow easy passage of the nozzle through the cut hole and allow sufficient return flow.

The next goal was to compare efficiency standards of this new system to other established abrasive cutting systems and to optimize the system's parameters. To do this two metrics were utilized- specific energy (SE) defined as the power required for a material removal rate (MRR) and an abrasive utilization efficiency term defined as mass of material removed per mass of abrasive utilized. Even though the abrasive and material debris can be processed and reused, this has a cost and was considered important to optimize the system.

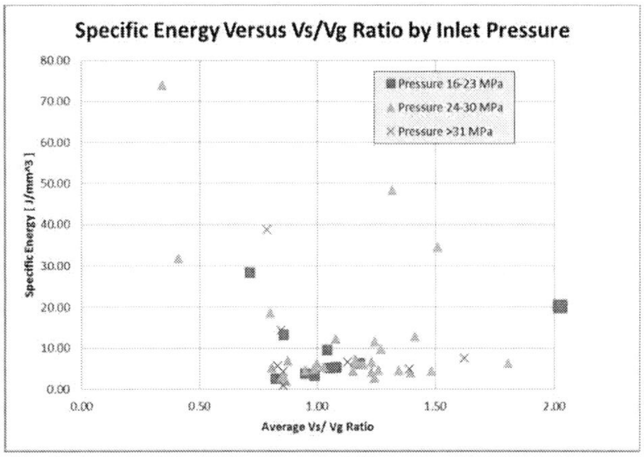

Figure 7. Specific Energy (J/mm^3) versus Slurry/Gas Volume Ratio Sorted by Inlet Pressure (MPa)

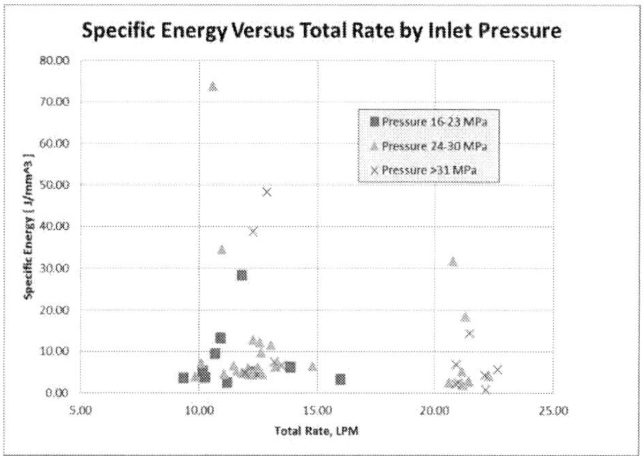

Figure 8. Specific Energy (J/mm^3) versus Total Rate (LPM), Sorted by Inlet Pressure (MPa)

Table 2. Impact Test Data using Carbon Dioxide on Rock Targets

Test No.	Nozzle	Type	Rock Cut (mm) Width	Rock Cut (mm) Depth	Volume	40-60 mesh Abrasive Used (kg)	Cutting Time min	CO_2 Temp °C	Inlet Pressure Mpa	Slurry Vol: Gas Vol Ratio	Total Rate LPM	Abrasive Efficiency R-kg/A-kg	Specific Energy J/mm^3
12	500S-2	Grey Granite	30.70	22.64	169,688	23.90	12.0	25.5	29.68	1.81	13.23	0.029	6.30
13	500S-2	Grey Granite	25.34	40.68	248,395	52.80	26.3	26.4	37.96	1.13	13.49	0.042	6.68
14	500S-2	Grey Granite	24.57	58.62	339,687	34.84	14.5	29.4	32.96	1.62	13.22	0.028	7.52
15	500S-2	Grey Granite	26.91	57.77	338,442	25.99	18.3	27.7	27.79	1.16	10.07	0.028	7.32
16	500S-2	Grey Granite	32.61	71.60	516,520	20.37	12.8	23.9	24.86	1.39	9.82	0.042	3.96
17	500S-2	Grey Granite	19.78	16.07	72,988	19.96	20.3	32.8	28.75	0.34	10.54	0.006	73.98
18	500S-2	Grey Granite	19.59	55.43	246,183	10.75	14.0	32.8	19.17	10.00	9.31	0.020	3.73
19	500S-1	Grey Granite	20.25	11.79	48,194	46.67	24.3	33.6	30.06	1.51	10.95	0.003	34.60
20	500S-1	Grey Granite	28.31	8.27	65,368	33.25	8.8	36.3	33.44	0.79	12.29	0.004	38.77
21	500S-1	Grey Granite	24.39	30.76	151,203	30.39	9.7	5.2	28.89	1.24	13.05	0.009	11.51
22	500S-2	Grey Granite	22.95	46.97	238,618	39.75	9.5	30.9	23.52	1.19	12.10	0.015	6.05
23	500S-2	Grey Granite	23.89	47.42	261,497	33.03	18.2	34.2	24.34	1.00	12.51	0.016	6.22
24	500S-2	Grey Granite	23.82	10.12	65,327	27.82	8.7	37.8	23.23	0.71	11.80	0.004	28.49
28	500S-2	Grey Granite	24.78	41.40	240,432	57.66	15.8	43.5	24.95	1.16	14.81	0.028	6.45
29	500S-2	Grey Granite	23.59	33.70	196,940	139.22	32.5	40.3	20.25	1.18	13.86	0.023	6.34
30	500S-2	Grey Granite	25.27	36.71	211,375	89.67	28.5	45.1	16.82	1.05	12.34	0.025	5.18
31	500S-1	Grey Granite	23.27	59.59	325,129	20.48	15.5	24.5	25.59	1.48	12.28	0.038	4.40
32	500S-1	Grey Granite	28.68	51.49	366,989	30.15	12.5	4.4	31.67	1.39	12.00	0.043	4.95
33	500S-1	Grey Granite	27.68	28.78	201,910	28.78	14.5	45.0	17.02	1.08	10.12	0.024	5.41
34	500S-1	Grey Granite	28.36	34.04	240,059	19.68	5.2	41.0	23.67	0.95	11.06	0.040	4.69
35	R100-3"	Grey Granite	19.02	39.02	172,812	13.37	8.0	37.4	27.36	1.23	11.47	0.029	6.66
36	500S-1	Grey Granite	25.50	41.80	257,067	13.39	8.2	34.1	28.95	1.26	12.70	0.043	4.68
37	500S-1	Grey Granite	29.37	33.43	255,662	17.16	9.5	29.8	26.66	0.99	11.82	0.043	4.87
38	500S-1	Grey Granite	24.82	39.55	230,315	15.11	7.8	10.0	26.94	1.03	11.60	0.039	5.35
39	500S-1	Grey Granite	58.12	73.74	87,039	14.00	7.7	-7.3	28.07	1.41	12.29	0.015	12.77
40	500S-1	Grey Granite	60.45	242.09	231,612	168.62	64.2	36.6	31.21	1.32	12.86	0.004	48.36
41	500S-1	Grey Granite	21.96	26.65	126,254	5.54	9.3	48.0	29.87	1.27	12.64	0.060	9.80
42	500S-1	Grey Granite	23.51	16.24	83,111	6.96	9.8	50.2	17.42	1.04	10.65	0.040	9.53
43	500S-1	Grey Granite	21.67	24.22	109,461	4.88	6.8	46.0	29.85	1.08	12.60	0.052	12.18
44	500S-1	Grey Granite	22.06	14.75	69,639	6.28	7.2	49.9	18.38	0.86	10.90	0.033	13.24
45B	500S-1	Grey Granite	36.47	163.45	85,392	129.21	62.3	47.8	25.83	1.04	11.77	0.007	30.21
46	500S-1	G Granite-500oF	25.85	59.11	243,976	13.89	4.5	43.6	27.45	1.34	12.08	0.041	4.55
47	500S-1	G Granite-500oF	29.92	64.51	289,142	16.04	5.7	44.9	30.16	1.15	12.24	0.049	4.52
48	500S-1	G Granite-500oF	31.44	41.97	235,258	20.47	6.8	43.0	19.00	0.95	10.24	0.040	3.85
49	500S-1	Grey Granite	30.93	94.27	434,915	22.03	7.3	45.0	21.90	0.83	11.17	0.073	2.58
50B	500S-1	Sandstone	39.42	200.20	162,899	131.75	50.8	35.7	24.92	1.29	12.34	0.003	55.37
51	500S-1a	Grey Granite	42.55	57.26	661,517	28.74	7.0	46.0	30.11	1.23	22.27	0.051	4.21
52	500S-1a	Grey Granite	45.75	35.33	451,703	35.41	8.3	58.9	15.02	0.99	15.99	0.035	3.41
53	500S-1a	Grey Granite	41.98	58.38	619,287	26.51	6.8	40.4	31.81	0.83	22.63	0.047	5.77
54	500S-1a	Grey Granite	30.26	16.97	144,491	65.61	13.2	38.2	28.82	0.41	20.77	0.006	31.85
55A	500S-1a	Grey Granite	52.43	63.97	1,265,576	58.05	8.5	36.2	30.82	0.86	21.01	0.056	2.61
56	500S-1a	Grey Granite	47.01	60.25	840,429	54.57	7.2	2.0	26.69	1.24	21.43	0.037	2.84
57	500S-1a	Grey Granite	49.57	35.00	610,439	63.42	9.2	48.1	29.30	0.81	21.15	0.027	5.30
58	500S-1a	Grey Granite	48.59	10.07	180,910	61.40	8.0	53.0	30.07	0.80	21.27	0.008	18.47
59	500S-1a	Grey Granite	75.37	59.18	1,056,521	61.51	8.0	33.6	29.07	0.85	21.42	0.047	2.96
60	500S-1a	Grey Granite	47.35	25.40	1,506,725	0.00	0.0	31.4	30.58	0.87	20.88	0.066	2.16
61	500S-1a	Grey Granite	43.69	-	1,190,913	0.00	0.0	14.2	29.17	0.86	20.61	0.052	2.62
62	500S-1a	Grey Granite	46.25	-	469,096	2.72	0.0	33.9	30.72	0.88	20.91	0.020	6.93
63	500S-1a	Red Granite	50.11	30.39	1,097,425	0.00	2.0	29.9	28.38	0.87	21.18	0.047	2.06
64B	500S-1a	Red Granite	35.36	59.18	173,867	61.51	8.0	31.6	27.30	0.84	20.79	0.007	68.22
T65-68B	500S-1a	Sandstone	107.95	-	3,408,509	0.86	1.7	31.7	29.82	0.91	21.10	0.147	0.00
T69-70B	500S-1a	Sandstone	84.33	-	3,052,910	1.11	2.5	31.8	30.52	0.89	21.28	0.132	0.00
71	500S-1a	Red Granite	50.52	-	296,884	0.45	0.5	32.9	31.53	0.85	21.46	0.013	14.31
72	500S-1a	Red Granite	49.61	-	526,041	0.91	2.0	30.1	34.01	0.86	22.12	0.023	4.32
73	TriNozzle	Sandstone	64.16	-	5,000,021	0.00	0.0	34.5	32.11	0.86	22.15	0.214	0.86
74B	TriNozzle	Sandstone	42.19	-	849,997	0.00	0.0	33.0	29.57	0.83	20.82	0.036	4.73

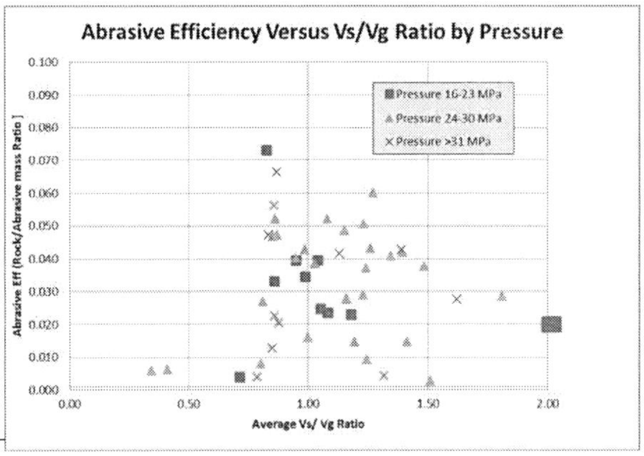

Figure 9. Abrasive Efficiency (Rock Removed/Abrasive Used, kg) versus Slurry/Gas Ratio, Log Scale on Efficiency axis

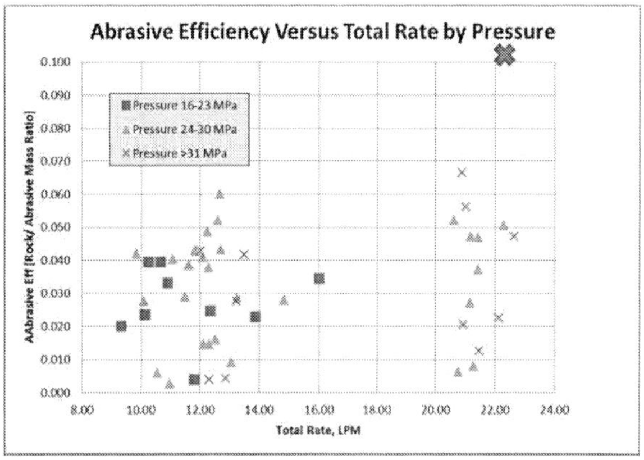

Figure 10. Abrasive Efficiency (Rock Removed/abrasive Used, kg) versus Total Rate (LPM), Sorted by Inlet Pressure (MPa)

The variability of the data was considered a function of the size of the system, the number of different nozzles tested and the limitations on the accuracy of controlling the various parameters during the tests. It can be seen from the data that SE is not a function of total flow rate. Gas volumetric content greater than the slurry volume provides the best SE values. Surprisingly, no strong influence of inlet pressure was found on SE values. Abrasive efficiency was relatively low across all conditions tested, ranging from 0.5 to 7% and averaged about 4%.

2.6 Evaluations at The University of Tulsa

Computer simulations of the supercritical gas abrasive cutting process were made and a model was developed for predicting performance [25]. Fluent™ was used to make a discrete phase simulation of the flow of particles through the nozzle. The particles of the discrete phase were assumed to be spherical in shape with a diameter of 300 μm and a density of 4100 kg/m3 (corresponding to garnet particle) and were injected from the inlet face. Figure 11 shows the results of the simulations, the plot shows the velocity of the particles, which are colored according to their velocity magnitude at that location. It can be observed that like the carrier fluid, the particles are accelerated through the nozzle. It is apparent that the velocity of the particles is also conserved similar to the fluid. It was observed that the particle velocity near the wall is significantly higher with this nozzle design. In particular, the velocity of the particles in the region away from the centerline is also high implying that this higher velocity will contribute towards more efficient material removal along the outer diameter of the wellbore.

2.7 Comparison of Specific Energies (SE) in the Literature

SE is not the only factor to compare in evaluating various cutting systems, but it provides a relative comparison of the efficiency of various processes and for optimization within a given process. From Tables 1 and 2 for the MS&T and Impact FLASH tests, the SE for cutting ranged from 0.133 to 74 J/mm^3 and averaged 10.5 J/mm^3. Ucun et. al. reported 4.5-10 J/mm^3 for a diamond circular saw [34]. Summers in Waterjetting Technology [1] reported water jetting (WJ) SEs at 5,000-25,000 J/mm^3. Darling [35], from data sourced by Rostamie et. al. 1994 [36], reported SEs based on mean particle size of the detrital – Bit Drilling averaged 0.21 J/mm^3 (50-70 Hp*hr/yd^3) and Tunnel boring averaged 0.035 J/mm^3 (10 Hp*hr/yd^3). Kolle [37] reported - Ultra-high pressure waterjetting (WJ) at 5-100 J/mm^3, abrasive water jetting (AWJ) at 3-100 J/mm^3, abrasive slurry water jetting (ASJ) at 70-500 J/mm^3, and Diamond rotary bit at 1-10 J/mm^3. Yao reported 10-80 J/mm^3 for a mechanical bit [38].

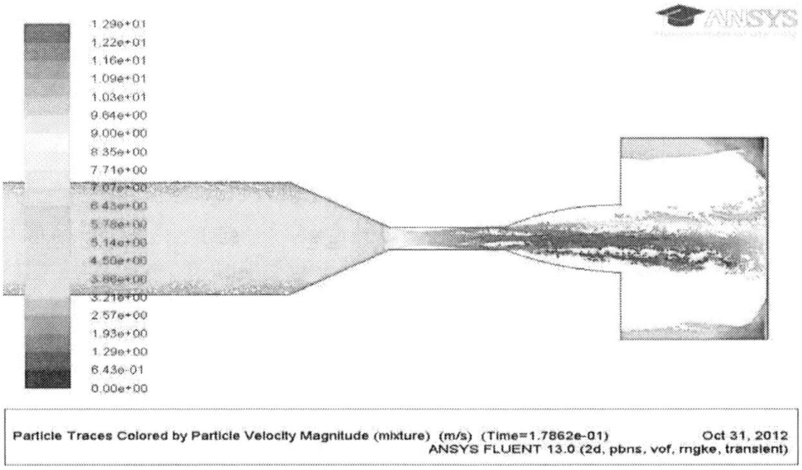

Figure 11. Particle Velocity Magnitude Plot

From this comparison of SE values in the literature, FLASH processes are not as low as mining or tunneling values, but are significantly lower or on the lower end of the range reported for waterjetting (WJ), abrasive waterjetting (AWJ) and even standard abrasive slurry jetting (ASJ) SE values. They are within the range of diamond saws and lower than mechanical rotary bits.

3. CONCLUSIONS

1. High Pressure Slurry Pumps (HPSP) are effective for pumping abrasive slurries to high pressures, but such pumps must have larger bores and slower stroke speed, due to the inertia of suspended dense solids. Otherwise, a viscous carrier and clean fluid must be used that require a pressurized suction. HPSPs may be used to directly pump the supercritical abrasive systems, but this was not directly tested.
2. Supercritical abrasive systems are very efficient, based on SE values, when compared to all water jet and water abrasive systems, and even some diamond cutting systems, but it is not as efficient as reported for mining and tunneling processes, including explosives.
3. Supercritical abrasive systems can cut wide bores sufficient for nozzle entry and drilling of microbores. Rotation of the nozzle can provide even wider created bores.
4. Optimal conditions for the supercritical abrasive system were found to be broad and stable, with only a Vs/Vg ratio above about 1.0 found to be important.
5. Such supercritical systems can be applied in utility trenching, tunneling (finished wall cut), shallow drilling for geothermal heat sink bores and instrumented bores.

4. ACKNOWLEDGMENTS

Stripper Well Consortium at Penn State University; Oklahoma Center for the Advancement of Science and Technology (OCAST); Department of Energy- Fossil Fuels and Geothermal Energy programs; The University of Tulsa - Dr. Ovadia Shoham, Missouri University of Science and Technology- Dr. D. Summers, Dr. G. Galecki, Dr. P. Nambiath; Lawrence Berkeley National Laboratory- Dr. S. Finsterle, Dr. Y. Zhang, Dr. L. Pan, Dr. P. Dobson; Dr. B. Felber; Dr. D. Rychel; and Impact Technologies LLC- B. McCollam, B. Oglesby, T. Luscoito, and N. Novak.

5. REFERENCES

1. Summers D.A. Waterjetting Technology, 1st ed, E & FN Spon, London, 1995.
2. Maurer W.C., Advanced Drilling Techniques, 1st ed.; Petroleum Publishing Co., ISBN-13: 978-0878141173, ISBN-10: 0878141170.
3. Fairhust R.M. "Abrasive Water jet Cutting", M.Sc. Thesis, Cranfield Inst. of Tech., Jan. 1982.
4. Fairhurst R. M., R.A. Heron and D. H. Saunders, "DIAjet – A New Abrasive Waterjet Cutting Technique", Proceedings of 8th International Symposium on Jet Cutting Technology, Durham England, pp 395 – 402, September 1986.
5. Hashish, M., " Steel Cutting with Abrasive Waterjets", Proceedings of the 6th International Symposium on Jet Cutting Technology, University of Surrey, UK, BHRA Fluid Eng., 1982.
6. Leach S.J. & Walker G.L. "Some Aspects of Rock Cutting by High Speed Water jets," Phil Trans Royal Society, London, Vol 260A, 1966, pp.295 – 308.
7. Selberg B.P. and Barker C.R. "Dual-Orifice Waterjet Predictions and Experiments," Erosion: Prevention and Useful Applications, ASTM STP 664, pp. 493–511.
8. Barker C.R. and Selberg B.P. "Water Jet Nozzle Performance Tests,"Paper A1, I 4th International Symposium on Jet Cutting Technology, Canterbury, UK, April 1978.
9. Oglesby K.D., Summers D.A., Galecki G., Rychel D., Felber B., "Advanced Mud System for Microhole Coiled Tubing Drilling", US Department of Energy, DE-FC26-04NT15476, Final Technical Report, 16 March 2009.
10. Hollinger, R.H. Perry, W.D., and Swanson R.K., "Precision Cutting with a Low Pressure Coherent Abrasive Suspension Jet", Proceedings of the 5th American Water Jet Conference, Toronto, Canada, U.S. Water Jet Technology Association, 1989.
11. Hollinger, R.H. and Mannheimer, R.J., "Reheological Investigation of the Abrasive Suspension Jet", Proceedings of the 6th American Water Jet Conference, pp. 515-528, 1991.
12. Hashish, M., "Cutting with High-Pressure Abrasive Suspension Jets", Proceedings of the 6th American Water Jet Conference, pp. 439-455, 1991.
13. Resnick, R., "Abrasive Suspension Jet Machining", Proceeding of the 13th International Conference on Jetting Technology, pp786-799, 1996.
14. D.A. Summers, R.D. Fossey, J.W. Newkirk, G. Galecki, UMR; M. Johnson, NSWC; G. Olson, DAC, " Results of Comparative Nozzle Testing Using Abrasive Waterjet Cutting," 2001 American Waterjet Conference, Minneapolis, MN, August 18-21, 2001.
15. Meng, H. S. and Ludema, K. C.: "Wear Models and Predictive Equations: Their Form and Content", Wear, Vol. 181 – 183, pp. 443-457, 1995.
16. Finnie, I.: "Erosion of Surface by Solid Particles", Wear, Vol. 3, Issue 2, pp. 87–103, 1960.

17. Finnie I.: "The Mechanisms of Erosive Wear in Ductile Metals, Corrosion-Erosion Behavior of Materials", Spec. Vol. published by Metallurgical Society of AIME, pp 118-126, 1980.
18. Levy A., "The Platelet Mechanism of Erosion of Ductile Metals", Wear, Vol.108, 1986.
19. Finnie I., and McFadden, D.: "On the Velocity Dependence of the Erosion of Ductile Metals by Solid Particles at Low Angles of Incidence", Wear, Vol. 48, pp. 181-190, 1978.
20. Lindsley, B. A. and Marder, A. R.: "The Effect of Velocity on the Solid Particle Erosion Rate of Alloys", Wear, Vol. 225 – 29 part 1, pp. 510-516, 1999.
21. Finnie, I.: "Some Observations on the Erosion of Ductile Materials", Vol. 19, 1972.
22. Winter, R. E. and Hutchings, I. M.: "Particle Erosion of Ductile Materials: A Mechanism of Material Removal", Wear, Vol. 27, pp. 121-128, 1974.
24. Palsamudram, S. L. and Bahadur, S.: "Particle Characterization of Angularity and the Effects of Particle Size and Angularity on Erosion in a Fluidized Bed Environment", Wear, Vol. 203-204, pp. 455 – 463, 1997.
25. Padsalgikar, A., "Solid Particle Transport in Gas-Liquid Stratified Slurry Flow", Ph.D. dissertation, Mechanical Engineering at The University of Tulsa, 2015.
26. Oglesby, K., Finsterle, S., Zhang, Y., Pan L., Dobson P., Mohan R., Shoham O., Felber B. and Rychel D., "Microhole Arrays Drilled with Advanced Abrasive Slurry Jet Technology to Efficiently Exploit Enhanced Geothermal Systems", DOE DE-EE0002783, March 2013.
27. Oglesby, K., Summers, D., "Advanced Mud System for Microhole Coiled Tubing Drilling", DE-FC26-04NT15476, Final Technical Report, March 2009
28. Oglesby, K., "High Pressure Slurry Piston Pump", US11/033,615 and international patents
29. Oglesby, K., "High Pressure Slurry Plunger Pump with Clean Fluid Valve Arrangement", US11/705,222 and international patents
30. Oglesby, K., "High Pressure Slurry Pump for Cutting and Drilling", Oklahoma Center for the Advancement of Science and Technology (OCAST), AR082-065 #7121 & #7373, July 2010.
31. Nambiath, P., "Design and Optimization of an Abrasive Feed System, Control, Circuit and Jet Drilling Tool for Mining Applications", Ph.D. dissertation, Missouri University of Science and Technology, 2008.
32. Oglesby, K., "Method and Apparatus for Jet-Assisted Drilling and Cutting", US12/400,507 and international patents for FLASH ASJ™ methods and process.
33. Oglesby, K., "Method and Apparatus for Jet-Assisted Drilling and Cutting", US13/589,626 and international patents for FLASH ASJ™ nozzles.
34. Ucun I., Buyuksagis I.S., Tasgetiren S., "Determination of specific energy in cutting process using diamond saw blade of natural stone", Energy Science and Research, 2012, Vol 28,641-648.
35. Darling, P., SME Mining Engineering Handbook, ISBN 978-0-87335-264-2.
37. Kolle, J., "A Comparison of Water Jet, Abrasive Jet and Rotary Diamond Drilling in Hard Rock", Tempress Technologies, 1999.
38. Yao, Q., "An Investigation of Rock Cutting: Towards a Novel Design of Cutting Bits", Doctor of Philosophy (PhD) dissertation, University of New South Wales, 2012.

6. NOMENCLATURE

A = Area [m2]
An = Angularity Parameter
At = Area of Throat of the Nozzle [m^2]
CS = Slurry Concentration
CT = Target Concentration
CV = Volume Fraction
ER = Erosion Rate
Fs = Shape Factor
m_a = Total Mass of Abrasive Particles Used [kg]
n = Velocity Exponent

American WJTA-IMCA Conference and Expo
November 2-4 • New Orleans, Louisiana

Paper

NOVEL PROCESSES FOR IMPROVING PRECISION OF ABRASIVE WATERJET MACHINING

(Peter) H.-T. Liu§
OMAX Corporation
Kent, WA 98032
www.omax.com

ABSTRACT

Abrasive waterjet (AWJ) possesses several technological and manufacturing merits unmatched by most machine tools. As a "floppy" machine tool, AWJ has inherent characteristics that tend to cause certain blemishes in various degrees on the finished parts. For precision machining, such blemishes must be minimized. For example, blemishes include (1) edge taper as the result of the spread of the AWJ upon exiting the mixing tube; (2) damage induced by AWJ during the initial piercing of delicate materials including composites, laminates, and brittle materials such as glass and silicon wafers; and (3) the frosting and edge rounding on the jet entry surface and the chipping and burr on the jet exit surface. This paper is devoted to demonstrating several novel methods to minimize or mitigate the above blemishes to qualify AWJ as a precision machine tool.

Organized and Sponsored by WJTA® - IMCA®

§ Senior Scientist

1. Introduction

Abrasive waterjet (AWJ) possesses several technological and manufacturing merits unmatched by most machine tools. As a "floppy" machine tool, however, AWJ inherently has a number of characteristics that could induce blemishes in various degrees on the finished parts. Such blemishes must be minimized in order for the AWJ to become a truly precision machine tool. For example, blemishes include (1) edge taper due to the spread of the AWJ after exiting the nozzle; (2) damage induced by AWJ during the initial piercing of delicate materials including composites, laminates, and brittle materials such as glass and silicon wafers; (3) the frosting and edge rounding on the jet entry surface and the burr and edge chipping on the cut edges of the jet exit surface. There are other blemishes that can be controlled by optimizing the process parameters. For example, a striation pattern is often formed on the AWJ-cut edge when the AWJ travels too fast. Slowing down the AWJ traverse speed reduces the amplitude of the striation pattern. Reducing the size of abrasives would also reduce the amplitude and wavelength of the striation pattern (Liu, 2014). Also, overcuts at the intersection between the lead-in and tool path result when the dwell time of piercing is too long. Controlling the dwell time during piercing is the key to mitigate overcutting.

Edge taper is the direct consequence of the spread of the AWJ after exiting the nozzle. Piercing damage takes place when the stagnating pressure building up inside the blind hole exceeds the tensile or adhesive strength of delicate materials (Liu, 2006a). Frosting and edge rounding result from the strayed particles outside the high-speed core of the AWJ (Liu, 2014). Burr and chipping on the exit edge are induced when the AWJ breaks through the bottom surface of the material, with chipping occurs mostly for brittle materials.

Extensive investigations have been conducted to understand the above characteristics. Several interim and permanent remedies have been developed to minimize the AWJ-induced blemishes. For example, an abrasive cryogenic jet (ACJ) using liquefied nitrogen as the working fluid was investigated to minimize the stagnation pressure inside the blind hole (Liu, 2006b). Based on the operational principle, a superheated AWJ or FAWJ was subsequently developed to emulate the ACJ (Liu et al., 2010). The success of the ACJ and FAWJ, however, has led to the development of two viable processes to mitigate piercing damage; Turbo Piercer and Mini Piercer are for cutting large and small internal features, respectively (patented). The Turbo Piercer manipulates the AWJ to reduce the stagnating pressure to be below the tensile strength of target delicate materials. The Mini Piercer applies low-pressure piercing together with vacuum assist.

In this paper, we will describe several processes and apparatus to minimize or mitigate the blemishes described above. Examples of the test parts cut with these processes and apparatus were presented to demonstrate their effects in improving the part accuracy.

2. Technical Objectives

Abrasive waterjet as a "floppy" machine tool has the tendency to introduce several undesirable features that degrade its performance for precision machining. The technical objectives of the R&D activities described in this paper are therefore to examine the performance characteristics of AWJ machining and develop novel remedies to mitigate those features. Our goal is to implement these remedies to quality AWJ as a precision machine tool for a wide range of materials and part sizes.

3. Technical Approach and Facilities

3.1 Technical Approach

AWJ is a versatile machine tool capable machining most materials from macro to micro scales. As a "floppy" machine tool, however, there are inherent behaviors of the AWJ process that could lead to certain blemishes on AWJ-machined parts. Such behaviors and the resulting blemishes include but are not limited to:

- AWJ spreading - edge taper
- Development of large stagnation pressure in blind holes – piercing damage in delicate materials
- Strayed abrasives – edge rounding and surface frosting
- Challenges in AWJ micromachining (feeding of fine abrasive in micro nozzle) – varying kerf width and skip cutting

Our approach was to conduct laboratory experiments to investigate in detail AWJ machining processes. Based on the understanding of these processes, novel remedies were developed to minimize or mitigate blemishes on AWJ-machined parts to achieve the goals described in Section 2.

3.2 Test facilities

3.2.1 MicroMAX® JetMachining® Centers and nozzles

OMAX has two product lines of equipment for AWJ precision machining, the OMAX JetMachining® Center and the MAXIEM JetMachining Center, to meet a wild range of machining needs. Under the current support of an NSF SBIR Phase II/IIB grant, OMAX has developed and commercialized micro abrasive waterjet (μAWJ) technology for precision meso-micro machining. One of the key objectives was to downsize the AWJ nozzle capable of machining features around 100 to 50 μm. Several micro nozzles together with novel accessories were developed to achieve the above objective. At present, in addition to our standard nozzles, there are three miniature nozzles at various developmental stages:

Nozzle	**Orifice ID in/mm**	**Mixing Tube ID in/mm**	**Optimum Garnet Size**	**Status**
7/15 Mini MAXJET5	0.007/0.18	0.015/0.38	≲ 150 mesh	In production
5/10 Mini MAXJET5	0.005/0.13	0.010/0.25	≲ 220 mesh	In beta testing
3/8 Nozzle	0.003/0.076	0.008/0.20	≲ 240 mesh	R&D

There were several challenges in downsizing the AWJ nozzle (5/10 and smaller).

- Difficulty in fabricating mixing tubes (ID < 0.010" or 0.25 mm) with good circularity and sufficiently large aspect ratio required for creating tightly collimated AWJ.
- It is important to size the abrasive such that its average size should be ≲ three times the mixing tube diameter to prevent nozzle clogging through bridging of two large abrasive particles. For very fine powdery abrasives, it is well known that they are difficult to feed consistently under the gravity feed mode.
- As the nozzle being further downsized, the 3-phase slurry flow through the mixing tube bore would transition from gravity-dominated to capillary-dominated microfluidic flow. According to the Hagen–Poiseuille equation, a 2-phase microfluidic flow resistance is proportional to the -

4 power of the diameter of the conduit. We anticipate that the 3-phase flow resistance of the microfluidic slurry flow through the mixing tube would also overcome the driving force, the high pressure, such that micro AWJ would no longer be effective for cutting.

Several novel processes, apparatus, and accessories, leading to five relevant patents, were developed to meet the above challenges. The μAWJ technology was subsequently commercialized by introducing a new product, the MicroMAX® JetMachining® Center, which debuted in September 2013 at the EMO (Hannover, Germany) and MD&D (Chicago) trade shows. Figure 1 is a photograph of the MicroMAX that was specially designed to more than match the requirements for precision meso-micro machining. With the addition of the MicroMAX, OMAX has extended the capability of AWJ machining from macro to micro scales. Refer to the link for a detail description of MicroMAX's specifications (https://www.omax.com/waterjet-cutting/machine/model/micromax).

3.2.2 Accessories

Accessories were developed to improve the performance of the μAWJ technology. They could be divided into several categories based on their functionalities. One of the accessories is a refocuser (patented – Liu and Shubert, 2014).

Refocusers and Stencil-Aided Waterjet System (SAWS)

These accessories were designed to further reduce the diameter for hole drilling and slot cutting. The refocuser is a short mixing tube, with ID smaller than that of the mixing tube, attached to the tip of the primary mixing tube. Only the core of the AWJ exiting the primary mixing tube would be allowed to enter into the blind hole and the abrasives outside of the core will be discarded. As such, the hole drilled with the refocuser would be considerably smaller than that drilled without the refocuser.

A second accessory under this category is Stencil-Aided Waterjet System (SAWS). The SAWS is formed by two pieces of carbide plates with the provision of a narrow slit along the edges where the two carbide plates meet. With the AWJ traversing along the axis of the narrow slit, the kerf widths of the slots cut with the SAWS setup would be limited by the width of the sacrificial slit.

Vacuum Assist Kit

The vacuum assist kit was developed for low-pressure piercing when the vacuum created by the Venturi effect was too low to entrain all the abrasive fed from the hopper. The additional vacuum issued by the vacuum assist kit would help remove the excessive abrasives through the vacuum discharging line into a waste tank. As soon as high-pressure cutting resumes, the abrasives are fully entrained by the vacuum generated by the full-power AWJ. The above operation is carried out automatically with software control of the on-off valves.

Tilt-A-Jet

The AWJ is a "floppy" machine tool. The diameter of the AWJ spreads and meanders after it exits the nozzle. Therefore, there is a natural taper on the edge of the AWJ-cut parts. The Tilt-A-Jet (TAJ) was designed to remove taper from the AWJ-cut part through dynamically tilting during cutting (https://www.omax.com/waterjet-cutting-accessories/tilt-a-jet/61). The TAJ was also adapted for the MicroMAX.

3.3 CAD/CAM programs

OMAX's PC-based CAD program, LAYOUT, was designed to either draw from scratch or import from AUTOCAD or other CAD software tool paths of parts for AWJ machining. The tool paths were subsequently passed onto the CAM program, MAKE, where various parameters pertaining to the workpiece (material types and thickness), the abrasive waterjet (pressure and nozzles), the abrasives (type, mesh, and mass flow rate), and cut qualities (from Q1 through Q5) were specified. The part according to the tool paths was cut automatically by MAKE, which controls operations of a suite of hardware accessories such as the TAJ for taper compensation, Rotary Axis and A-Jet for 2D and 3D machining.

4. Results

4.1 Mitigating edge taper

In the absence of the TAJ, the edge cut with the AWJ exhibits a taper due to the spreading of the AWJ as it travels downstream. The TAJ tilts dynamically to compensate for the edge taper. Before cutting a part that requires a minimum taper, a taper compensation test was first conducted by cutting a series of coupons with the same thickness the part to be cut. A taper compensation program was used to dial in the tool and taper offsets to achieve the minimum edge taper.

As an example to demonstrate a part that requires a minimum taper, Figure 2 illustrates an interlocking part made of 0.125" thick titanium. It consists of an inner and outer components that fit snuggly with a very narrow gap, say around 20 μm. The part was cut on the MicroMAX with a 5/10 beta nozzle using 240 mesh garnet at a flow rate of 0.11 lb/min (50 gram/min). In the presence of an edge taper, the two components could only be locked into position from one direction by flipping one of them. As a result, one component has a positive taper angle and the other a negative taper angle.

By activating the TAJ, the part was cut nearly taperless. The two components could then be locked into position from both directions. The part has adequate precision such that the two components could be locked into position by rotating one of the components any multiples of 12 degrees (i.e., there are a total of 30 teeth).

4.2 Gear Making

AWJ is an excellent gear maker. OMAX LAYOUT has a gear generation package which is a subset of the "Parametric Shapes Tool". The gear generator supports internal and external gears, racks, sprockets, in both U.S. and metric standards, as illustrated in Figure 3. Resulting geometry can then be further edited using LAYOUT tools to combine with other shapes. Gear cutting could by executed (1) without using the TAJ but flipping gears to partly cancel taper or (2) with the use of TAJ to make them taper free. For an in-depth discussion of AWJ gear making and examples, refer to Olsen (2015).

In the early developmental stage of the μAWJ technology, a planetary gear set consisting of five planetary gears was fabricated, assembled, and operating driven by a micro motor operating with two AAA batteries (Liu and Schubert, 2012). The gear set was cut with a 5/10 nozzle and 320 mesh garnet at 0.1 lb/min (45 gram/min). No TAJ was used and the gear components were flipped to match the taper.

Most recently, in demonstrating the performance of the MicroMAX®, in terms of its cutting precision and vibration isolation capability three sets of cycloidal gears with different sizes were cut from 2.9 mm thick titanium using the 5/10 nozzle and 240 mesh garnet at 0.11 lb/min (0.5

gram/min). Cycloidal gears are traditionally designed for making watches and clocks. Figure 3 illustrates the photographs of the three sets of cycloidal gears. The large gears were cut according to the tooth paths. Smaller gears were cut by nesting several of them together. A narrow tab was designed into one of the tips of each gear to keep it from falling into the catcher tank. Individual gears were then extracted by machining away the tabs. The three sets of gears were mounted on pins that were fixed on supporting titanium plates. Each set of gears could be rotated by hand with ease.

4.3 Delicate Materials

Delicate materials include composites, laminates, and brittle materials such as glass and silicon wafer are difficult to machine by conventional AWJ processes. During the initial piercing of these materials, a large stagnating pressure is developed, as the high-speed waterjet decelerates, stops, and reverses near the bottom of the blind hole, before breakthrough takes place. These materials would result in delamination or internal cracking provided the stagnating pressure exceeds their tensile/adhesive strengths (Liu, et al., 1998). In other words, conventional AWJ is simply is incapable of preserving the structural integrity of parent materials. Considerable efforts were made to understand the AWJ piercing process and several interim remedies were developed to mitigate AWJ piercing damage. One of the remedies used an abrasive cryogenic jet (ACJ) with liquefied nitrogen as the working fluid (Liu, 2007). The liquefied nitrogen remains a liquid under pressure but begins to evaporate as soon as it exits the mixing tube. Therefore only abrasive goes into the blind hole without developing stagnant pressure and therefore no piercing damage results.

Another remedy was a super-heated AWJ or FAWJ to emulate the ACJ. The water was superheated to beyond the boiling temperature. It evaporates as soon as the waterjet exits the mixing tube and only abrasives enter into the blind hole without inducing stagnating pressure and piercing damage. The FAWJ was used to machine a variety of delicate materials to demonstrate the success in mitigating piercing damage (Liu et al., 2010).

In practice, both the ACJ and FAWJ are not suitable for operating in industrial environments because of their complexity and severe operating temperatures. The experiences gained in these two investigations, however, have helped understand the AWJ piercing process leading to piercing damage of delicate materials. Based on such understanding, two proprietary processes for mitigating piercing damage in delicate materials were developed. These processes utilize air-assisted abrasive feeding (Turbo Piercer) and low-pressure piercing with vacuum assist (Mini Piercer), respectively. The vacuum assist helps to remove excess abrasives accumulated in the mixing chamber at low-pressure piercing where the Venturi vacuum is too weak to entrain and accelerate all the abrasives fed from the hopper. Otherwise, nozzle is prone to be clogged by excessive abrasives left in the mixing chamber/mixing tube. Since the air-assisted feeding process, Turbo Piercer, tends to increase the spread of the AWJ, it is most suitable for machining relatively large features. On the other hand, low-pressure piercing with vacuum assist with the Mini Piercer does not affect the geometry of the AWJ and is suitable for machining small features especially when miniature nozzles are used.

Figure 4 illustrates photographs of an aluminum-laminate sample (BAC1534-63F) pierced with the Turbo Piercer. The laminate consisted of 19 aluminum sheets 0.076 mm thick with an overall thickness of 1.6 mm. The 14/42 nozzle with 80 mesh garnet were used. Piercing was carried out by slightly pressurizing the abrasive hopper. Cutting was performed at p = 380 MPa. The photographs on the left and right correspond to the samples machined before and after Turbo Piercer was optimized, respectively. The effectiveness in mitigating delamination of the optimize Turbo Piercer is evident. Figure 5 illustrates photographs of internal patterns machined with Mini Piercer on the aluminum laminate and a fiber-reinforced plastic (0.85 mm thick), respectively. The 7/15 nozzle with the 240 mesh garnet at 0.68 gram/min was used to cut these parts. Piercing and cutting were carried out at p = 41 MPa and 380 MPa, respectively. No delamination is present on any of the four identical patterns machined on each of the finished part.

One of the premier composite materials is PEEK (polyetheretherketone). It has among the best chemical resistance of any thermoplastic material and is capable of maintaining stiffness at elevated temperatures. As a result, it is has been often specified for high temperature applications in aerospace, semiconductor, medical, and food processing. PEEK has found to be superior to titanium for manufacturing implants as PEEK is not only biocompatible but also much more bone-friendly and is much more compatible with diagnostic imaging than metal implants. Sample materials of unfilled and carbon fiber reinforced PEEK were furnished by Victrex to conduct test cutting with AWJ. The planetary gear set described in Section 4.2 was cut using both PEEK materials. Figure 6 illustrates two sets of nested components and assemblies of two planetary gear sets made from the two materials. The components were machined without the TAJ activated. The assemblies were as cut without any secondary processes except the planetary gears were flipped to account for the edge taper. For the carbon reinforced PEEK, the carbon fibers were cut cleanly without fray. The embedded carbon fibers could present difficulties to cut with other machine tools.

The Mini Piercer with a 5/10 nozzle was used to cut small internal features on the aluminum laminate described above. In this case, pressures of 41 MPa and lower were required to mitigate delamination. For such low pressures, the Bernoulli vacuum developed after the waterjet exits the orifice is too weak to entrain all the abrasives fed from the hopper. Vacuum assist is required to remove excessive abrasives accumulated in the mixing chamber. Otherwise, the mixing tube will be clogged by the excessive abrasives. As soon as piercing is through, normal high-pressure cutting at 380 MPa resumes to cut internal features. Figure 7 illustrates the top and bottom surfaces of the aluminum laminate.

For thin workpieces such as the 0.076 mm individual shims of the aluminum laminate, the best way to machine them with AWJ was through stack cutting. It is imperative that individual shims must be firmly adhered to one another to resist delamination during piercing and easily separated after cutting is complete. The top and bottom shims of the stack would serve as the sacrificial covers to protect the interior ones from frosting (top shim) and burr (bottom shim). After the stack is cut, the internal shims would be nearly identical with no frosting and burr, as illustrated in Figure 8.

4.4 AWJ Meso-Micro Machining

The MicroMAX was built to meet the requirements for precision meso-micro machining. With the use of the 5/10 nozzle and abrasives 240 mesh and finer, small parts with miniature features could be machined with high precision. For example, the smallest cycloidal gear set shown in Figure 3 could only be cut with nozzles 5/10 or smaller. Certain features such as the minimum diameter of holes and minimum kerf width of slots are determined by the diameter of AWJ. As a cold cutting process with no heat induced distortion and minimum force exerted onto the workpiece, AWJ is capable of cutting very thin webs if the AWJ is steady and minimum nozzle vibration. Proprietary processes, apparatus, and hardware components were incorporated into the design of the MicroMAX to achieve constant feed rate of fine abrasives and minimum nozzle vibration. Another criterion is that the target material must be sufficiently stiff to support the thin webs.

The beta 5/10 nozzle mounted on the precision MicroMAX equipped with the TAJ and vibration isolation is fully capable of machining micro-size features. An important example is the machining of nearly taperless thin webs used as a critical component - flexures – of mechanical load cells (Section 4.5). A test part consisting of five thin webs with widths of 0.5, 0.25, 0.1, 0.076, and 0.05 mm were machined to demonstrate the above capability. The 5/10 nozzle was operating at 413 MPa. The abrasives was 240-mesh garnet with a mass flow rate of 50 gram/min. Titanium with a thickness of 2.9 mm was chosen as the target material to take advantage of its large ratio of stiffness versus weight. Prior to machining the parts, the TAJ was dialed in to achieve a taper angle of less than 0.126 degree. Figure 9 illustrates the top and bottom surfaces of one of the test parts. Note that the webs appear to be thinner on the jet entry surface than on the jet exit surface because the induced frosting and edge rounding on the entry surface. The web with the width of 0.05 mm tended to be slightly warped most likely due to stress release or insufficient material stiffness. Note that the length-to-width and length-to-thickness ratios of the 0.05 mm wide web are 125 and 2.54, respectively. Titanium parts with such large aspect ratios would be difficult to cut with other machine tools. For example, the heat generated by CO_2 laser could lead to considerable warping/twisting or even breakage of the thin webs; a large amount of slag or recast could left behind as the molten metal runs down the web walls solidifies (Liu, 2015). It should be pointed out that, by reducing the aspect ratio or for materials with high stiffness in the annealed state, thin webs with widths less than 0.05 mm could be machined with the 5/10 nozzle on the MicroMAX. As a versatile tools, such micro-size webs could be machined from a wide range of materials, which is beyond the capability of most machine tools.

4.5 Case Studies

The first academic and industrial installations of the MicroMAX are at the Hobby Shop of MIT. Two case studies were presented in this section to demonstrate the real-world applications of the MicroMAX. One of the studies was a reengineering of a Rostock MAX 3D printer at MIT's Hobby Shop. The original printer had injection-molded linkages and laser-cut plastic frame, which proved to be unstable and brittle in some places. These linkages and plastic frame were subsequently replaced with aluminum counterparts cut with AWJ using the MicroMAX (Figure 10). The reengineered printer results in a lighter, stiffer, and more heat-resistant structure.

The second case study applied the MicroMAX equipped with the TAJ for machining nonlinear stiffening load cells (Kluger et al., 2015). These load cells, with high resolution (within 1% of the force value) that can function over a large force range (5 orders of magnitude), and with minimal hysteresis and intrinsic geometric protection from force overload, was developed (patent pending). The stiffening nature of the load cell causes its deflection and strain to be very sensitive to small forces and less sensitive to large forces. High stiffness at high forces prevents the load cell from over-straining. The detailed design and theory of the load cell was given elsewhere (Kluger et al., 2015). AWJ was chosen as the machine tool to manufacture the symmetric and monolithic load cell consisting of four 1 mm thick springs as the force sensing elements. The nonlinear springs with cantilever beams that increasingly contact rigid surfaces with carefully chosen curvatures as more force is applied was physically implemented. A 7/15 nozzle operating at 380 MPa and using 240 mesh garnet at 0.68 gram/min was use to cut the load cell. Figure 11 is the AWJ-machined load cell and the inserts used for an optional setup. Figure 12 illustrates the experimental setups for two configurations, one without, and the other with the inserts in place (Figure 12a and 12b). Figure 12c presents the test results for the two configurations. The nonlinearity of the response to small and large load is evident.

5. Summary and future work

5.1 Summary

Abrasive waterjet (AWJ) has several technological and manufacturing merits that cannot be matched by most other machine tools. However, as a "floppy" machine tool, AWJ presents several challenges for precision machining. OMAX has overcome most of the challenges in order to elevate the AWJ as a versatile precision machine tool. This paper examines several undesirable characteristic of AWJ and the resulting blemishes left on AWJ-cut parts. The blemishes include (1) edge taper, (2) AWJ piercing damage in delicate materials, (3) frosting and edge rounding on the entry surface and chipping and burr on the exit edge, and (4) challenges in downsizing of AWJ nozzle for micromachining.

The accessory Tilt-A-Jet (TAJ) that is capable of tilting on the fly the nozzle head +/- 9 degrees from the vertical was used to demonstrate nearly taper free AWJ machining. The TAJ tilts dynamically during cutting such that the part is nearly taper free while the scrap has twice the taper. An interlocking part that could be locked into position from both directions was machined to demonstrate the performance of the TAJ.

Two proprietary processes, Turbo Piercer and Mini Piercer, were developed for mitigating piercing damage of delicate materials. Turbo Piercer and Mini Piercer were most suitable for machining large and small internal features, respectively. Internal features on aluminum laminates, fiber reinforced plastics, PEEK were machined to demonstrate the ability of mitigating piercing damage using the two Piercers.

AWJ stack machining was applied to cut internal features on thin aluminum shims. Not only is stack machining an effective means to stiffen the thin shims but also significantly increases the productivity. The top and bottom shims serve as the sacrificial covers while the internal shims are free of frosting, edge rounding, edge chipping, and burr.

As the AWJ continues downsizing toward micromachining, the size of the abrasives must decrease to avoid nozzle clogging by large abrasives. It is well known that powdery fine abrasives 320 mesh and finer tend to clump together and forms worm holes and are difficult to feed under gravity. Inconsistent abrasive feeding would vary kerf widths and cause skip cutting (Liu, 2014). Proprietary processes and apparatus were developed to mitigate clumping and the formation of worm holes and therefore achieve constant abrasive feed rate. As a result, we are able to feed abrasives as fine as 350 mesh for meso-micro machining (Liu, 2014).

In addition, the excellent gear making capability of AWJ was demonstrated by cutting cycloidal and planetary gears made from titanium and PEEK composite, unfilled and carbon fiber reinforced. The cycloidal gears are often used in watch making.

Titanium test parts with webs as narrow as 0.05 mm were machined on the MicroMAX using the beta 5/10 nozzle with the TAJ activated. Such a combination ensures that all the webs were cut precisely. These are attributed to several technological merits of the AWJ in terms of cold cutting and minimum impact force on the workpiece. The incorporation of the vibration isolation mechanism ensures that nozzle vibration is minimized. The slight distortion of the 0.05 mm wide web shown in Figure 9 is likely caused either by stress relief or by inadequate material stiffness. Webs with widths < 0.05 mm could be machined with the same setup either by reducing the aspect ratio or by selecting materials of high stiffness.

Two case studies were presented by using the MicroMAX for real-work applications. The first study was performed at MIT by replacing injection molded linkages and laser-cut frame with AWJ-cut aluminum counterpart. The second case study applied the MicroMAX for machining a monolithic, stiffening nonlinear load cell capable of high resolution with a wide force range (Kluger et al., 2015).

5.2 Future Work

The result of commercialization of the μAWJ technology has led to the award of an NSF SBIR Phase IIB Supplemental funding for continuing refinement of the MicroMAX for meso-micro machining. One of the tasks was to continue downsizing the AWJ nozzle toward micromachining. Preliminary results showed that a 3/8 combination was successfully operating to cut miniature parts. Figure 13 illustrates the progress toward AWJ micromachining by showing photographs of several tweezers machined with nozzles of different sizes. Other tasks involved the upgrade of the MicroMAX for 3D machining and the improvement of ergonomics.

Acknowledgments

This work was partly supported by an OMAX IR&D fund and by an NSF SBIR Phase II/IIB Grant #1058278. Any opinions, findings, and conclusions or recommendations expressed in this material are those of the authors and do not necessarily reflect the views of the NSF. The author wishes to thank OMAX's technical staff members for their teamwork. They are responsible not only for the design and production release of the MicroMAX but also for the development and upgrade of a rich set of accessories which enhance the performance of the MicroMAX for precision meso-micro machining.

References

Liu, (Peter) H.-T. (2015) "Abrasive-waterjet machining of most materials from macro to micro scales," Proc 2015 TechConnect World Innovation Conference and Expo, June 14-17, Washington DC.

Liu, (Peter) H.-T. (2014) "Roles of Abrasives in AWJ Meso-Micro Machining," Proc 22nd International Conference on Water Jetting, the Netherlands, 3 – 5 September.

Liu, H.-T. (2006a) "Collateral Damage by Stagnation Pressure Buildup during Abrasive-Fluidjet Piercing," Proc. 18th Int. Conf. On Water Jetting, Gdansk, Poland, September 13-15.

Liu, H.-T. (2006b) "Hole Drilling with Abrasive Fluidjets," Int. J. of Adv. Manuf. Tech., Vol. 32, pp. 942-957. (DOI 10.1007/s00170-005-0398-x).

Liu, (Peter) H.-T. (2014) "Roles of Abrasives in AWJ Meso-Micro Machining," Proc 22nd International Conference on Water Jetting, the Netherlands, 3 – 5 September.

Liu, H.-T. and Schubert, E. (2012) "Micro Abrasive-Waterjet Technology (Chapter Title)", Micromachining Techniques for Fabrication of Micro and Nano Structures, Ed. Mojtaba Kahrizi, INTECH Open Access Publisher, ISBN 978-953-307-906-6, January, pp. 205-34

Liu, H.-T., Schubert, E., McNiel, D., and Soo K. (2010) "Applications of Abrasive-Waterjets for Precision Machining of Composites," Proc. SAMPE 2010 Conference and Exhibition, May 17-20, Seattle, WA.

Liu, H.-T., Miles, P., and Veenhuizen, S. D. (1998) "CFD and Physical Modeling of UHP AWJ Drilling" Proc. 14th Int. Conf. on Jetting Technology, Brugge, Belgium, September 21 – 23, pp. 15-24.

Kluger, J. M., Sapsis, T. P., and Slocum, A. H. (2005) "A High-Resolution and Large Force-Range Load Cell by Means of Nonlinear Cantilever Beams," Precision Engineering, http://dx.doi.org/10.1016/j.precisioneng.2015.08.003.

Olsen, C. (2015) "Manufacturing Gears with Water Machining," Feature Article in Gear Production, March.

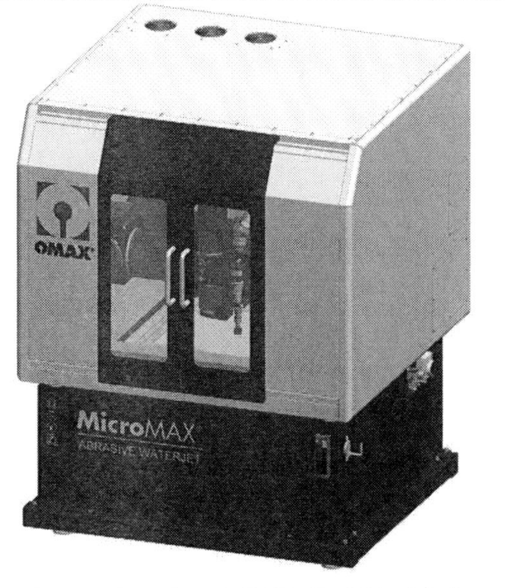

Figure 1. MicroMAX JetMachining Center

Figure 2. 2.9 mm thick well-matched interlocking part

Figure 3. Cycloidal gear sets made from 2 mm thick titanium

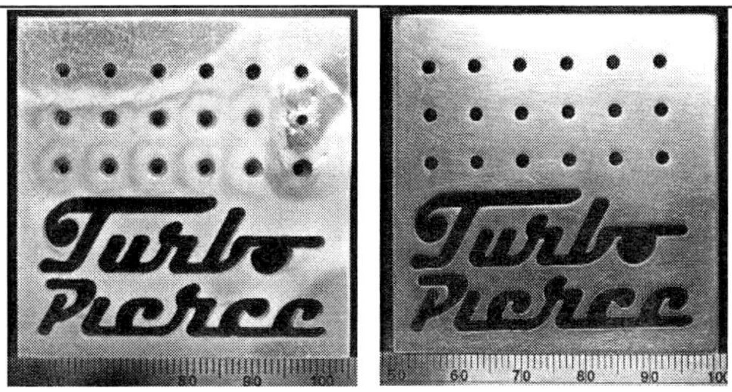

a. Before optimized b. After optimized

Figure 4. AWJ-machined internal features on aircraft aluminum laminate with Turbo Piercer

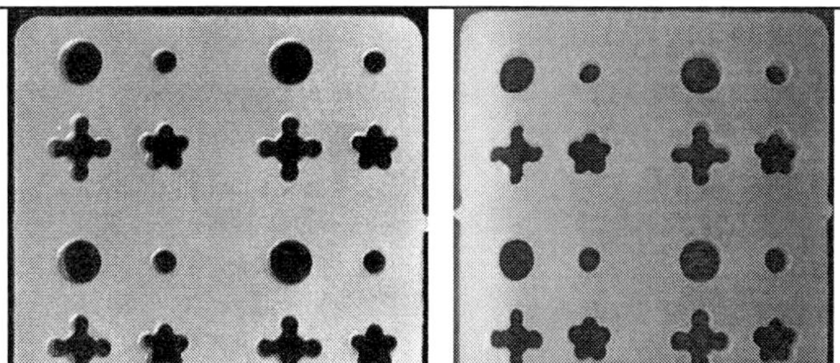

a. Aluminum laminate b. Fiber-reinforced plastic

Figure 5. AWJ-machined internal patterns on delicate materials with Mini Piercer (Scale: mm)

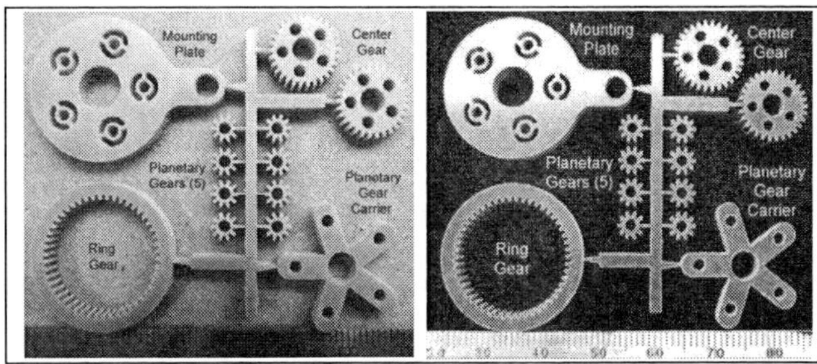

a. Unfilled b. Carbon fiber reinforced

c. Assembled planetary gear sets

Figure 6. Planetary gear sets made from PEEK

Aluminum Laminate
(19 layers of 0.076 mm thick shims)

a. Top view

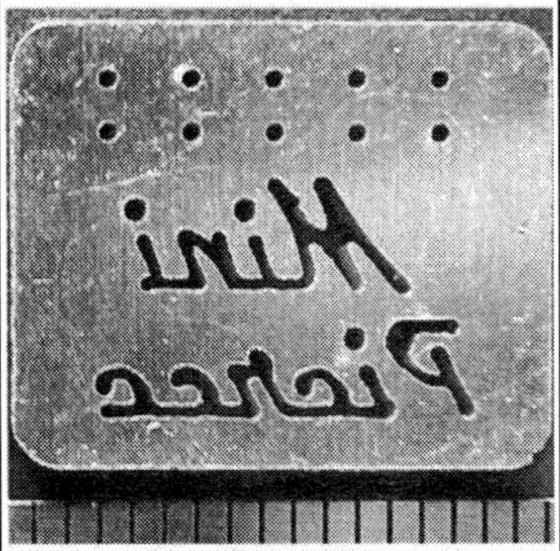

b. Bottom view

Figure 7. Internal features cut with the Mini piercer on aluminum laminate

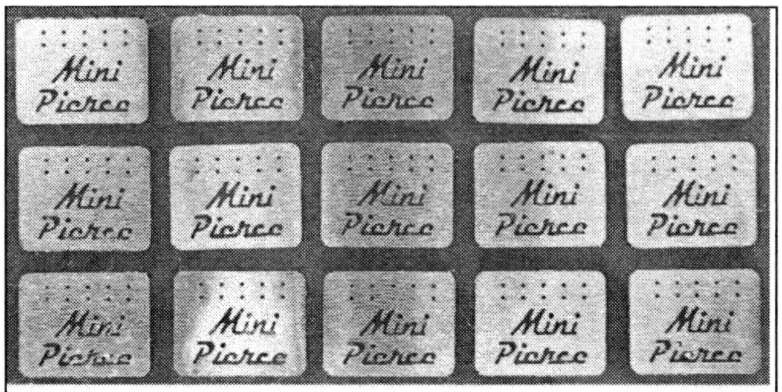

Figure 8. Interior 76 mm thick aluminum shims of AWJ-machined stack

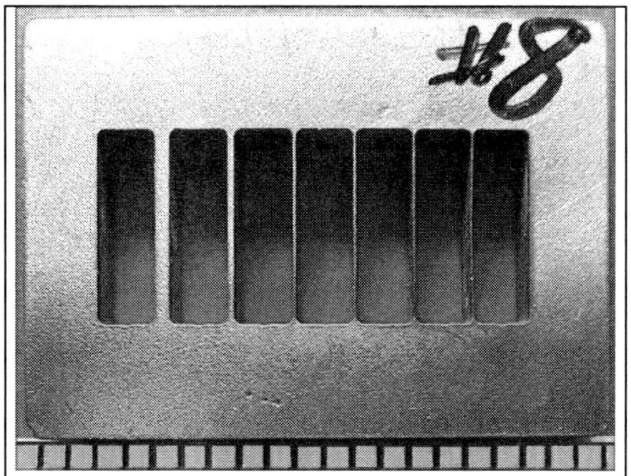

a. Top view

b. Bottom view

Figure 9. Test parts with thin webs

a. Before

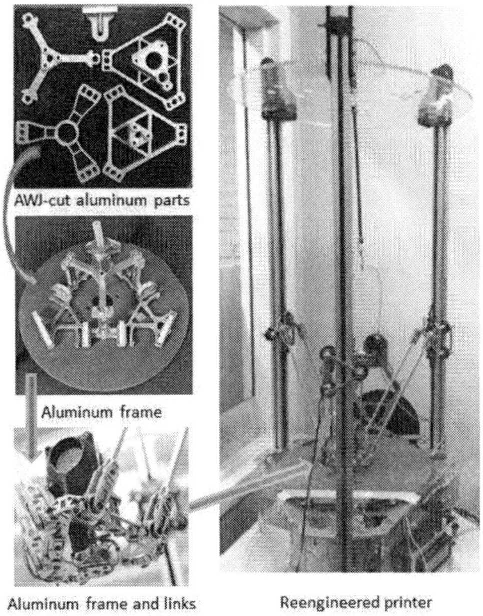

b. After

Figure 10. Reengineering of Rostock MAX 3D printer

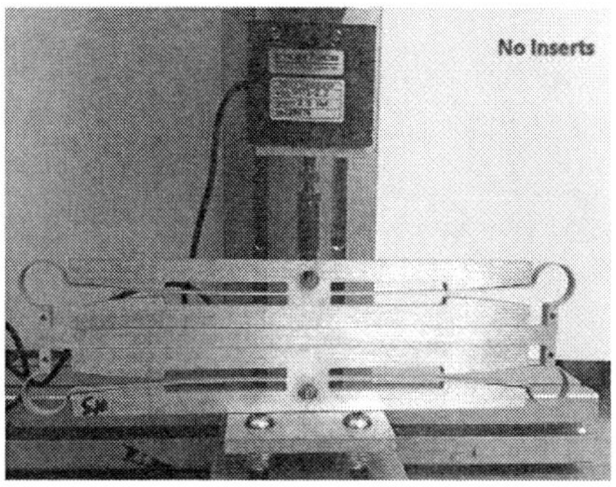

a. Setup #1

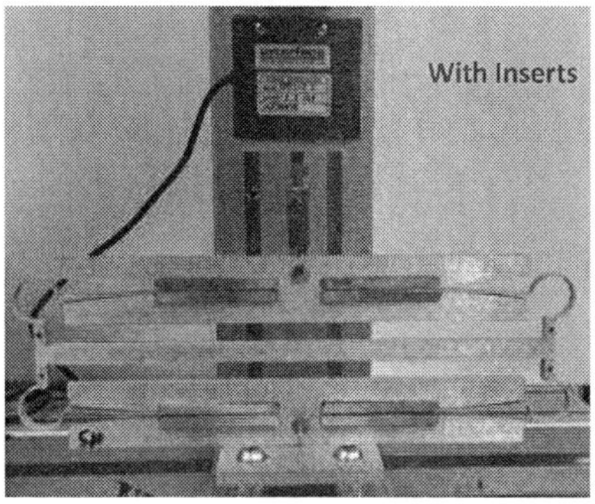

b. Setup #2

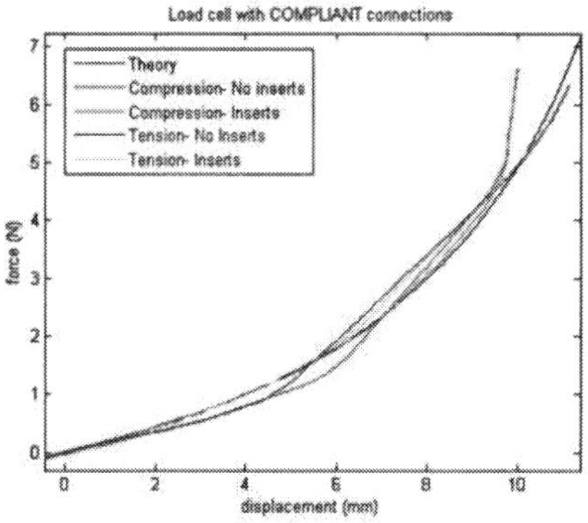

c. Test results

Figure 12. Setups of load cells and results of force measurements (Kluger et al., 2015)

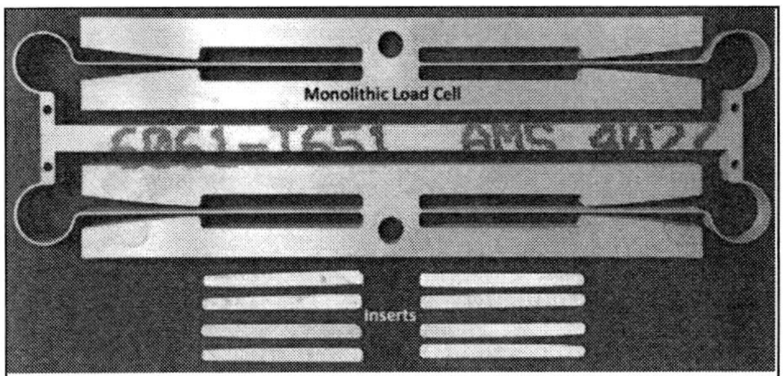

Figure 11. AWJ-cut nonlinear and monolithic load cell on the MicroMAX

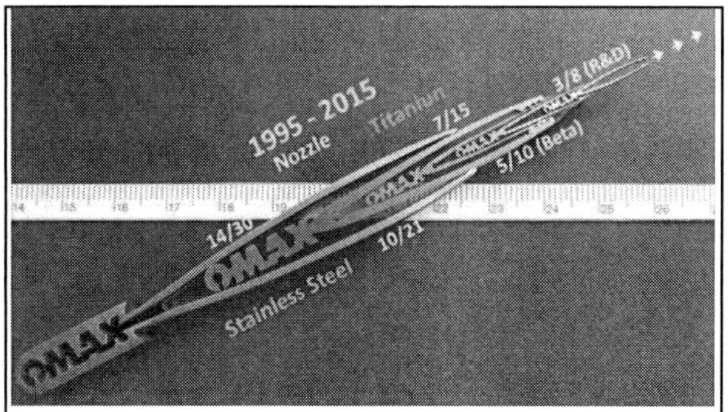

Figure 13. Progress in AWJ micromachining

2015 WJTA-IMCA Conference and Expo
November 2-4 • New Orleans, Louisiana

Paper

PARTICLE DISINTEGRATION IN THE ABRASIVE WATER INJECTION JET

D. Zaremba, P. Heese, M. Bauer, H. J. Maier, T. Hassel
Institut für Werkstoffkunde (Materials Science), Leibniz Universität Hannover, Germany

ABSTRACT

In the Abrasive Water Injection Jet (AWIJ) process, abrasive injection is pneumatically enabled through the generation of technical vacuum by a water jet. Fed particles are collected by the jet in a mixing chamber and are subsequently accelerated. Before impact on the work piece surface, the turbulent multiphase flow containing air, water and abrasive particles is focused at a certain distance, enabling cutting applications. During acceleration and focusing, the current flow conditions provoke numerous collisions of the abrasive particles with the focusing tube. Thus, highly wear resistant materials like tungsten carbide are necessary for this component. The abrasive particles themselves are also subject to wear during collision with each other or the harder focusing tube. As a result, the actual size of the ready-to-cut particles is smaller than the original particle size.

Manufacturers of currently available machines and cutting systems tend to increase the hydraulic power through higher-pressure levels (> 500 MPa), and to minimize jet diameters for micromachining. However, the size of the particles, which carry out the actual micro chipping process, is unknown.

In this study particle disintegration in the abrasive water injection jet process was investigated. The objective was to better understand the material removal procedure during AWIJ machining. Micro- and macro cutting systems corresponding to the actual state of the art were utilized in a pressure range between 250 MPa and 550 MPa. The particle disintegration was investigated for several abrasive load ratios.

Organized and Sponsored by WJTA®-IMCA®

1. INTRODUCTION

In the Abrasive Water Injection Jet (AWIJ) process, the initially formed jet includes air at its surface and at the surface of its droplets, which results in a technical vacuum formed in the mixing chamber. This allows for a pneumatic transport of abrasive particles into the jet. The high-speed waterjet collects the fed particles along with a substantial amount of air in a mixing chamber and subsequently accelerates them (Figure 1). During acceleration and focusing, the flow conditions provoke collisions of the abrasive particles among themselves or with the focusing tube.

Manufacturers of currently available machines and cutting systems tend to increase the hydraulic power through higher-pressure levels ($p > 500$ MPa), and to minimize jet diameters for micromachining. These parameters have a major influence on the transport process in the focusing tube. However, due to particle wear, the size of the particles, which carry out the actual micro chipping process, is unknown. Thus, the objective of the present study was to investigate particle disintegration and contribute to understanding of the material removal procedure.

2. STATE-OF-THE-ART

The utilization of the water jet's kinetic energy as an acceleration medium for abrasive particles has a major effect on the jet's material removal behavior. The accelerated, sharp-edged abrasive particles cause the micro-chipping process. Prior to impact on the work piece surface, the turbulent multiphase flow containing air, water and abrasive particles is focused at a distance l_f (30 mm $\leq l_f \leq$ 100 mm, dependent on orifice diameter d_o and focusing tube diameter d_f). During acceleration and focusing, the current flow conditions provoke numerous collisions. The particle acceleration direction finally results from the sum of the particles' eccentric and centric impacts among themselves, as well as with the focusing tube's cladding (1,2). Thus, highly wear resistant materials like tungsten carbide are necessary for this component. In addition to the tool wear, the abrasive particles are also subject to wear during collision between each other or with the harder focusing tube.

The accelerated particles either cause micro-chipping, micro-crenation or micro-cracking processes at the work piece. Velocity, shape and mass of the abrasive particles are important parameters that affect material removal. Criteria typically used to evaluate the material removal potential of abrasives are cutting ability, cutting quality and degree of fragmentation and particle size is a key parameter in this respect. A decrease of the particle size directly involves a decrease of cutting power because of the lower maximum momentum transfer. However, microscopic comparison of particles prior to and after the jetting process indicated a significant particle disintegration. Secondary disintegrations occur on work piece impact, as well as collisions with slowed-down particles after work-piece impact. On particle disintegration, the kinetic energy is split between the fragments. The direction of movement alters and disturbs the abrasive waterjet (2).

Since the beginning of its industrial applications, several authors have studied the Abrasive Water Injection Jet (AWIJ) process. Given the complexity of the process, most of the models developed are empirical in nature and assume given values for the particle mass or particle energy (1,3–8). In a related study, particle disintegration has been investigated with the aim of estimating the recycling potential (2). The grade of disintegration was analyzed for three different stages (jet formation, cutting process and residual energy conversion). It was shown that the predominant disintegration takes place within the jet formation process, whereas the actual cutting process has only a minor influence in this respect. The grade of fragmentation decreases when smaller particles are used in the same cutting head (d_o = 0.25 mm, d_f = 0.9 mm, l_f = 40 mm). The grade of fragmentation ranged from 20% (d_p = 40 μm) to 60% (d_p = 600 μm) (2). However, these data were acquired with nozzle combinations that are unusual in today's industrial applications and did not consider modern ultra-high pressure and micro-cutting applications. Thus, the present study was conducted to determine the resulting particle size under such conditions and help to better understand the material removal process.

3. MATERIALS AND METHODS

3.1 Materials

Abrasives used were GMA Garnet #120 (GMA Garnet Pty Ltd, Perth, Australia) for macro applications and Barton HPX #220 (Barton Mines Co LLC, Glen Falls, NY, USA) for micro applications, technical specifications are summarized in Table 1. Initially, both abrasives were subject to a sieve analysis. According to specifications from GMA and Barton (9,10), the analysis was carried out according to DIN ISO 565 (11), series R40/3. The results were used as reference values for the following investigations, but also for verification of the particle size distribution specified by the manufacturers. For the sieve analysis, an analytical sieve shaker (manufacturer Retsch, type VS 100) was used. According to DIN ISO 565 series R40/3, the installed sieves were: 250 μm, 212 μm, 180 μm, 150 μm, 106 μm, 90 μm, 75 μm, 63 μm, 53 μm, 45 μm and 38 μm. For the gravimetric analysis, a precision balance (manufacturer Shinko, type DJ-1506, accuracy of measurements 0.001 g) was used.

All jetting experiments were carried out on a lab machine with an Engelhardt C-55 control unit. The high pressure pump was a ThyssenKrupp Uhde HPS 6045 (max. operating pressure p_w = 600 MPa, max. volume flow rate Q_w = 2.8 l/min). A total of three cutting heads were used. To represent normal job shop operation, a standard cutting head commonly used in industry was employed. This was equipped with a common nozzle combination (manufacturer BFT, type TJ006070/591, d_o = 0.28 mm, d_f = 0.76 mm, l_f = 70 mm), in the following referred to as "Macro". To represent common micro-cutting applications, the nozzle combination used was: d_o = 0.125 mm, d_f = 0.3 mm, l_f = 32 mm. The use of two different cutting heads was necessary in order to provide the best possible nozzle centering on the one hand, but also providing the necessary sealing for pressures $p \geq 450$ MPa on the other. The cutting heads used were a self-centering one for operating pressures of $p \leq 350$ MPa (IW Hannover / Dick & Dick GmbH, referred to as "Micro A") and a statically centering one for operating pressures $p \geq 450$ MPa (manufacturer Allfi, type Centerline II, referred to as "Micro B"). All water orifices were sapphires (manufacturer Comadur); all focusing tubes were made of tungsten carbide (manufacturer Ceratizit, type WJNS/Standard).

3.2 Cutting performance benchmark

Prior to the jetting experiments, the cutting performance of the Micro and Macro cutting systems was benchmarked by kerf tests with wedge-shaped specimen. The feed rate was oriented from the summit into the specimen. The tests were carried out with five different abrasive mass flow load ratios (5 %, 10 %, 15 %, 20 % and 25 % by weight), which allowed for the determination of cutting power at different loading situations. In most industrial applications, an abrasive/water load ratio between 0.18 and 0.20 by weight has been evaluated as very economical. Dependent on the particle size mixture, a certain saturation level can already exist on lower abrasive/water proportions. The applied investigation method not only provides for a benchmark of the abrasives but only allows for an estimation of the optimum abrasive/water load ratio.

The tests were carried out with stainless steel specimen (DE 1.4301, similar to AISI 304) and a wedge angle of $\alpha = 15°$. The feed rate was kept constant at f = 100 mm/min, and two pressure levels were employed ($p_2 = 350$ MPa and $p_4 = 550$ MPa).

3.3 Jetting experiments

This test series was carried out in order to investigate the abrasive particle disintegration in the jet forming process; all experiments were carried out without a work piece. For the experiments, abrasive collecting vessels according to Figure 2 were constructed. Prior to the experiments, the vessels were completely filled up with water. The focusing tube of the cutting head was positioned slightly underneath water level before the experiment was started for a determined test duration. The different abrasive load ratios and nozzle sizes required varying test durations; for a required balancing weight of $m_b = 30$ g, the test duration was always set individually to receive a minimum of $m_a = 150$ g of abrasive in the vessel. After each experiment, the vessel was removed and the abrasive was left for sedimentation for a couple of days. The sedimentation was controlled optically through laterally inserted inspection glasses. After sedimentation, the excessive water was carefully decanted and the garnet specimens were subject to a further drying process in a heat treatment furnace at a temperature of $T_f = 200$ °C (392 °F) for a minimum duration of $t_f = 8$ h. According to DIN 66165 (12), the balancing weight was chosen as $m_b = 30$ g for a sieve diameter of $d_s = 200$ mm and an abrasive bulk mass of $m'_b = 2{,}3$ g/cm³. The sieving duration was experimentally determined to be $t_s = 20$ min, which also corresponds to the guide value in Ref. (13).

The experiments were carried out with the “Macro” and “Micro” cutting heads for the pressure levels of $p_1 = 250$ MPa, $p_2 = 350$ MPa, $p_3 = 450$ MPa and $p_4 = 550$ MPa and five different abrasive mass flow load ratios (5 %, 10 %, 15 %, 20 % and 25 % by weight).

A sieve analysis initially results in a cumulative distribution of the mesh sizes. A linear diagram mostly leads to an S-shaped curve progression, which cannot easily be adapted for a statistical analysis. According to (13,14), three distribution functions may be used for the graphic presentation of particle size distributions. These include the power function (15), the log-normal distribution (16,17) and the Weibull distribution (RRSB function) (18). For a sieve analysis with a wide particle size distribution and a high fine particle share, the log-normal distribution is considered most suitable.

Initially, the quantities are plotted as cumulative distribution Q_r. For every sieve run R, equation (3.3.1) applies:

$$R = \frac{subset\ (which\ passes\ a\ sieve\ with\ given\ mesh\ size\ w)}{total\ quantity\ (of\ the\ sieve\ job)} \qquad [-] \qquad (3.3.1)$$

The sieve grades represent an interval of particle sizes Δx_i, therefore the arithmetic mean of its interval width $\bar{x}_i$ is characteristic for this particle class.

$$\Delta x_i = x_i - x_{i-1} \qquad [\mu m] \qquad (3.3.2)$$

$$\bar{x}_i = \frac{x_i + x_{i-1}}{2} \qquad [\mu m] \qquad (3.3.3)$$

For each run R and each mean particle size of the intervals $\bar{x}_i$, the distribution density $q_{r,i}$ represents the mass fraction that remains between two sieves.

$$q_{r,i} = \frac{subset\ x_{i-1} \ldots x_i}{total\ quantity * interval\ width} \qquad [\mu m^{-1}] \qquad (3.3.4)$$

According to (14), the amount type r is set to r = 3 for mass and volume distributions. To analyze the size distribution of each sieve run R in a better way, the following mean values were used. The <u>median value</u> $x_{50,3}$ describes a particle size, which is bigger than 50% of the total particles. The <u>mode</u> $x_{h,3}$ is defined as the most common particle size. The <u>mean particle size</u> $\bar{x}_3$ indicates the weighted mean in relation to the occurring mass fraction.

$$\bar{x}_3 = \sum_{i=1}^{n} \bar{x}_i * R \qquad [\mu m] \qquad (3.3.5)$$

According to (13,14,16), the log-normal distribution is suitable for this application and was therefore used for the further analysis. In this distribution type, the logarithm of the particle size is normally distributed. As a 2-parameter approximation for the measured size distribution, the function gives a simple illustration of the distribution. The distribution density $q_3(x)$ results from equation (3.3.6):

$$q_3(x) = \frac{1}{\sigma_{ln}\sqrt{2\pi}} \cdot \frac{1}{x} \cdot exp\left[-\frac{1}{2}\left(\frac{\ln(x/x_{50,3})}{\sigma_{ln}}\right)^2\right] \qquad [\mu m^{-1}] \qquad (3.3.6)$$

According to equation (3.3.7), the standard deviation can be calculated (16) as

$$\sigma_{ln} = \frac{1}{2} \cdot ln\frac{x_{84}}{x_{16}} \qquad [-] \qquad (3.3.7)$$

where the values for x_{16}, x_{84} and $x_{50,3}$ are obtained from the cumulative distribution graph.

By utilization of equations (3.3.8) and (3.3.9), the specific surface S_V and the Sauter mean diameter d_{32} can be calculated. Hereby, f indicates the so-called Heywood factor for different grain shapes. Based on past investigations of abrasives for industrial customers, a mean value of f= 1.8 was used in this calculation. The Sauter mean diameter indicates the equivalent diameter of a ball with a similar surface compared to the particle.

$$S_V = \frac{6f}{x_{50,3}} \cdot exp\frac{\sigma_{ln}^2}{2} \qquad [\mu m^2] \qquad (3.3.8)$$

$$d_{32} = \frac{6}{S_V} \qquad [\mu m] \qquad (3.3.9)$$

The mode $x_{h,3}$ of the log-normal distribution is given by

$$x_{h,3} = x_{50,3} \cdot exp[-\sigma_{ln}^2] \qquad [\mu m] \qquad (3.3.10)$$

Fine particles can supersaturate the AWIJ without providing an increased cutting power. The fine particle fraction is determined with regard to analytical methods for soil samples derived from geoscience (19). The fine particle share ≤ 63 μm is considered separately in the results.

3.4 Cutting experiments

The procedure of the cutting experiments was similar to that of the previously described jetting experiments (3.3). The main difference comprises that the jet was not immediately dissipated after formation but was used for a defined cutting process. As test material, unalloyed construction steel (DE 1.0503, similar to AISI 1045) with a thickness of t_1 = 10 mm for macro applications and t_2 = 3 mm for micro applications was used. An unalloyed engineering steal was intentionally chosen in order to remove the metal chipping from the used abrasive magnetically. To ensure a similar grade of abrasive stress between the different loading situations, orifice sizes and pressure levels, the feed rates were calculated individually regarding to (5) for a cutting quality of q = 3. The calculation based on (5) was preferred as compared to (6), because the consideration of pressure-dependent water density is not possible with the latter. The specimens were mounted in the abrasive vessels according to Figure 2 (D) and covered with a water level of t_w = 10 mm prior to the experiment in order to avoid reflections. Data acquisition and analysis were carried out identically to the previously described jetting experiments (3.3).

4. RESULTS

4.1 Raw material sieve analysis

The sieve analysis results of the unused garnets are displayed in Figure 3. Both abrasives meet their manufacturer's specifications, the distribution curves are almost identical to ones given in the data sheets. The most common particle size for GMA #120 is 180 μm; the most common size for Barton #220 is 90 μm.

4.2 Cutting performance benchmark

The results of the carried-out kerf tests are displayed in Figure 4. According to (20), the cutting performance P_s is calculated from kerf depth k_s and feed rate f:

$$P_s = k_s \cdot f \qquad [mm/s] \qquad (4.2.1)$$

For the Macro system, optimum load ratios of $R_{macro,350}$ = 0.19 (p_2 = 350 MPa) and $R_{macro,350}$ = 0.21 (p_4 = 550 MPa) were determined. The load ratio values for the Micro system are slightly lower and amount to $R_{micro,350}$ = 0.17 (p_2 = 350 MPa) and $R_{micro,350}$ = 0.19 (p_4 = 550 MPa). The cutting performance of the micro system at high pressure approximately corresponds to the macro system at lower pressure. Hydraulic power P_w is however significantly different in both cases

($P_{w,micro,550}$ = 4,38 kW and $P_{w,macro,350}$ = 11,42 kW), therefore the effectivity of particle acceleration seems to be higher for the micro jet.

4.3 Jetting Experiments

A plot of log-normal distribution curves is displayed in Figure 5 for the case of macro cutting. In the table attached, the associated data (median value $x_{50,3}$, standard deviation σ_{ln}, mode $x_{h,3}$, mean particle size $\bar{x}_3$ and Sauter mean diameter $d_{3,2}$) are displayed. A decrease of the particle size for higher pressure levels can be discerned from in both cases (macro and micro cutting). Simultaneously, the fine particle share increases. The same behavior can be recognized for the 8 other test, with only slight differences between the experimental series. The maximum mean particle size occurs at the lowest pressure levels (p_1 = 250 MPa) while the highest pressure (p_4 = 550 MPa) effects the minimum mean particle sizes. For the micro cutting experiments, the standard deviation is slightly higher for cutting system "Micro B" (p_3 = 450 MPa, p_4 = 550 MPa), this can be seen by the flatter curve progession. This can be attributed to the better orifice - focusing tube alignment of the cutting system "Micro A".

4.4 Cutting Experiments

As stated in section 3.4, the test material (DE 1.0503, similar to AISI 1045) was intentionally chosen in order to remove the metal chipping from the used abrasive magnetically. However, different attempts to separate the micro chipping from the abrasive failed, even the installation of magnets in the sieve shaker did not separate the chipping from the abrasive grains. In Figure 6 A, several pictures of used abrasive with metal chipping with an approaching magnet are displayed. A higher magnification image is shown in Figure 6 B. The metal chipping aligns along the magnetic field and entrains the abrasive grains. The further analysis therefore had to be carried out with the attached particles. For abrasive grains with agglomerating chipping, the apparent grain size increases. In Figure 7, SEM- and photomicroscopic micrographs of single particles after the cutting experiment are displayed. The potential error regarding the chipping is considered to be low, because the microscopic pictures indicate considerably smaller chip sizes compared to the grain sizes. This error therefore mainly affects the fine fraction.

A representative plot of log-normal distribution curves is displayed in Figure 8 for micro cutting. In the attached table, the associated data (median value $x_{50,3}$, standard deviation σ_{ln}, mode $x_{h,3}$, mean particle size $\bar{x}_3$ and Sauter mean diameter $d_{3,2}$) are summarized. Overall, the results correspond to those from the jetting experiments. The maximum mean particle size occurs at the lowest pressure levels (p_1 = 250 MPa), while higher pressure effects the minimum mean particle sizes. However, compared to the jetting experiments, some deviations between the experimental series become obvious, which possibly can be attributed to the mentioned effect through chipping adhesion. The mean values, however, do not indicate an additional significant particle disintegration through the cutting process.

5. DISCUSSION

A possible explanation for the effect of metal chipping agglomeration on the abrasive grains is an increasing electrostatic chargeability for small particle sizes. According to (13), smaller grain sizes tend to agglomeration and wall adhesion. Compared to the jetting experiments, some deviations between the log-normal distribution curves of some experimental series became apparent. This can probably be attributed to the mentioned effect through chipping adhesion.

The plots of the mean particle size $\bar{x}_3$ of the particles > 63 µm from the jetting experiments are displayed in Figure 9 (macro cutting) and Figure 10 (micro cutting). In the case of micro cutting, the graph is split in the middle in order to make clear that two cutting heads with different centering mechanisms have been used. Considering the experimental data displayed in the figures, the influence of pressure clearly becomes apparent for both micro and macro systems. Higher pressure directly reduces the particle size. This tendency has been previously reported (2). The load ratio has less influence on particle disintegration. While for the macro system, a decrease of particle size for an increasing load ratio is still clearly recognizable, this effect is not big enough to become apparent in the experimental data of the micro system. A further, significant particle disintegration through the cutting process could not be detected, this corresponds to the results of the related investigation in Ref. (2).

The fine particle fraction (share of particles ≤ 63 µm) is plotted in Figure 11 (macro cutting) and Figure 12 (micro cutting). A distinct tendency to an increasing fine particle share for increasing pressure can be recognized. The influence of an increasing abrasive load ratio is only minor, but it can also be found to increase the fine particle share.

The decrease of the mean particle size $\bar{x}_3$ is calculated from the mean particle size $\bar{x}_{3,s}$ of the experimental series s and the mean particle size $\bar{x}_{3,n}$ of the new condition grains. This can also be referred to as "degree of particle disintegration", D.

$$D = \frac{\bar{x}_{3,s}}{\bar{x}_{3,n}} \qquad [\%] \qquad (5.1)$$

In Figure 13, the degree of particle disintegration D is plotted for several pressures. The influence of pressure clearly becomes apparent here for both systems. An important finding is that D is consistently lower for micro cutting systems. Noticeable are the similar experimental values for $D_{micro,550}$ and $D_{macro,350}$. Considering Figure 4, the curve progression of the cutting performance is also almost identical. It seems likely that D is generally dependent on the particle momentum, equally if the particle velocity is provided by big orifices or high pressure. A higher degree of particle disintegration does not only have an effect on cutting performance, but also on lifetime of the focusing tube. The most significant amount of particle disintegration takes place in the jet forming process. Disintegration can take place either in mixing chamber and the focusing tube inlet guiding cone or the focusing tube cylinder. The ratio of disintegration in these places is, however, hard to determine. Disintegration in the mixing tube and inlet cone can possibly be reduced by delivery of the abrasive in jet direction or an in-process deflection of the abrasive delivery, maybe also through optimized inlet cones.

6. ACKNOWLEDGEMENTS

The authors thank the German Working Group of Waterjet Technology (AWT) for fruitful discussion.

7. REFERENCES

1. Henning, A. Modelling of the kerf geometry with abrasive waterjet cutting: Modellierung der Schnittgeometrie beim Schneiden mit dem Wasserabrasivstrahl; Jost-Jetter: Heimsheim, 2008.
2. Ohlsen, J. Recycling von Feststoffen beim Wasserabrasivstrahlverfahren; VDI-Verl.: Düsseldorf, 1997.
3. Hashish, M. A Modeling Study of Metal Cutting With Abrasive Waterjets. J. Eng. Mater. Technol. 1984, 106, 88.
4. Hashish, M. A Model for Abrasive-Waterjet (AWJ) Machining. J. Eng. Mater. Technol. 1989, 111, 154.
5. Zeng, J.; Kim, T. Parameter Prediction and Cost Analysis in Abrasive Waterjet Cutting Operations; Proceedings of the 7th American WJTA Conference, August 19-21, Seattle, Washington., 1993.
6. Zeng, J. Determination of machinability and abrasive cutting properties in AWJ cutting; Proceedings of the 2007 American WJTA Conference and Expo, August 19-21, Houston, Texas., 2007.
7. Lichtarowicz, A., Ed. Jet Cutting Technology; Springer Netherlands: Dordrecht, 1992.
8. Hoogstrate, A.M.; Karpuschewski, B.; van Luttervelt, C.A.; Kals, H. Modelling of high velocity, loose abrasive machining processes. CIRP Annals - Manufacturing Technology 2002, 51, 263–266.
9. GMA Garnet Group. GMA GARNET. Das natürliche Wasserstrahlschneidemittel. Produktdatenblatt, 2009.
10. Barton International. HPX® Fine Waterjet Abrasives Typical Particle Size Distribution. Graph and Table: Glen Falls, NY, USA, 2011.
11. DIN ISO 565. DIN Deutsches Institut für Normung e.V. Test sieves - Metal wire cloth, perforated metal plate and electroformed sheet - Nominal sizes of openings, 1998.
12. DIN 66165. DIN Deutsches Institut für Normung e.V. Partikelgrößenanalyse, Siebanalyse. Teil 1: Grundlagen; Teil 2: Durchführung, 1987.
13. Stiess, M. Mechanische Verfahrenstechnik: Partikeltechnologie, 3rd ed; Springer Berlin Heidelberg: Berlin, Heidelberg, 2009.
14. DIN ISO 9276. DIN Deutsches Institut für Normung e.V. Representation of results of particle size analysis - Part 1: Graphical representation; Part 2: Calculation of average particle sizes/diameters and moments from particle size distributions; Part 4: Characterization of a classification process; Representation of results of particle size analysis, 2004-2012.
15. DIN 66143. DIN Deutsches Institut für Normung e.V. Darstellung von Korn-(Teilchen-) größenverteilungen. Potenznetz, 1974.
16. DIN 66144. DIN Deutsches Institut für Normung e.V. Darstellung von Korn-(Teilchen-) größenverteilungen. Logarithmisches Normalverteilungsnetz, 1974.
17. ISO 9276-5. ISO - International Organization for Standardization. Representation of results of particle size analysis - Part 5: Methods of calculation relating to particle size analyses using logarithmic normal probability distribution, 2005.
18. DIN 66145. DIN Deutsches Institut für Normung e.V. Darstellung von Korn-(Teilchen-) größenverteilungen. RRSB-Netz, 1976.

19. DIN 18123. DIN Deutsches Institut für Normung e.V. Soil, investigation and testing - Determination of grain-size distribution, 2011.
20. Peter, D. Untersuchungen zu Einsatzmöglichkeiten von Wasserabrasivstrahlen für Rückbaumaßnahmen; PZH, Produktionstechn. Zentrum: Garbsen, 2007.
21. Barton International. Technical Data & Physical Characteristics for Hard Rock Garnet Abrasives: Glen Falls, NY, USA, 2012.

8. TABLES

Table 1: Garnet product data (9,21)

Product labeling	GMA Garnet #120	Barton Garnet HPX #220
Relative density	4.1	3.9
Mohs hardness	7.5 – 8.0 Mohs	7.5 – 8.5 Mohs
Melting point	1250 °C (2282 °F)	1315 °C (2399 °F)
Electric conductivity (max.)	25 ms/m	25 ms/m
Almadine content	97 % - 98 %	92 % - 96 %

9. FIGURES

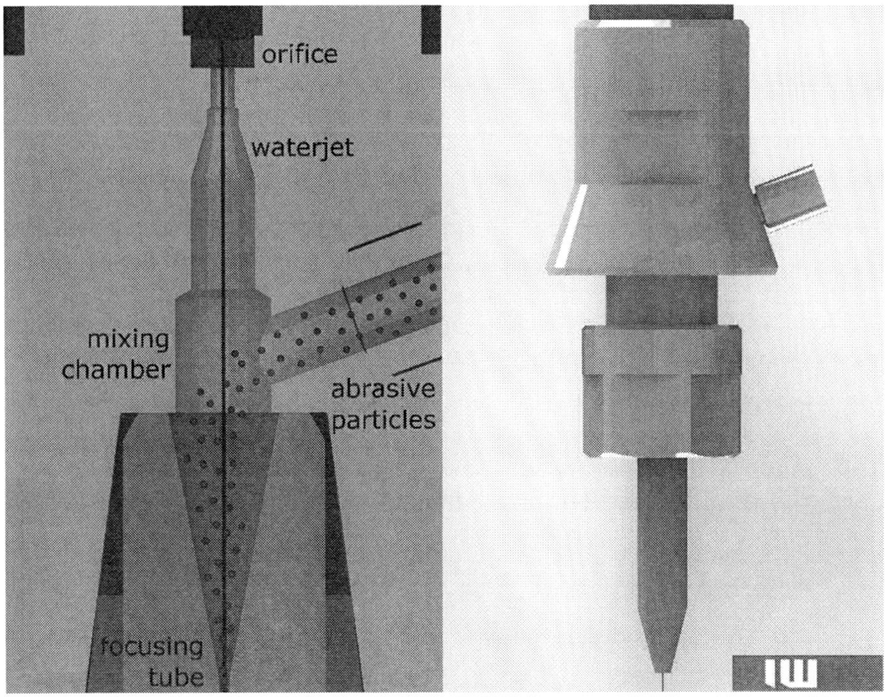

Figure 1: Schematic AWIJ cutting head

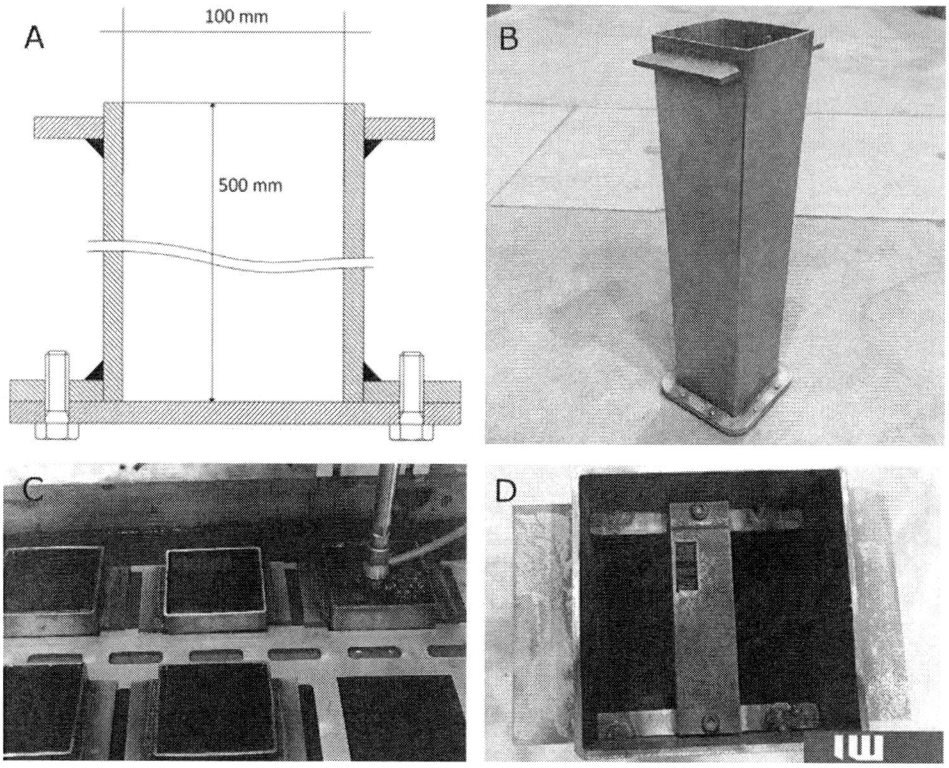

Figure 2: Abrasive collecting vessel (A-B), Jetting experiment with mounted vessels (C), mounted specimen for cutting experiments (D)

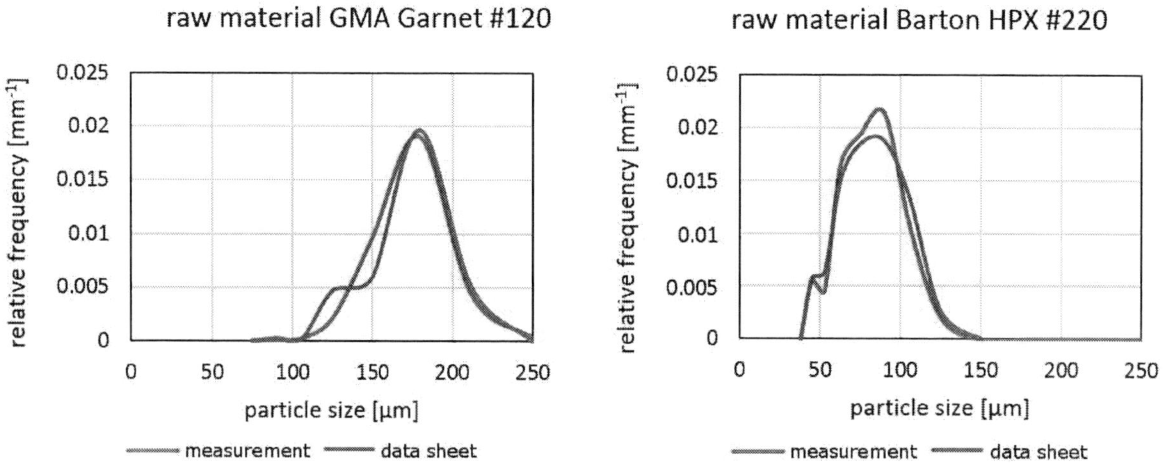

Figure 3: Abrasive sieve analysis according to ISO 565, R40/3 (11) prior to the jet formation (red), and particle size distribution as given by the manufacturers (blue) (9,10)

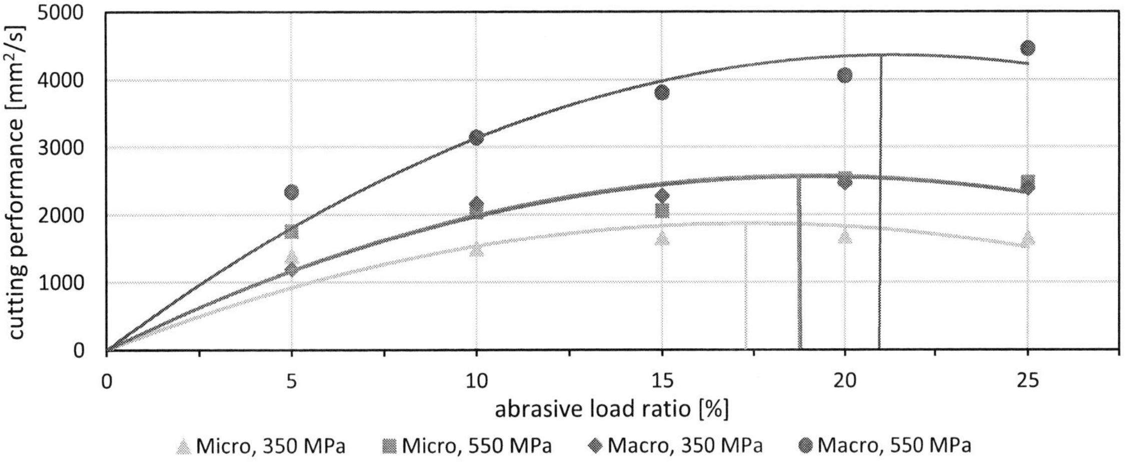

Figure 4: Cutting performance for different abrasive load ratios and operating pressures

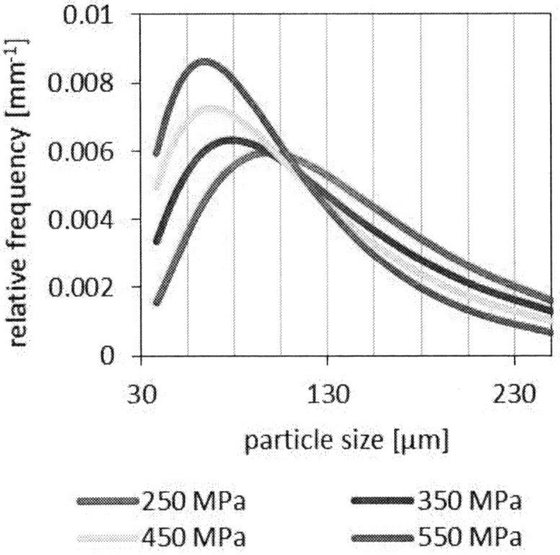

p [MPa]	250	350	450	550
$x_{50,3}$ [µm]	137	120	104	92
σ_{ln}	0,58	0,65	0,66	0,60
$x_{h,3}$ [µm]	98	79	68	64
$\bar{x}_3$ [µm]	129	115	107	100
$d_{3,2}$ [µm]	64	54	47	43
Fine fraction: < 63 µm [%]	14	20	23	23
$\bar{x}_3$ > 63 µm [µm]	144	135	130	120

Figure 5: Exemplary log-normal distribution curves for a jetting experiment with the macro system. d_o = 0,28 mm, d_f = 0,76 mm, R = 15 %

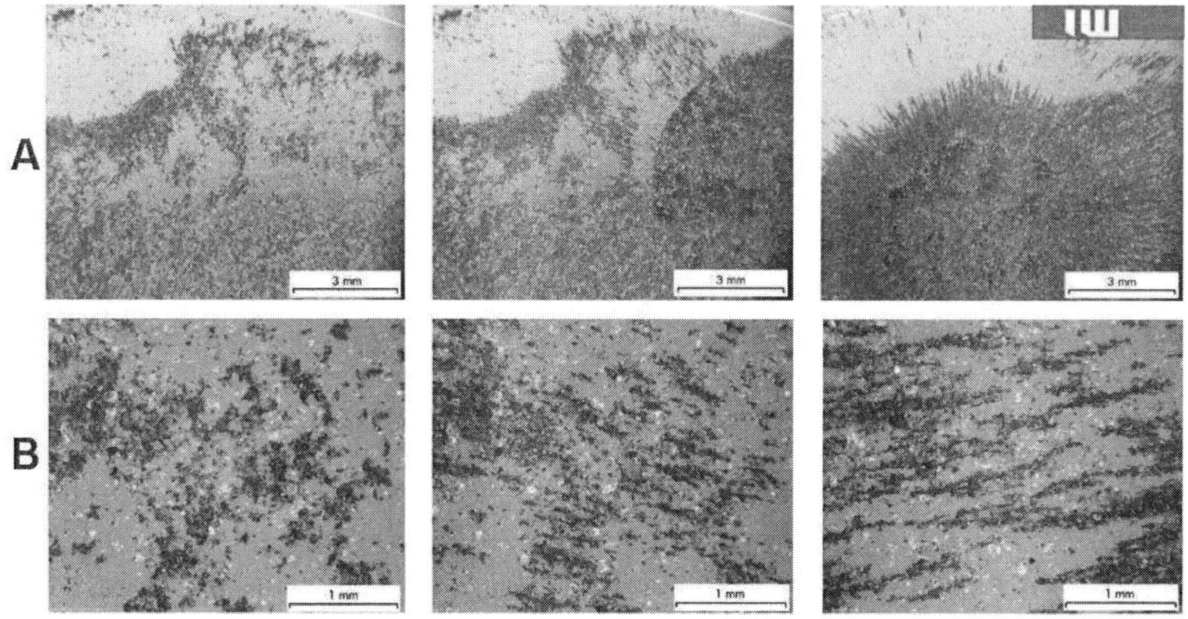

Figure 6: Photomicrographs of abrasive with attached, magnetic micro chipping

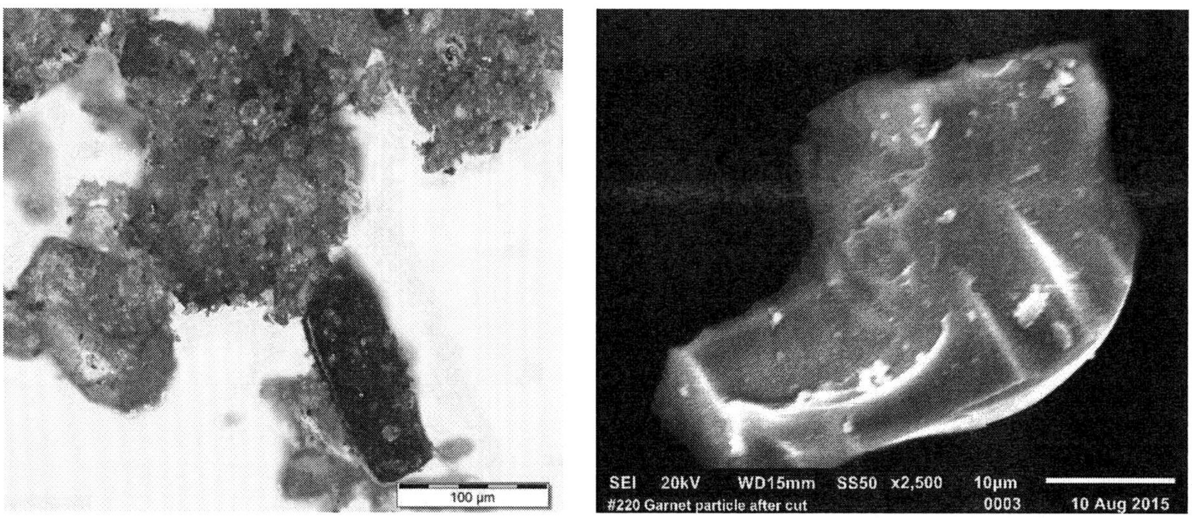

Figure 7: Photomicrograph and SEM-micrograph of disrupted abrasive particles (Barton HPX #220 after cutting experiment)

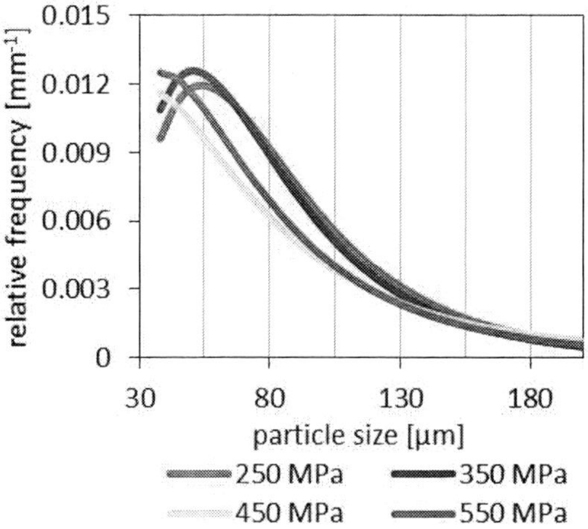

p [MPa]	250	350	450	550
$x_{50,3}$ [µm]	72	68	60	60
σ_{ln}	0,53	0,54	0,75	0,66
$x_{h,3}$ [µm]	54	51	34	39
$\bar{x}_3$ [µm]	69	64	58	58
$d_{3,2}$ [µm]	35	33	25	27
Fine fraction: < 63 µm [%]	29	37	45	46
$\bar{x}_3$ > 63 µm [µm]	85	84	82	82

Figure 8: Exemplary log-normal distribution curves for a cutting experiment with the micro system. d_o = 0,125 mm, d_f = 0,3 mm, R = 20 %

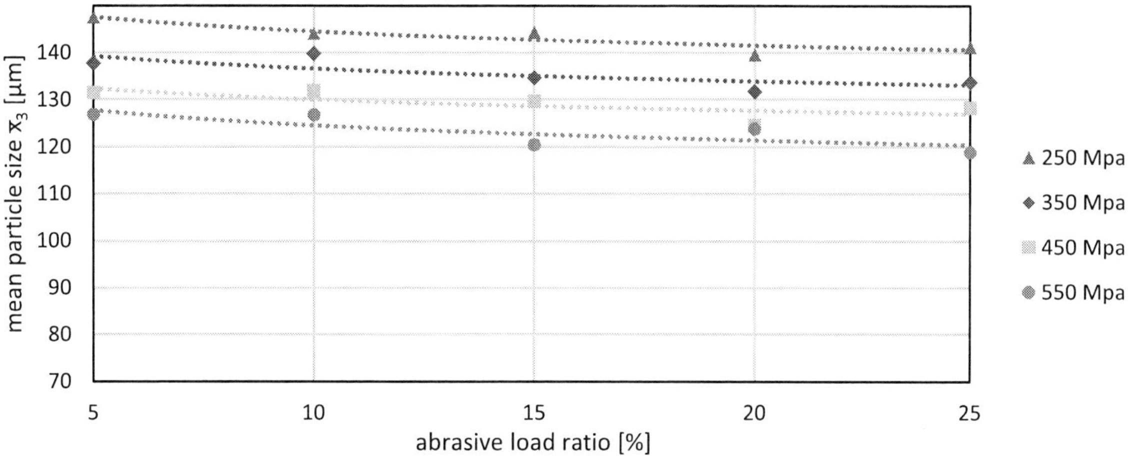

Figure 9: Plot of the mean particle size (> 63 µm) after jet formation (macro cutting)

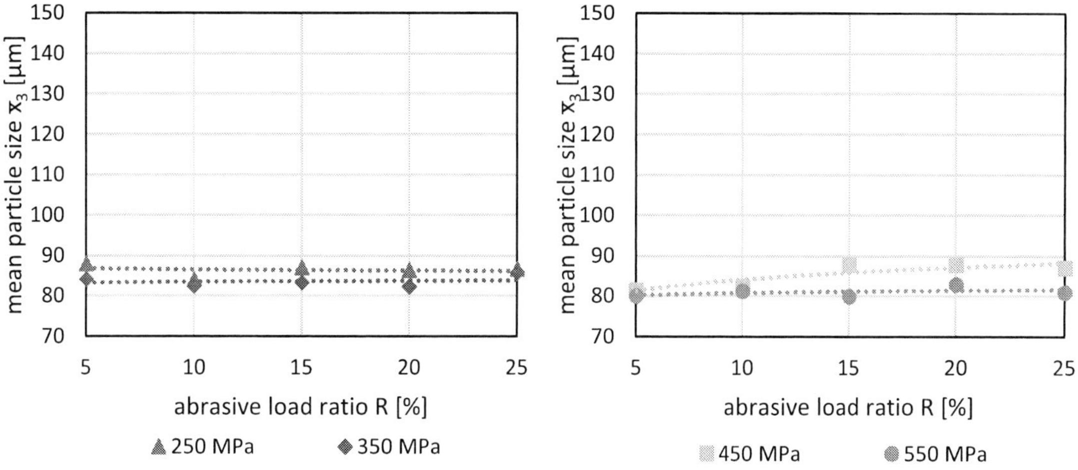

Figure 10: Plots of the mean particle size (> 63 µm) after jet formation (micro cutting)

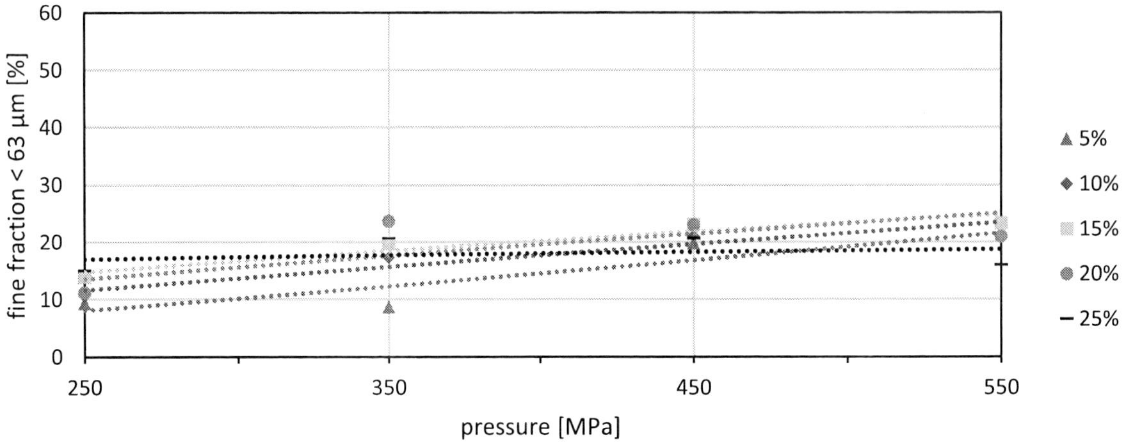

Figure 11: Plot of the fine particle fraction, macro cutting

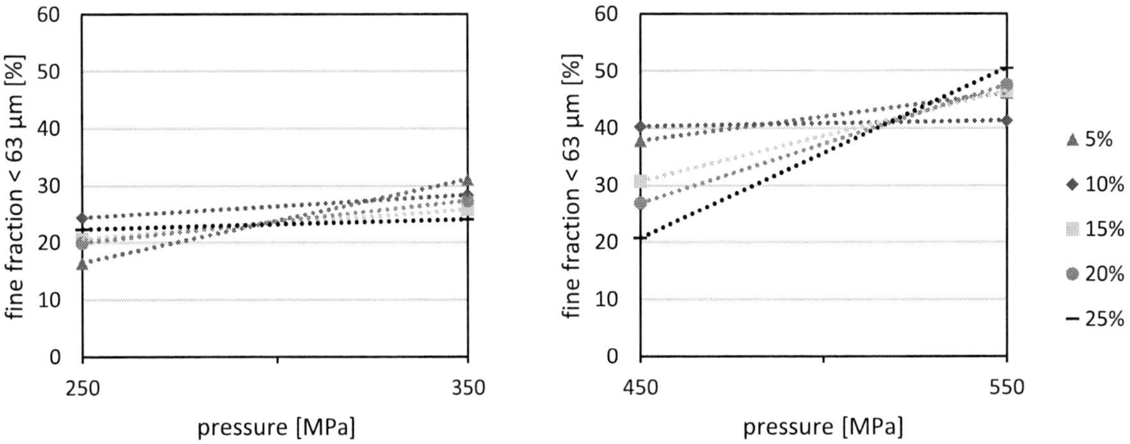

Figure 12: Plots of the fine particle fraction, micro cutting

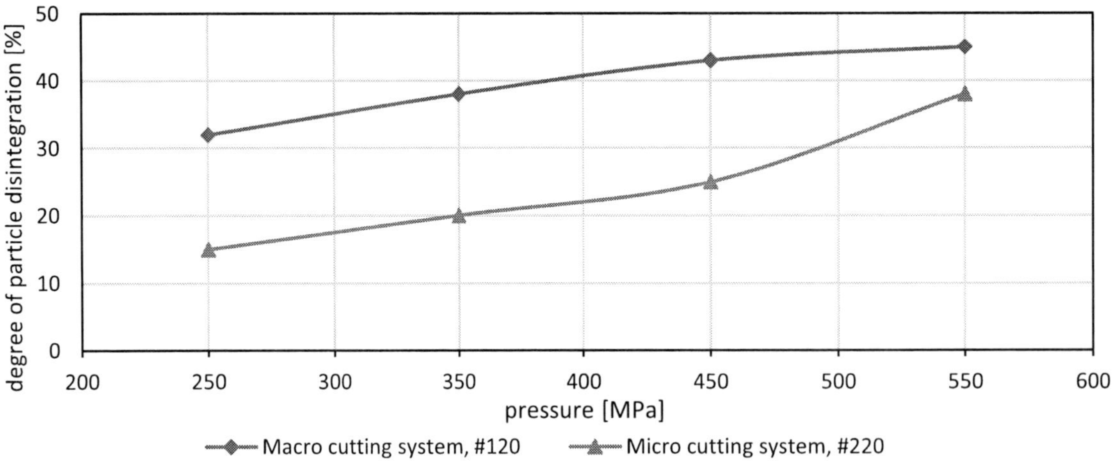

Figure 13: Degree of particle disintegration D for different pressures

2015 WJTA-IMCA Conference and Expo
November 2-4 • New Orleans, Louisiana

Paper

ENHANCING THE ADHESION STRENGTH OF COLD GAS DYNAMIC SPRAYED COATINGS BY PREPARING THE SUBSTRATES WITH THE HIGH-FREQUENCY PULSED WATERJET

M. Vijay, A. Tieu, W.Yan, B. Daniels, M. Xu
VLN Advanced Technologies
Ottawa, Ontario, Canada

B. Jodoin, T. Samson, D. MacDonald, R. Fernández. M. Yandouzi
University of Ottawa
Ottawa, Ontario, Canada

ABSTRACT

It is known that surface preparation can affect performance of thermal spray coatings more than any other variables. The objective of this study is to gain an understanding of the effect of different surface preparations on the adhesion strength of coatings applied by the cold gas dynamic spraying technique. Preparation methods under consideration are: polishing, grit blasting and high-frequency pulsed waterjet. The results demonstrate that adhesion strength of the grit blasted substrates is lower than the one achieved with the pulsed waterjet substrates. Furthermore, the results indicate that pulsed water jet provides a very simple method to control surface roughness by simply varying the standoff distance.

Organized and Sponsored by WJTA®-IMCA®

1. INTRODUCTION

As highlighted in the JPCL article (Bamhart et al.), the surface preparation of the substrate is one important part of a coating system. That is, the surface preparation can deeply affect the performance of the coating. The effectiveness of a coating depends on its bond to the substrate which is highly dependent on how the surface has been prepared. The ubiquitous method of preparing a surface is grit blasting (Helsel; Groff). However, for many coatings, the imbedding of grit particles is highly undesirable. In order to avoid this problem, use of continuous waterjet has been attempted for preparing surfaces, primarily for epoxy coatings (Groff).

The adhesion strength level of thermal spray coatings in general and more precisely in the case of cold gas dynamic sprayed coatings, is highly dependent on surface preparation and surface roughness level (Varacalle et al.). Conventional methods for surface preparation using grit blasting are aimed to improve adhesion strength of coatings by primarily enhancing the substrates surface area by increasing roughness profiles. The main disadvantage of using grit for creating such profiles are: a) grit particle embedment and b) limited roughness attainability. Grit contamination poses a considerable health issue for biomedical applications where costly revision surgeries have been performed to replace damaged joint implants. Even if the health effects of contamination are ignored, it is not possible to markedly increase the coating adhesion strength as particle embedment leads to early crack propagation, thus reducing considerably fatigue life of the coated part.

A novel technique used to increase surface roughness by using pulsed water jet has shown dramatic improvements in coating adhesion, superior to that of grit blasting, without the potential disadvantage of particle embedment at the interface. The main advantages of using water only to roughen surfaces is to i) eliminate grit embedment that can cause contamination and reduce fatigue life. ii) increase roughness beyond what grit blasting technique can produce to further promote adhesion strength.

This paper shows that using pulsed water jet to prepare surfaces to be coated using the cold gas dynamic spray process increases the coating adhesion strength by a factor of 4 over that of grit blasting methods. This increase in adhesion strength appears to be the result of unique nanometer to micrometer sized surface features.

2. EXPERIMENTAL PROCEDURE

2.1 Pulsed Water Jet – PWJ

PWJ has been in the industrial scene for past 20 years. It has mainly been used as an alternative technology to ultra-high pressure water jet for removing coatings such as epoxy based paint and much harder coatings such as thermal barrier coatings typically found on aircraft engine hot-section components. However, recent studies have shown that PWJ can also deliver embedment-free surface preparation that is superior to conventional grit blast method (Samson et al.)[m1]. This patented process can be applied on any material and water is the only required medium (Vijay et al.).

With PWJ, substrate material such as aluminum and high strength steel alloys all can be roughened with operating water pressures as low as 5000 psi. The fundamental physics behind the workings of pulse jet has been presented previously (Vijay et al.; Vijay). To completely explain how pulsed water jet can easily alter the surface characteristics without using any abrasives is beyond the scope of the paper. In general, pulsed jet generates and directly deposits cavitation bubbles via a water jet to implode on the targeted area. As intense cavitation bubbles collapses on one another, a stochastic surface profile is produced.

2.2 Cold Gas Dynamic Spraying – CGDS

CGDS is becoming a more common thermal spray technique used to deposit metallic coatings for a variety of applications (Irissou et al.). The value of this process resides in the fact that it allows coatings to be deposited without exposing the feedstock particles to high in-flight temperatures that may cause notable oxidation or phase change (Davis; Grujicic et al.; Richer et al.). This is accomplished by injecting particles into a flow of compressed gas (often nitrogen, although helium or air can be used) accelerated through a converging-diverging nozzle. Feedstock particles in this stream are accelerated to high velocities before impacting the substrate. Upon impact with the substrate, both the particles and substrate experience adiabatic shear instabilities (Schmidt et al.; Assadi et al.; Wu et al.; Hussain et al.; Grujicic et al.), resulting in jetting of material. This jetting often leads to the removal of oxide layers allowing for metal to metal contact, localised melting (at the nanometer scale) due to friction forces, as well as producing mechanical anchoring points which are enveloped by the impacting particles. This leads to a mix of two types of bonding mechanisms that have been identified in CGDS: metallurgical and mechanical. The parameters used in the current work for the CGDS deposition can be found in Table 1.

Table 1: CGDS Parameters used for Deposition

Gas Temperature	300°C
Gas Pressure	3.45 MPa
Carrier Gas	Nitrogen
Traverse Speed	20 mm/s
Step Size	1 mm
Passes	1
Feed Rate	14 g/min
Feed Wheel Type	320 Small Hole
Powder Feeder Gas Flow Rate	0.85 NCMH
Powder Feeder Carrier Gas	Nitrogen
Standoff Distance	15 mm

2.3 Substrate and Coating Material Selection

To evaluate the efficiency of PWJ prepared surfaces, coatings adhesion strength are to be compared to grit blasted and polished surfaces that have been coated using the same spray process and parameters. The deposition of pure aluminium powder onto 6061 aluminum alloy substrates has been used successfully accomplished in the past and can serve as a reliable reference point. The substrate material used in this study was 6061-T6511 aluminium alloy. The substrates were machined into 25.4 mm diameter cylinders in accordance to the ASTM C633 adhesion strength standard. To ensure consistency between sprays, the samples were thermally insulated during deposition using a high temperature polymer (polybenzimidazole) holder. The powder used in this investigation was pure aluminium powder (SST-A5001) [Centerline Ltd., Windsor, Ontario, Canada].

2.4 Substrate Preparation

Three different surface preparing techniques were used for comparisons namely, PWJ, grit blasting and mirror-finish polishing. A minimum of three samples were used for each group to assess process variances. The parameters for polishing and grit blasting can be found in Tables 2 and 3, respectively.

Table 2: Polishing Disks Used In Order of Decreasing Roughness Used To Prepare Polished Samples

Grinding / Polishing Disk	Expected Surface Finish
MD - Piano 220 solution	~ 68 μm
MD - Largo with Largo suspension solution	~ 3-9 μm
MD - Mol with Mol suspension solution	≤ 3 μm

Table 3: Surface Preparation Parameters for Grit Blasting

Abrasive	Aluminium Oxide
Size	177 μm
Process Gas	Filtered / Dry Air
Spraying Pressure	~ 400 kPa
Spray Angle	45°
Standoff Distance	51 mm

To evaluate the effect of substrate surface roughness on coating adhesion strength, PWJ samples were prepared with low, intermediate, high, and very high surface roughness. Low level is considered to be equivalent to surface roughness values obtained using grit blasting, whereas intermediate and above levels are typically not attainable by the grit blasting process. PWJ roughness can be controlled via standoff distance, pressure, flowrate and/or dwell time. Process parameters for the PWJ process are presented in Table 4.

Table 4: Parameters Held Constant For All PWJ Surface Preparations

Traverse Velocity	455 mm/s
Water Pressure	69 MPa
Nozzle Diameter	1.0 mm
Pulse Frequency	20,000 Hz
Step Index	0.1 mm/pass

2.5 Adhesion Strength Testing

The adhesion strength between the coatings and the substrates was determined in compliance with ASTM C-633. A thermally cured elastomeric adhesive with a tensile strength of 76.9 MPa was used. The samples were placed in a vertical fixture and then placed in a 177°C oven for two hours to ensure the adhesive had properly cured. The testing was performed using an Instron universal tensile testing machine.

2.6 Evaluation of Substrate Roughness

3D digital imaging by depth composition was performed with a digital microscope (VHX – 1000, Keyence Corporation, Osaka, Japan) on the substrates prior to coating deposition. These images were then used to perform roughness measurements and perform surface area calculations.

3. RESULTS & DISCUSSIONS

3.1 Substrate Roughness

In the experiment, surface roughness for the PWJ samples was altered by changing the standoff distance of the jet. Table 5 presents the roughness values determined for each substrate preparation method used in this work. It also presents the calculated increase in projected surface area compared to an ideally flat surface.

Table 5: Roughness results for substrates prepared by polishing, grit blasting, and PWJ

Sample	Standoff Distance [mm]	Measured Roughness, R_a [µm]	Increase in surface area [%]
Polished	N/A	0.5 ± 0.1	1.4 ± 0.1
Grit Blasted	51	1.7 ± 0.3	3.7 ± 0.6
Waterjet A	56	2.1 ± 0.3	6.4 ± 4.2
Waterjet B	51	4.6 ± 0.4	13.1 ± 4.6
Waterjet C	46	15.9 ± 1.3	49.3 ± 8.1
Waterjet D	42	21.7 ± 2.3	60.8 ± 20.2

3.2 Evaluation of Substrate Surfaces

Figure 1 shows that the grit blasted surface contained traces of embedded aluminum oxide. This feature is important, as it may result in reduced adhesion strength due to crack propagation. Figure 2 shows a PWJ surface at the same magnification. Unlike roughened surfaces created by grit blasting or traditional machining methods, pulsed water jet roughened surfaces are teeming with intricate networks of voids where the coating can find a firm grip to produce strong mechanical bond.

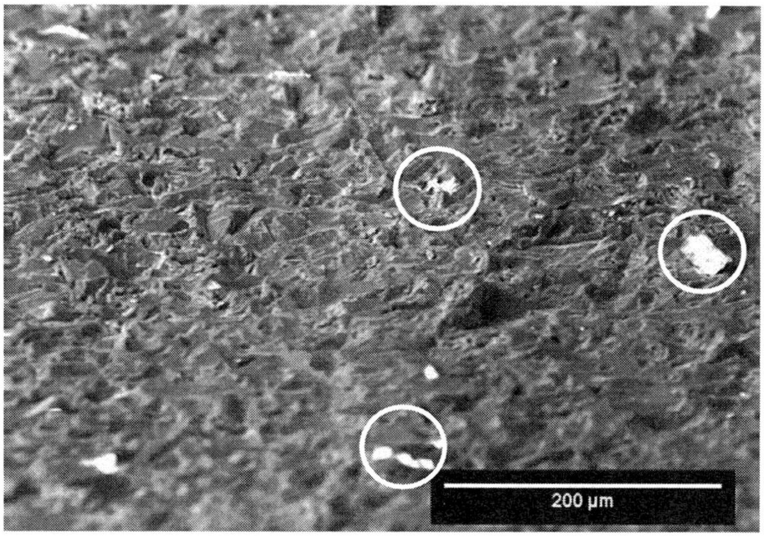

Figure 1: SEM image of grit blasted aluminum showing traces of alumina oxide grit embedment at 200X magnification at angle.

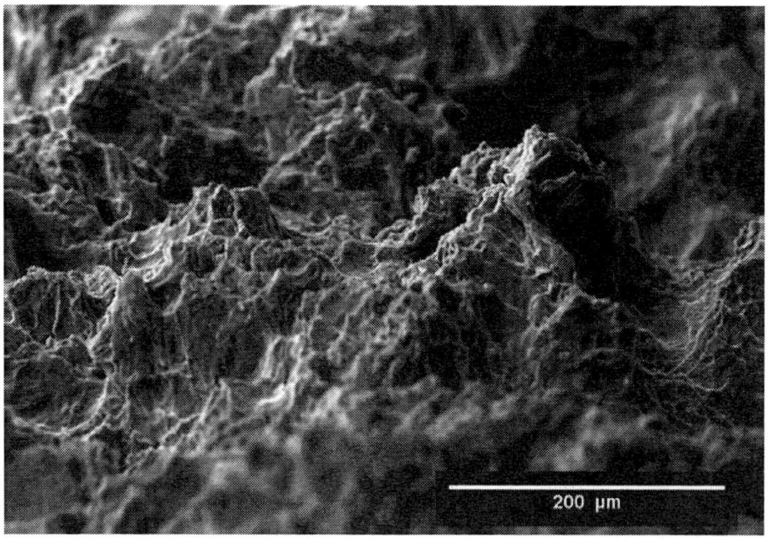

Figure 2. SEM image (200X) showing PWJ roughened surface free of grit contamination.

Figure 3 shows the top view of a 20-30 micrometer (Ra) surface produced by PWJ. This picture demonstrates that there are not only features at the micrometer scale, but also many features at the nanometer scale. These images lead to a clearer understanding into the way that the coating particles could entrench themselves into the cavities generated by the PWJ process.

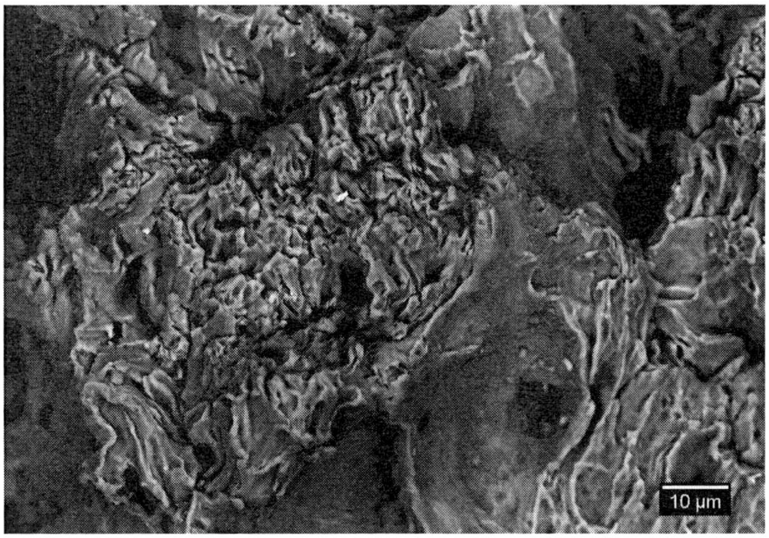

Figure 3. SEM image (1000X) of PWJ roughened surface topography. Note the density of random pits and crevices created by cavitation.

3.3 Evaluation of Adhesion Strength

The results of adhesion strength vs substrate roughness values can be found in Figure 4. It reveals a direct correlation between surface roughness and adhesion strength i.e, the rougher the surface the higher the adhesion strength. However, it can be observed that the adhesion strength levels off at the higher end of substrate roughness and suggests that increasing roughness beyond these levels will not translate into higher adhesion values. It is hypothesized that this is mainly due to poor coating quality; such that as the surface gets rougher than a specific value, the coating integrity diminishes. More work needs to be performed to clearly assess the physics behind these observations and hypothesis.

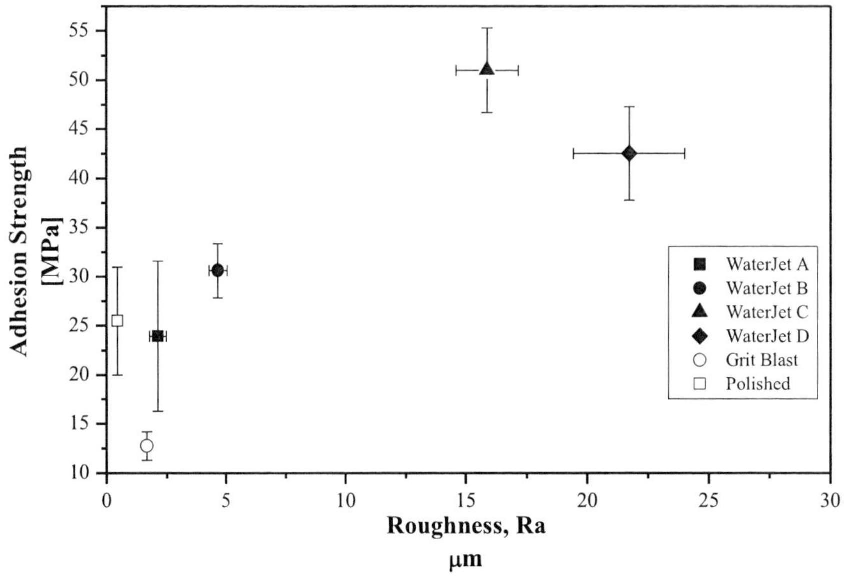

Figure 4. Plot showing the effect of surface roughness (surface preparation) on the adhesion strength of CGDS coatings.

Figure 4 also shows that the adhesion strength is higher on polished surface over grit blasted surface. This is counter intuitive in terms of increased surface area as discussed previously. However, the CGDS process is known to act as grit blasting for the first few particles impinging the substrate. As such, a polished surface would present some level of surface roughness due to that effect without foreign grit particle embedment that reduces the effectiveness of coating adhesion strength. The coating cross section of Figure 5 reveals that grit remained embedded at the substrate/coating interface for the grit blasted samples, which would substantially decrease the adhesion strength of the coatings.

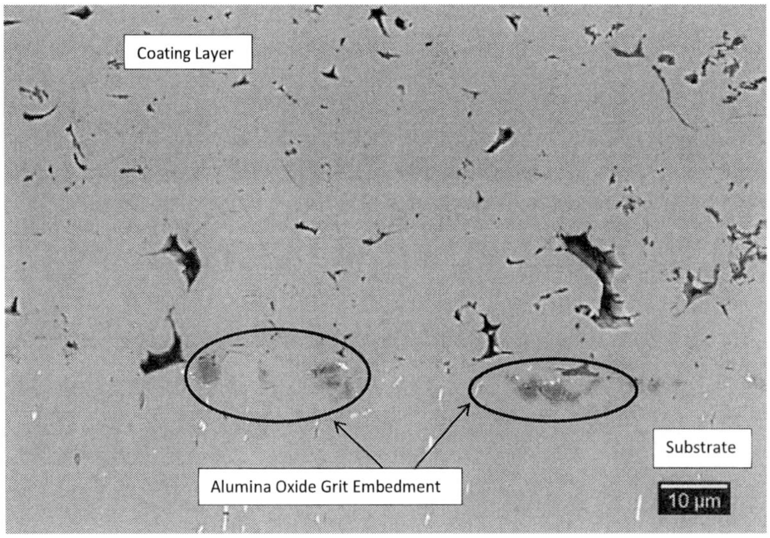

Figure 5. SEM image (500X) showing a cross sectional view of the coating and substrate interface. Note the grit particle contamination.

The high adhesion strength of the PWJ samples can be attributed to abundance of 3-dimensional roughness features that allow more mechanical anchoring and an increased surface area. Such high adhesion strength leads to cohesion failure which can be observed during the pull test where the tensile strength of the coating material itself is lower than the adhesion strength. A clear example of mechanical anchoring can be observed in Figure 6.

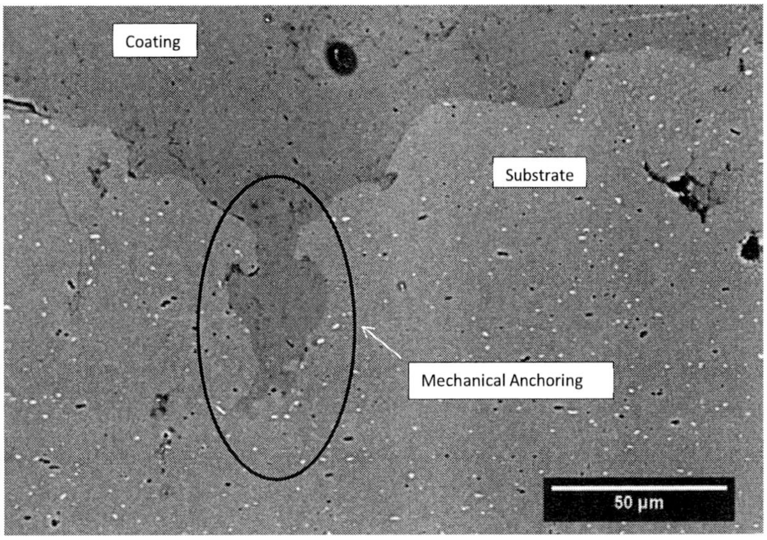

Figure 6. SEM image (500X) showing a cross sectional view of the coating and substrate interface. Note the interlocking effect offering strong mechanical anchoring.

Given that the coating spray parameters are kept constant throughout the various roughening techniques, the only variable that remains is the change in surface texture. These results demonstrate two very the important aspects when it comes to understanding and improving coating adhesion:

1) Consideration of the complex nature of the surface topography and not just measured values from a roughness meter.

2) Grit embedment is to be avoided.

4. CONCLUSIONS

This investigation highlighted the relationship between substrate surface roughness and coating adhesion strength. Substrate samples were prepared by polishing, grit blasting, and PWJ processing. Substrates were then coated with using cold spray of pure aluminium powder. The results presented suggest that increased adhesion strength is a result of increased surface area and increased mechanical anchoring features. It is hypothesized that these features allow incoming particles to conform and eventually encapsulate acting as a mechanical interlock.

PWJ was also evaluated as a novel method of surface preparation for cold gas dynamic spray operations. The system provides the ability to increase the surface roughness beyond what can be expected from grit blasting. It also provides the additional benefit of not using entrained particles. This ensures that embedded particles, often an issue with grit blasted substrates, are not present which removes the possibility of bond weakening caused by embedded grit.

5. ACKNOWLEDGEMENTS

VLN authors acknowledge the technical support provided by Dr. Bertrand Jodoin, Tyler Samson, Daniel MacDonald, Mohamed Yandouzi and Rubén Fernández from Cold Spray Research Laboratory at the University of Ottawa in Ottawa, Ontario, Canada.

6. REFERENCES

Assadi, Hamid et al. “Bonding Mechanism in Cold Gas Spraying.” Acta Materialia 51.15 (2003): 4379–4394. Web. 19 Feb. 2014.

Bamhart, Robert et al. “Why Surface Preperation Is Important.” Journal of Protective Coatings and Linings May (2013): n. pag. Print.

Davis, J R. Handbook of Thermal Spray Technology. Ed. J R Davis. ASM International, 2004. Print.

Groff, Pamela, ed. “Surface Preparation: Waterjetting.” SSPC Pocket Guide to Coating Information. 1st ed. Pittsburgh, PA: The Society for Protective Coatings, 2009. 70. Print.

Grujicic, M et al. “Adiabatic Shear Instability Based Mechanism for Particles/substrate Bonding in the Cold-Gas Dynamic-Spray Process.” Materials & Design 25.8 (2004): 681–688. Web. 19 Feb. 2014.

Helsel, Jayson L. “Surface Preparation: A Primer.” Journal of Architectural Coatings October/No (2008): n. pag. Print.

Hussain, T. et al. “Bonding Mechanisms in Cold Spraying: The Contributions of Metallurgical and Mechanical Components.” Journal of Thermal Spray Technology 18.3 (2009): 364–379. Web. 4 Feb. 2014.

Irissou, Eric et al. “Review on Cold Spray Process and Technology: Part I—Intellectual Property.” Journal of Thermal Spray Technology 17.4 (2008): 495–516. Web. 19 Feb. 2014.

Richer, P. et al. “Substrate Roughness and Thickness Effects on Cold Spray Nanocrystalline Al-Mg Coatings.” Journal of Thermal Spray Technology 15.2 (2006): 246–254. Web. 19 Feb. 2014.

Samson, T. et al. “Effect of Pulsed Waterjet Surface Preparation on the Adhesion Strength of Cold Gas Dynamic Sprayed Aluminum Coatings.” Journal of Thermal Spray Technology 24.August (2015): 984–993. Web.

Schmidt, Tobias et al. “Development of a Generalized Parameter Window for Cold Spray Deposition.” Acta Materialia 54.3 (2006): 729–742. Web. 19 Feb. 2014.

Varacalle, Dominic J. et al. “Effect of Grit-Blasting on Substrate Roughness and Coating Adhesion.” Journal of Thermal Spray Technology 15.3 (2006): 348–355. Web. 19 Feb. 2014.

Vijay, Mohan. “Apparatus and Method For Prepping A Surface Using a Coating Particle Entrained In A Pulsed Waterjet or Airjet.” 2013: n. pag. Print.

---. “Method and Apparatus for Prepping Surfaces with a High-Frequency Forced Pulsed Waterjet.” 2010: 1–57. Print.

Wu, Jingwei et al. "The Rebound Phenomenon in Kinetic Spraying Deposition." Scripta Materialia 54.4 (2006): 665–669. Web. 19 Feb. 2014.

7. NOMENCLATURE

CGDS: Cold Gas Dynamic Spray
PWJ: Pulsed Water Jet
Ra: Arithmetic Average of Roughness
SEM: Scanning Electron Microscope

2015 WJTA-IMCA Conference and Expo
November 2-4 • New Orleans, Louisiana

Paper

THE EFFECTS OF TIER 4 EMISSION STANDARDS ON THE WATERJET INDUSTRY

J. Van Dam Director of Product Development
NLB Corporation
Wixom, Michigan, USA

ABSTRACT

This paper documents the impact of Tier 4 final engines on the water jetting industry. Diesel engines, which have been the most common means of providing power for portable water jetting pumps , have been subject to the EPA's tightening standards for the emission of harmful exhaust gases. The final stage of the reduction process is now at a point that both the manufacturers and users of water jetting pumps are learning what it is like to operate these new diesel engines.

While the cost to procure a Tier 4 Final engine is typically greater than that of a Tier 3 or Tier 4i, the cost savings during the life of the engine can be significant. In addition they greatly reduce harmful exhaust gases like nitrogen oxides, carbon dioxide and carbon monoxide. Owning a Tier 4 Final engine also ensures that owners/operators will be able to take their diesel powered pump anywhere in the U.S. and operate it in the future.

Organized and Sponsored by WJTA®-IMCA®

1. EPA EMISSION STANDARDS - A BRIEF HISTORY

For nearly two decades, the Environmental Protection Agency (EPA) has been tightening standards for the emission of harmful exhaust gases from diesel-powered equipment, such as high-pressure water jet units. Until January, 2015, these regulations applied only to new engines under 750 hp, but the latest rules (known as Tier 4) mandate lower emission levels from larger engines as well.

Until recently, the engines driving water jet units were not designed to meet these standards, but today engine manufacturers are adding new technology. While this adds cost to the engine (and, therefore, to the water jet unit it powers), it also improves fuel economy, which reduces operating expense. These savings offset a substantial portion of the investment.

2. EMISSIONS REDUCTIONS

2.1 EPA goal

Tier 4 standards are part of a national program by the EPA to reduce emissions from non-road diesel engines, a category that includes water jet pump units as well as excavators, generators, compressors, and a host of other equipment. These reductions are to be accomplished using the Best Available Control Technology, or BACT. Full details can be found at the EPA website, http://www.epa.gov/otaq/nonroad-diesel.htm. Essentially, the EPA wants non-road engines to have the same sort of advanced emission control technologies required of trucks and buses.

In 1996 the EPA announced that future emissions of nitrogen oxides (NOx) had to be limited to 10% of then-current levels and particulate matter (PM) emissions to 5%. (PM is the black soot or smoke in the exhaust of a diesel engine.) The agency estimated that its new standards would reduce NOx emissions by about a million tons per year by 2010… the equivalent of taking 35 million cars off the road. Reductions in carbon monoxide (CO) and non-methane hydrocarbon emissions (NMHC) were also mandated.

Recognizing that achieving these changes would not be simple (or inexpensive) for manufacturers, the EPA decided to phase them in. Diesel engines of 11 hp to 750 hp would have to reduce emission levels to a certain level (called Tier 1) by 2000, followed by progressively lower levels in 2006 (Tier 2), and 2008 (Tier 3).

An even stricter set of standards — Tier 4 — was established for engines above 750 hp. This was implemented in two stages, with Interim (Tier 4i) taking effect in 2011 and Final (Tier 4F) in January, 2015.

2.2 The importance of compliance

Compliance with Tier 4F standards is now mandatory in areas that have not yet attained current EPA air quality standards, and in some areas that have. It is required by the California Air

Resources Board (CARB) and written into bid specifications and site permits for DOT and public works projects in various states, including New Jersey, Pennsylvania and Massachusetts.

In addition, a growing number of companies, especially large, publicly-owned firms — are making environmental goals key elements of their business strategies. Dow Chemical Company and3M, for example, now require that all diesel-powered equipment brought into their facilities (wherever they are) be Tier 4F-certified.

This puts water jet contractors whose equipment is not Tier 4F-compliant at risk of being shut out of projects. Furthermore, should they be caught using older, non-compliant equipment where it is not allowed, they can be liable for fines of $37,500 per violation, as well as the costs of litigation and lost time. Water jet users who want to be competitive in the future — or get ahead of competitors today — need to consider investing in new equipment.

3. THE ENGINES

3.1 Advanced engine technology

The two primary technologies that enable diesel engines to achieve Tier 4F emission levels are exhaust gas recirculation (EGR) and selective catalytic reduction (SCR). Some engines use a combination of the two.

EGR refers to cooled exhaust gas recirculation, which recirculates some of the exhaust gases back to the combustion chamber. This has the dual effect of reducing the combustion temperature and reducing the formation of NOx. Large engines (like those used on stationary generator sets) may also have an exhaust pre-heater and Diesel Particulate Filter (DPF) to jumpstart the NOx conversion process. Together, they can heat the exhaust to 450°F in as little as nine minutes.

SCR a technology widely used in Europe, is becoming more common in North America. It sprays diesel exhaust fluid (DEF), or aqueous urea, into the exhaust stream, where the urea reacts with the exhaust NOx. This reaction reduces NOx emissions to an average of less than 0.67 g/kW-hr. The DEF dosing system, supply and return tubing, and control and monitoring functions are all integrated into the engine's electronic controls. Additional emissions controls are added to the exhaust system, where the traditional muffler is typically replaced by a catalytic converter or particulate filter.

Despite the additional technology, water jet units with Tier 4F-compliant engines operate much the same as older units. Displacement is more efficient (more torque at lower horsepower), and some users report better cold weather starting.

3.2 Fluids, fuel efficiency and operating costs

- Users should be aware of some differences, however. Tier 4F engines are designed to run on environmentally-friendly fluids, including:

- Ultra-low sulfur diesel fuel (ULSD), with sulfur content of no more than 15 ppm

- Low-ash oil meeting CES 20081 CJ-4, typically
- 15W-40 (this can also be used in older engines)

- Fully-formulated coolant, 50/50, meeting
- ASTM D6210

- Urea meeting ISO 22241-1, such as AdBlue
- DEF 32.5%

While these fluids cost more than those typically used in Tier 3 or Tier 4i engines, this expense is more than offset by the Tier 4F engine's greater fuel efficiency.

Payback periods depend on the user's particular circumstances, but 1,000 hours a year is a conservative figure (for many water jet users, 2,000 hours is more common) and units used more frequently will pay for themselves sooner.

4. CONCLUSION - SUMMARY: THE CASE FOR TIER 4 COMPLIANCE

Advancements in technology, including increased pump horsepower and automated traversing devices have made the cleaning of heat transfer tubes in power plants easier and safer.

Water jet units that meet Tier 4F requirements deliver the cleanest water jet power ever, minimizing pollution and enhancing public health. But even without these "big picture" advantages, there are solid business reasons to use them:

1. Better fuel economy and lower operating costs

2. Opportunities to do jobs virtually anywhere

3. Avoiding fines and legal costs for non-compliance

4. Customer and public goodwill for reducing air pollution

2015 WJTA-IMCA Conference and Expo
November 2-4 • New Orleans, Louisiana

Paper

DEVELOPMENT OF A NEW METHOD IN ORDER TO DO SILENT ABRASIVE WATER JET TURNING PROCESS APPARATUS

F. Kartal
University of Kastamonu,
Kastamonu, Turkey

H. Gökkaya
Karabuk University,
Kastamonu, Turkey

Z. YERLIKAYA
University of Kastamonu,
Kastamonu, Turkey

ABSTRACT

This study aims to reduce the high level of noise generated during the machining process using abrasive water jet turning from a range of 100~130 dB(A) to a range of 70~85 dB(A). Moreover, this study also focus on a new method to be developed and applied in order to remove the ripple effect occurring in the water jet pool and to eliminate the abrasive water jet splash. This new method developed in the current study proved successful for the machining of polymer based plastics using abrasive water jet below the surface of water levels.

Keywords: Abrasive water jet machining, abrasive water jet turning, worker health and safety awareness, Noise.

Organized and Sponsored by WJTA®-IMCA®

1. INTRODUCTIONS

Abrasive Water Jet (AWJ) cutting has been used as an efficient method for machining of any material, especially the ones hard to cut, in line with the developments in the materials used for industrial purposes. Water jet technology with its structural differences from traditional machining techniques has found a place for itself in developed countries for the operational and economic advantages it offers [1,2].

It is possible to machine rotating workpieces using water jet along with machining and cutting flat surfaces. In this respect, water jet serves as a tool bit. However, there is an essential difference between water jet and the traditional tool bit. It utilizes the manufacturing of rotating symmetry out of hard-to-machine metals such as complex metal composites, tough alloys, and ceramics. Operation of the water jet involves the movement of the jet in x, y, and z coordinates while the workpiece rotates. The system is shown in Figure 6. An important advantage brought in by AWJ machining is that it offers very low cutting power which allows for machining of brittle bars with very small diameters and made of hard-to-machine materials [3].

The literature scan showed that there are several experimental studies which involve cutting and erosion using a milling device [4-13]. However, it was found that all the milling mechanisms used in these studies were similar and that the mechanism most of the times did not involve an enclosure in order to protect against pressure water and abrasive particles while the process was conducted above water level. Using a lathe chuck without any enclosure makes the machining process even harder to perform. The noise produced by the abrasive water jet nozzle when used above water level is rather high and may reach up to 110 dB(A) . Available two-dimensional abrasive water jet cutters machine the workpiece below the water level in order to reduce the noise produced and the splatter of the materials (Figure 1). Exposure to loud noises (> 80 dB(A)) for an extended period (months and years) can cause irreversible hearing loss . This hearing loss may be permanent or temporary, or both. Temporary hearing loss, also known as auditory fatigue, loses its effect when the subject leaves the environment with loud noise (often within 2 hours) [14-16].
Permanent hearing loss depends on the following;

- Personal Sensitivity,
- Level of noise (sound energy level),
- Frequency distribution of the noise-sound,
- Daily total exposure duration,
- The length of the duration one is exposed to loud sound,
- The nature of the sound, i.e. continuous, intermittent, or pulsed.

Figure 1. Two-dimensional cutting process below the water level of abrasive water jet,

Being exposed to continuous sound above 85 dB(A) causes significant hearing loss for many people and the damage is increased with the increasing sound level. The permitted exposure duration for unprotected ears should be decreased by half for every 5 dB(A) increase in the noise level. For example, the exposure duration for 90 dB(A) should be limited with 8 hours while this number should be 4 hours for 95 dB(A) and 2 hours for 100 dB(A). The permissible maximum noise level for the protected ear is 115 dB(A) for 15 minutes.

2. MATERIALS AND METHODS

Machining parameters of the abrasive water jet cutter used in the experimental study are shown in Table 1.

Table 1. The parameters used in the turning operation by abrasive water jet.

PARAMETERS	VALUE
Pressure, MPa	300
Traverse rate, mm/min	30
Abrasive mass flow rate, g/min	55
Orifice diameter, mm	0.33
Focusing tube, mm	0.76
Focusing tube length, mm	72
Stand-off distance, mm	2
Number of passes	1
Cutting head angle. Degree °	90°
Abrasive	Barton Garnet
MESH	80

Figure 2 shows the experimental turning mechanism used for abrasive water jet cutting below water level and the water jet lathe used. Figure 3 shows the computer assisted design of the turning mechanism used for abrasive water jet cutting below water level in this study.

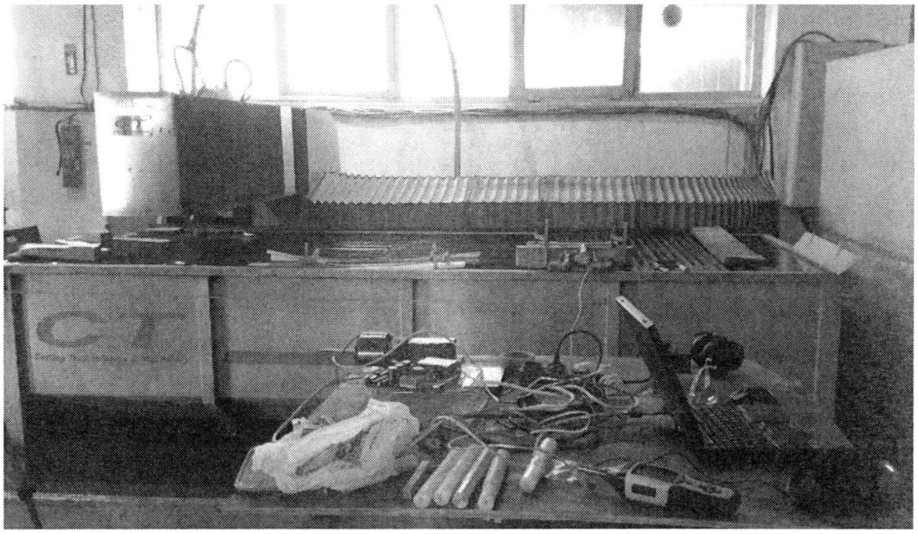

Figure 2. The water jet machining machine used in the experiment.

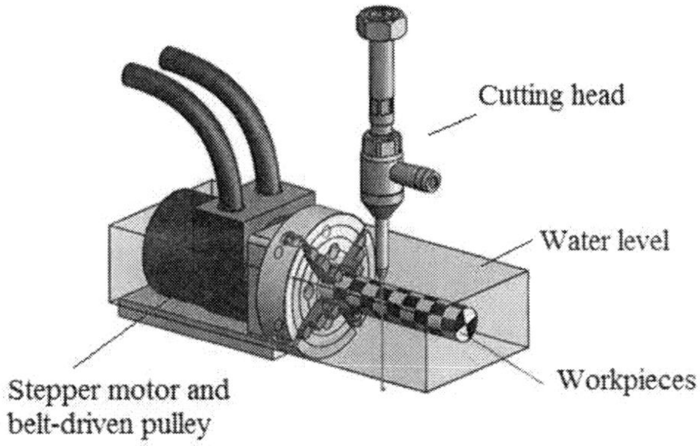

Figure 3. The three-dimensional image of the experimental setup model that is designed to make turning with abrasive water jet under the water level.

Custom experimental machining workpiece holder was tried for speeds under 50 rpm. It was only aimed to give circular motion to the workpiece in the same direction with the abrasive water jet. As low speed was preferred, lathe chunk did not cause ripples in the water. Nevertheless, water chuck and bearings were made of waterproof stainless material and polymer plastics. Figure 4 shows the prototype model developed experimental setup to make turning with abrasive water jet under the water level. DT-8852 digital sound / noise level meter data logger used in this study is shown in the Figure 5 below. Specifications of this device are given in Table 2.

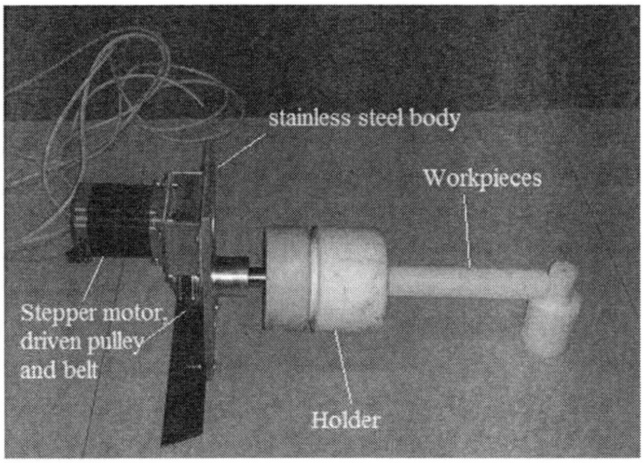

Figure 4. The image of the prototype model developed experimental setup to make turning with abrasive water jet under the water level.

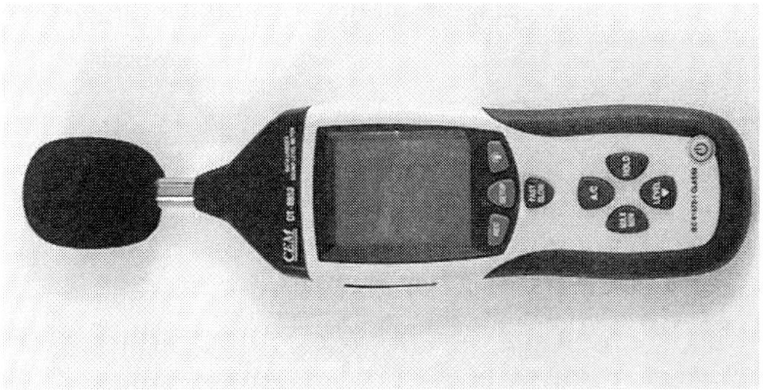

Figure 5. DT-8852 sound level meter [17].

Table 2. DT-8852 sound level meter characteristics [17].

Standard applied	IEC61672-1 Type 2, ANSI
Accuracy	±1.4 dB(A)
Level ranges	Auto:30 dB(A) ~130 dB(A)
Frequency range	31.5 HZ ~ 8 KHZ
Dynamic range	50 dB(A)
Time weighting:	125 mS,
Microphone	1/2 inch electret condenser microphone
Resolution	0.1 dB(A)
Display Update	2 times/sec
Analog output	AC/DC outputs, AC=1Vrms, DC=10mV/ dB(A)

3. RESULTS AND DISCUSSION

Abrasive water jet processes in which the water jet nozzle operates below and above the water level have resulted in different outcomes. The noise level obtained from the nozzle which was placed 2mm above water level where the meter was placed 1 meter away from the mechanism was recorded as 108,9 dB(A) (see Figure 6 (a)). The noise level obtained from the nozzle which was placed 2 mm below the water level, on the other hand, was recorded as 85,4 dB(A) (see Figure 6 (b)). Noise level decreases by 27.51% when the machining is performed when the nozzle is located 2mm under water. This is a significant difference which cannot be ignored.

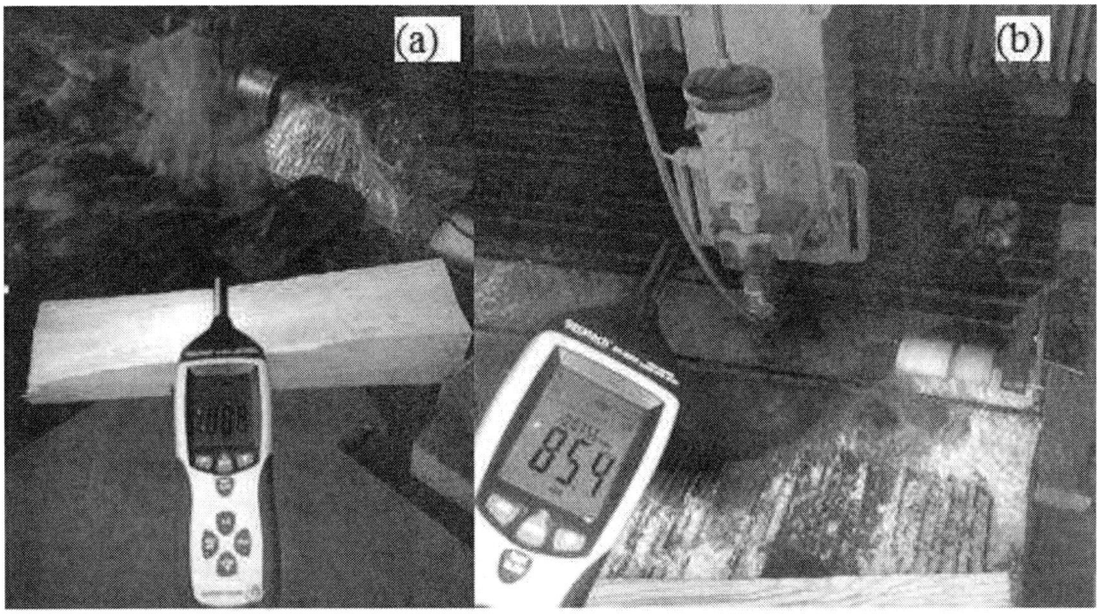

Figure 6. (a) Turning with abrasive water jet over the water level, (b) turning with abrasive water jet under the water level.

4. CONCULISONS

It is possible to reduce the noise level at its source, to reduce it on the way the energy of sound diffuses, and to protect the exposed personnel in order to reduce the noise produced by the water jet and the pump during the abrasive water jet machining process. This study found that it was possible to reduce the noise level by 27% by machining the workpiece under water where nozzle is also located underwater. According to the occupational health and safety standards, a noise level of 85.4 dB(A) is in the acceptable range for 8 hours daily work. This is a significant number in terms of occupational health and safety. Nevertheless, machining where workpiece and nozzle are placed under water eliminates the unfavorable conditions such as ripples in the water jet pool and water splash. It allows for a safer and orderly working environment. It is expected that the findings of this study will contribute to the future research in the field of abrasive water jet cutting.

5. ACKNOWLEDGMENT

The Authors are deeply grateful for the financial support of Kastamonu University Scientific Research Fund, Grant No: KU-BAP-04/2015-1

6. REFERENCES

[1] K.G. Swift and J.D. Booker, Chapter 7 - Non-traditional Machining Processes, In Manufacturing Process Selection Handbook, edited by K.G. SwiftJ.D. Booker, Butterworth-Heinemann, Oxford, 2013, Pages 205-226

[2] R. Kovacevic, Surface texture in abrasive waterjet cutting, Journal of Manufacturing Systems, Volume 10, Issue 1, 1991, Pages 32-40

[3] Momber, A., Kovacevic, R. "Principles of Abrasive Waterjet Machining" Springer-Verlag, London, 1998.

[4] D. A. Axinte, J. P. Stepanian, M. C. Kong, J. Mc Gourlay, Abrasive water jet turning-An efficient method to profile and dress grinding wheels, International Journal of Machine Tools and Manufacture, Volume 49, Issues 3–4, March 2009, Pages 351-356.

[5] Zhong, Z. W., and Han, Z. Z., "Turning of Glass with Abrasive Waterjet," Materials and Manufacturing Processes, Vol. 17, No.3, 17 June 2002, pp. 339-349.

[6] M. Hashish, "A modeling study of metal cutting with abrasive waterjets," ASME J. of Eng. Mat. And Tech., vol. 106, no. 1, pp. 88–100, 1984.

[7] Kartal, F. Gokkaya, H. "Aşındırıcı Su Jeti İle Tornalama Deney Düzeneği Tasarımı" International Iron& Steel Symposium, 02-04 Nisan 2012, Karabük, Türkiye.

[8] Kartal, F. Gokkaya H., "Turning with abrasive water jet machining - a Review", Engineering Science & Technology, an International Journal , 2013, Vol. 16 Issue 3, p113-122

[9] F. Kartal, H. Gökkaya, M. Nalbant, "Turning of (Cu-Cr-Zr) alloy with abrasive water jet" 21st International Conference on Water Jetting, Ottawa, Canada. September 2012

[10] Kartal, F. Gokkaya H., "The Effect of Process Parameters on Machining Volume and Depth of Cut in Turning Operation of AISI 1040 Steel with Abrasive Water Jet". Pamukkale J Eng Sci. 2014; 20 (1): 20-24

[11] F. Kartal, M. H. Cetin, H. Gökkaya, Z. Yerlikaya, "Optimization of Abrasive Water Jet Turning Parameters for Machining of Low Density Polyethylene Material Based on Experımental Design Method" International Polymer Processing, vol. 29 (4), pp. 535-544, (2014).

[12] M. Hashish "Macro Characteristics of AWJ Turned Surfaces" 2001 WJTA American Waterjet Conference August 18-21, 2001 o Minneapolis, Minnesota

[13] I. Ansari A. M. Hashish "Effect of Abrasive Waterjet Parameters on Volume Removal Trends in Turning" Journal of Engineering for Industry, November 1995, Vol. 117 / 475.

[14] http://www.ccohs.ca/oshanswers/phys_agents/exposure_can.html.

[15] http://www.labour.gov.on.ca/english/hs/pubs/liveperformance/gl_live_sound.php.

[16] http://www.isguvenligi.net/gurultu/.

[17] http://www.cem-instruments.com/en/pro/pro-385.html.

2015 WJTA-IMCA Conference and Expo
November 2-4 • New Orleans, Louisiana
Paper

NAHAD HOSE SAFETY INSTITUTE WHITEPAPER:

EYE ON SAFETY AND RELIABILITY FOR LIQUID VACUUM OPERATIONS

Editor: D. Mitchell (see Participating Companies in this paper for a list of contributors)
Hose Safety Institute©, NAHAD - the Association for Hose & Accessories Distribution, Inc.
Annapolis, Maryland.

ABSTRACT

This whitepaper represents an on-going effort to document best practices for ensuring safety and reliability of Liquid Vacuum hose assembly applications. It is intended to represent the consensus of contributing companies and organizations as to the recommended best practices in the design, fabrication, and on-going maintenance of hose assemblies used in Vacuum applications, specifically Liquid Vacuum work. Liquid Vacuum service has the potential to be a high risk craft if not managed properly. Environmental and Safety considerations hold an equally high prominence. This white paper is part of a comprehensive approach to ensure worker safety and environmental stewardship, one which requires equal effort in three critical areas: behavior based programs, standard equipment design and maintenance, and employee training to the standard. It addresses the need to improve the level of safety for the sourcing, fabrication, and service life issues of Liquid Vacuum hose applications, largely in industrial cleaning applications. Safety processes increase costs, but for these applications, they are critical.

Organized and Sponsored by WJTA®-IMCA®

1. INTRODUCTION

Liquid Vacuum work is a staple of the pulp, paper, petrochemical, refining, automotive, steel, etc. industries, in applications ranging from routine to time critical. Vacuum trucks have become standard operating equipment for refineries, chemical plants, paper mills, and power plants. Providing services such as moving waste, evacuating blast water, covering line breaks, skimming ponds, removing catalyst, vacuuming sands and media, transporting acids or pond bugs, lowering oil sumps, maintaining water levels during flooding, and managing spills are now common service lines for industrial cleaning companies to offer.

Liquid Vacuum service has the potential to be a high risk craft if not managed properly. Environmental and Safety considerations hold an equally high prominence. This white paper is part of a comprehensive approach to ensure worker safety and environmental stewardship, one which requires equal effort in three critical areas: behavior based programs, standard equipment design and maintenance, and employee training to the standard.

While each area is of equal importance, this paper focuses on best practices in standard equipment design relating to hose assemblies based on input from Vacuum service contractors and hose and equipment manufacturers and distributors.

Services have grown over the last 25 years, while policies and procedures vary across services providers, often with a lack of a comprehensive hose management policy, including marking, inspection, and testing, which can lead to hose failure, injury or environmental incident. In addition, customers, such as major petrochemical companies, have elevated their expectations and now seek proof of a hose management policy in order to perform work in their facilities. Operational consistency across Liquid Vacuum companies, their sites and divisions is required, but few programs offer a complete hose life cycle solution; therefore service provider's costs vary greatly. And finally, no common industry standards exist for Liquid Vacuum hose, relating to inspection, testing, service life, or common definitions of terms.

Intended benefits of this paper include:

- Providing a common lexicon – ensuring that the hose industry and the end user understand what the other is referring to
- helping end users get the right hose for the application – and understand the variables involved; hose compatibility with the material being transported is crucial
- Understanding the criticality of knowing what is being "sucked up" – investment up front in purchasing the right hose for the job is cheaper than the costs downstream of cleaning up a spill or dealing with an accident; regulatory issues/costs are hugely important and relevant

This paper is limited to hoses used for liquid vacuum operations with chemicals, or petroleum products; it will not address dry vacuum operations or water applications, nor will it explicitly address tank truck industry applications.

This paper delineates recommended best practices to provide guidance to end-users, distributors, and others involved in this business.

- **Strategic business model**
  Liquid Vacuum hoses are one of the most critical components of Liquid Vacuum Operations. Hose reliability and safety directly affect a provider's ability to deliver safe, environmentally reliable, productive service. Following practices that increase reliability and safe performance of these components is essential for the continued health and vitality of the business.
- **Top line impact**
  Clients are attracted to vendors who can provide service in the most safe, environmentally sound and cost efficient manner. Providers who can demonstrate processes and strategies which insure the highest standards of hose integrity will gain more business opportunities and consequently grow their top line revenue.
- **Bottom Line Impact**
  Injuries, environmental impacts and downtime due to hose failures lead to costly direct and indirect expenses which will erode any top line revenue gains. Hoses that have higher integrity protect profits.
- **Environmental Impact**
  Liquid Vacuum hose failures can lead to environmental losses which can range from relatively minor to reportable quantity (RQ) level. RQ losses can significantly impact the environment and lead to EPA or DEQ inspections, regulatory penalties, fines and work delays.
- **Human cost**
  Employees who work for companies with strong hose management programs enjoy higher confidence in their equipment which leads to higher morale and productivity. Liquid Vacuum hose failures can lead to significant injuries, restricted or lost workdays and large medical costs.

This whitepaper can be added to over time, may include sections on pre-sale considerations, post-sale in-service issues/recommendations, and whatever else the team deems appropriate.

1.1 Primary audience:
- Risk Management Managers
- EH&S Managers
- Senior Management at Liquid Vacuum Services companies
- Client Maintenance Managers
- WJTA-IMCA members (Water Jet Technology Association - Industrial and Municipal Cleaning Association)
- Hose and Accessories Distributors

1.2 Call to action**:** This white paper is a first step towards documenting best practices for liquid vacuum service for the hose and accessory industry.

2. MAIN BODY OF PAPER

This paper focuses on three main aspects of the vacuum hose life cycle: design and specification of the hose assembly used for vacuum hose applications (actions taken before the sale), fabrication of the hose assembly, and guidance around the use and management of these hose assemblies once they are in service.

2.1 Pre-Sale Considerations: Specification and Design – STAMPED

An effective way to identify application factors that need reviewing prior to defining the proper specifications of a hose assembly is to remember the simple acronym STAMPED.

The STAMPED acronym stands for the 7 major information areas required to provide a quality hose assembly for the customer, as follows:

2.1.1 S stands for **SIZE**; I.D. and length; any O.D. constraints

- overall length should be specified to include fittings
- tolerances need to be specified if special requirements exist

I.D., O.D. and overall length of the assembly

- To determine the replacement hose I.D., read the layline printing on the side of the original hose. If the original hose layline is painted over or worn off, the original hose must be cut and inside diameter measured for size.
- The inside diameter of the hose must be adequate to keep pressure loss to a minimum, maintain adequate flow, and avoid damage to the hose due to heat generation or excessive turbulence.
- Length tolerances should be considered for all types of hose assemblies.
- **Pressure Loss -** The flow rate of the system in conjunction with the inside diameter of the hose will dictate the pressure loss through the hose. Please consult your hose supplier for specific recommended flow rates.
- Loading and unloading flow rate is impacted by the inside diameter of the hose.
- Note: hoses with a large OD will likely require special handling.

2.1.2 T stands for **TEMPERATURE** of the material conveyed and environmental conditions

- How hot is the material being conveyed
- Are there factors such as heat sources in the environment in which the hose will be used
- Continuous (average) and minimum and maximum temperatures have to be specified for the environment
- Care must be taken when routing hoses near hot manifolds and in extreme cases a heat shield may be necessary.
- In subfreezing temperatures, care must be taken to keep water flowing through hoses. All hoses must be drained on completion of the project. In starting in subfreezing conditions, hoses must be flushed to remove ice crystals prior to installing the tip.
- Other things to consider: maximum intermittent ambient temperature, fluid temperature, ambient temperature and maximum temperature.

2.1.3 A stands for the **APPLICATION**, the conditions of use

- Configuration/routing (add a sketch or drawing if applicable)
  - is the hose hanging, laying horizontally, supported, unsupported (orientation and aspect of the hose); routing up or down hills – requirements for pressure calculations (feet of incline etc.) – head pressure requirements; pull forces if vertical routing is

included
    - what else is attached to the hose, any external load on the hose
    - bend radius requirements, flexibility
- Immersion in the material being conveyed
- Quantify anticipated movement and geometry of use requirements
- Intermittent or continuous service
- Indoor and outdoor use
- Unusual mechanical loads (vehicle traffic, etc.)
- Excessive abrasion
- External conditions – abrasion, oil (specify type), solvents (specify type), acid (specify type and concentration), ozone, salt water, ultraviolet (UV) radiation (sunlight), geographic temperatures (e.g., Alaska vs. Louisiana)
- Hose now in use
    - Type of hose
    - Service life being obtained and description of failure or source of customer dissatisfaction
- strength and frequency of impulsing or pressure spikes
- non-flexing applications (static), flexing applications (dynamic)

2.1.4 M stands for the **MATERIAL or MEDIA** being conveyed, type and concentration

- Are there special requirements for this hose tube
    - Any special specifications (or agency requirements) that need to be considered (e.g., FDA, API)
    - Will the material be continuously flowing, or sit in the hose for long periods of time (specify)
- Media velocity, flow rate
- Weight of media (specific gravity)
- Chemical name/concentration (MSDS)
- Solids, description and size
- Fluid Compatibility - Some applications require specialized oils or chemicals to be conveyed through the system. Hose selection must assure compatibility of the hose tube. In addition to the hose materials, all other components, which make up the hose assembly (hose ends, o-rings, etc…), must also be compatible with fluid being used. Depending on the fluid, your hose supplier may lower the maximum temperature or pressure rating of the assembly. When selecting any hose assembly, always consult your hose supplier's recommendations.
- Temperature of product
- Corrosively of product; potential corrosiveness of mixed media (resulting from improperly cleaned hoses)

2.1.5 P stands for the **PRESSURE** to which the assembly will be exposed (or vacuum for negative pressure or inches of mercury)

- System pressure, including pressure spikes. Hose assembly working pressures must be equal to or greater than the system pressure. Pressure spikes greater than the maximum working pressure will shorten hose life and must be taken into consideration.

- Temperature implications
- **Maximum Operating Pressure -** This is the maximum pressure that the system should be exposed to in normal operating conditions. This pressure should be dictated by the relief setting of the system. Both the hose and hose end should not be rated to a pressure less than the maximum operating pressure of the system.
  Pressure Spikes - When a system is subjected to a large load in a short period of time, the system pressure can overshoot the relief setting and exceed the maximum operating pressure. Frequent pressure spikes can reduce the life of hose assemblies.
- Vacuum as measured in inches of mercury (HG)
- Hose routing (will the hose be straight or bent for the application) impacts the hose requirements; some hoses will hold a vacuum while straight, but will collapse if bent.

2.1.6 E stands for **ENDS**; style, type, orientation, attachment methods, etc.
- Specify end style
- Materials and dimensions (steel, stainless, etc.)
- Conductivity requirements
- Specify attachment method – banding, crimped
- Consideration should be given to the lowest rated component (hose, fitting, and attachment) in determining overall MWP of the entire hose system.
- Clarify sleeves and ferrules (add to fabrication portion)
- Specify impact of any pull forces, if vertical routing of hose assembly is anticipated

2.1.7 D stands for **DELIVERY**
- Specific to customer requirements
- Testing and certification requirements
- Any special packaging requirements
- Any special shipping requirements
- Tagging requirements

When analyzing hose failures, note there can be multiple root causes based on the above parameters from STAMPED.

2.2 Hose Assembly Fabrication Considerations

Fabrication of hose assemblies should be done in accordance with industry best practices such as NAHAD's Hose Assembly Guidelines, supplemented by specific manufacturer instructions. As part of the fabrication process, assemblies should be tested to ensure the integrity of the product when first made; this is usually done through hydrostatic pressure testing, but may include vacuum testing if required by customer. Inspection and re-testing of in-service assemblies should be conducted annually. Field inspection should be conducted prior to each use as well as end of job use. (See the section on In-Service Considerations in this paper for more details.) For assembly handling, storage and shipping best practices, please refer to the appropriate sections found in NAHAD's Hose Safety Institute Handbook©.

Documentation of assemblies should be thorough and trackable in conformance with a formal hose management program including tagging of assemblies through a formal marking scheme, and a formal tracking process. See In-Service Considerations for more details. Specific remediation of hoses which have been taken out of service is beyond the scope of this document. Repair and Remediation of hoses should be done in accordance with industry best practices.

Crimping and banding are both acceptable methods of attaching fittings/couplings to "Vacuum Hose" (chemical, petroleum transfer, material suction/discharge hoses). When crimping, the best practice is that all components (fittings/couplings, ferrules, sleeves) should be from the same manufacturer; if not, the Distributor should consult with the appropriate manufacturers, and verify the viability of the assembly via proper testing. Further, the procedures, process, and crimp specs should be based on manufacturer requirements. Components should be tested and verified by the manufacturer, as related to pressure and temperature requirements and should carry safety factors associated with the entire hose system; the full hose assembly should be properly tested by the Distributor.

Banding is also an acceptable attachment method. (Check with manufacturer for recommendations). Band clamps are typically placed behind a raised part of the barb, such that if the hose were to try and slip off, the clamp would have to expand to fit over the raised part, thus keeping the hose attached to the barb (see the Industrial Hose Fabrication guide for further details). When banding with a hand tool it's imperative to properly install the clamp (bring the clamp material to its yield prior to completing the clamp installation) to achieve repeatability. (Contact your Hose Safety Institute distributor or manufacturer for methods).

NOTE: The overall Maximum Working Pressure of the entire hose system is determined by the component (hose, fitting, and attachment) with the lowest pressure rating, regardless of attachment method used. Repeatability and consistency of the fabrication process are crucial in yielding safe, reliable hose assemblies.

Note: when specifying the appropriate hose assembly to meet vacuum hose application requirements, it is important to consider pull forces on the coupling/connections based on any vertical elevation pull. Hose weight and media weight both need to be taken into consideration if the application includes vertical routing of the hose assembly. If this is the case, end pull forces need to be re-stated as equivalent psi requirements, depending on hose size. Appropriate fittings and connections need to be specified.

2.3 In-Service Considerations

2.3.1 Hose Management Program

Hoses should be managed as an individual asset from fabrication of hose until end of useful life. Such management should include the following:

- An asset management program should include the following
  - Methods to track hose from initial fabrication until end of hose life
  - Methods to alert end user companies of scheduled inspections

    - Methods to alert end user of end of life so hose can be removed from service
    - A policy of hose life expectancy
    - Records confirming results of inspection program and asset certification
    - Techniques for proper marking of assets (e.g., mechanically etched ferrules)
- A proper use and care program should include the following
    - Manufacturer and Distributor recommendations
    - Proper handling of hose
    - Proper storage of hose
    - Do's and Don'ts for hose handling while in service.
- An inspection program should contain
    - Field visual inspection criteria
    - Periodic pressure test and integrity protocols
- A failure inspection program should contain
    - Definition and Types of hose failures
    - Methods for reporting hose failures
    - Methods for investigation hose failures
    - Methods for implementing lessons learned across the organization
- A training program should contain
    - All of the above programs
    - Documented hands on competency verification

Hoses should have documentation or markings on site which allows the user to easily confirm

A. What the hose is designed to load or unload
B. Temperature rating
C. Vacuum and Pressure rating
D. Current Test Data

2.3.2 Service Life

The recommended service life of Vacuum hose is dependent on the care it receives, the application it is being used in, the material being transferred, cleaning processes, and a thorough inspection program.

The human element is critical and can be mitigated only partially by good training programs, clear processes, etc. Variations in equipment (e.g., trucks), stress points, and other variables will all impact service life of the hose assembly. Organizations should focus on best practices in hose tracking: identifying what their experience has been with hose assemblies, capturing data from failures, and noting what trends emerge over time, allowing them to better forecast what to expect in terms of hose service life.

The object of the following testing and inspection procedures is to detect any weakness in a hose assembly before the weakness causes failure of a hose in service.

It is imperative that hose, while in storage or in service, not be subjected to any form of abuse such as kinking, exposure to an environment involving extremes of temperature, corrosive or oxidizing fumes or liquids, oils and solvents, ozone, etc.

For chemical hose, care must be taken when different chemicals are conveyed in the same hose; the chemicals may react and shorten the service life of the hose. When it is impractical to disconnect the hose line after use, drain any remaining chemical from the hose.

2.3.3 Hose Use and Care

Handling - Crushing or kinking of the hose can cause severe damage to the reinforcement. Care should be exercised to prevent mishandling. Do not drag the hose or lift large bore hose from the middle of its length with the ends hanging down. Limit the curvature of the hose to that recommended by the manufacturer and avoid sharp bends at the end fittings and at manifold connections.

Operation - Important: Personnel involved in an operation using chemical hose must use safety precautions such as wearing eye or face protection, rubber gloves, boots, and other types of protective ·clothing. Pressures and temperatures are to be monitored to see that the hose is not exposed to conditions above specified limits. Exceeding specified limits could injure the hose and result in damage to property and serious bodily harm.

Never allow chemicals to drip on the exterior of a hose or allow hose to lay in a pool of chemicals as the hose cover may not have the corrosion resistance of the tube. Should a corrosive material come in contact with the reinforcing material, early failure could result. If kinking or crushing occurs, examine the hose carefully, and, if the outside diameter is reduced 20% or less, the hose must be immediately subjected to a Hydrostatic Pressure test. If the reduction in diameter is more than 20%, retire the hose from service.
Care must be taken when different chemicals are conveyed in the same hose; the chemicals may react and shorten the service life of the hose. When it is impractical to disconnect the hose line after use, drain any remaining chemical from the hose.

Storage - Before placing chemical hose in storage, the hose must be completely drained and any potentially explosive vapors or corrosive residues flushed out. NOTE: Extreme care must be taken when flushing out a chemical hose with water; some chemicals, such as concentrated acids, may react with water and cause spattering. When flushing a hose, disposal of the effluent must be made in such a manner that environmental problems are not created.

Chemical hose should be stored so that air can circulate through it—laid straight on a solid support. This procedure helps extend the life of the hose. Ideally, hose should be stored in a cool, dark, dry place.

Inspections - A hose and fitting maintenance program is critical to maintaining a safe functional hose system and will reduce equipment down time and maintain peak operating performance. Hoses should be inspected and pressure tested in house or by a third party prior to placement into service and whenever a defect is suspected or at a minimum of annually. A visual inspection should be conducted prior to each use as well as end of job use. Retesting hose is a recommended procedure to insure the safe continued serviceability of hose assemblies. Retesting is recommended every 12 months.

Inspections should include

- Visual inspection
- Pressure testing
    - New hoses and hoses for re-certification should be pressure tested at 1.5 times maximum allowable working pressure for a minimum of 5 minutes; consult manufacturer for specific requirements. Note: the maximum allowable working pressure for a hose assembly is dictated by the component of the assembly (hose, fitting, etc.) with the LOWEST pressure rating.
    - Hose assemblies shall be inspected and tested immediately after the hose is subjected to abnormal abuse such as: severe end pull, flattening or crushing or sharp kinking. Any hose that has been recoupled shall be proof tested and inspected before being placed in service.
- Inspection Results Documentation

Visual Inspections - the operators of the hose are a company's last line of defense against spills resulting from damaged hose. Companies should have training programs that educate the operators on what to look for before the hose is placed into service. The operators should be trained to always err on the side of safety and "When in doubt, remove the hose from service."

When inspecting hoses in service, attention should be paid to:

- Removable protective covers
    - Any cuts, gouges or tears in the cover which do not expose the reinforcement should be repaired before the hose is returned to service. If the reinforcement is exposed, retire the hose from service.
    - Covers may show surface cracking or crazing due to prolonged exposure to sunlight or to ozone. Such deterioration, which does not expose reinforcing material, is not cause for retirement.
- Carcass
    - Look for any indication of kinking or broken reinforcement as evidenced by any permanent distortion, longitudinal ridges, or bulges. Crushed or kinked spots where the outside diameter of the hose is reduced by 20% or more of the normal outside diameter shall require that the hose be retired from any service. Hose containing kinked or crushed spots where the outside diameter is reduced less than 20% may be used if the hose passes the hydrostatic tests.

- Fittings
    - All metals are subject to attack by various chemicals. Check with the manufacturer to make sure that suitable end fittings, appropriate to both the hose and the chemical being handled, are being used.
    - Exposed surfaces of couplings, flanges and nipples shall be examined for cracks or excessive corrosion. Either condition shall cause the hose to be retired from service. Any evidence of coupling or nipple slippage on the hose is cause for removing the hose from service.

The hose and fittings shall be visually inspected for:

- Leaks at the hose fittings or in the hose.
- Damaged, separated or pulled back covers
- Cracked, damaged, deformed or badly corroded fittings.
- Other signs of significant deterioration such as blisters.
- Compromised reinforcement where the wires are exposed and show signs such as unwrapped, broken or corroded.
- Dents, twists, or kinks
- Discoloration of color coded hose cover
- Verify test date and pressure is in conformity with requirements for the application.
- Fitting Thread and seat condition
- Verify condition and seating of gaskets

In addition to the above conditions, do not place any Liquid Vacuum Hose in service that has visible or exposed reinforcing wires (braids) and immediately remove from service if there are compromised reinforcing wires.
Visual Inspection Checklist

Visually inspect hose before each use. Immediately remove Hoses from service upon finding any of the following conditions:

External inspection:

☐ Evidence of leaks at the hose fittings or in the hose.
☐ Damaged, separated or pulled back covers (de-lamination)

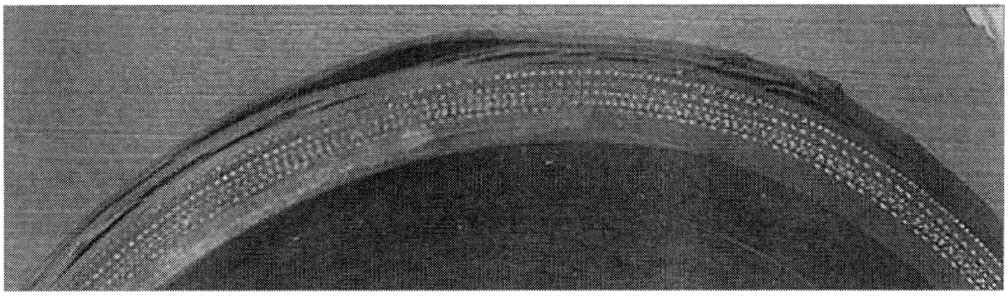

☐ Cracked, damaged, deformed or badly corroded fittings

- ☐ Poorly crimped fittings – can't see the hose through the crimp; ferrule and hose are not fully engaged

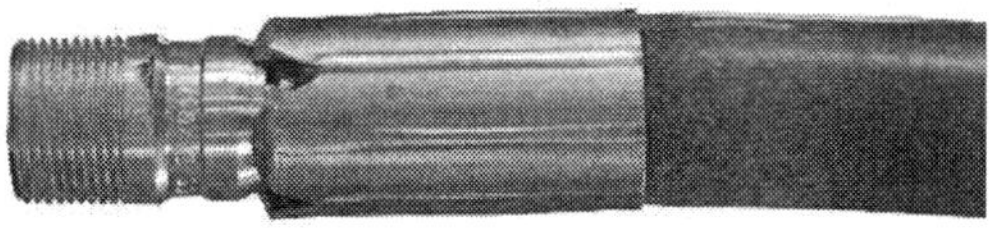
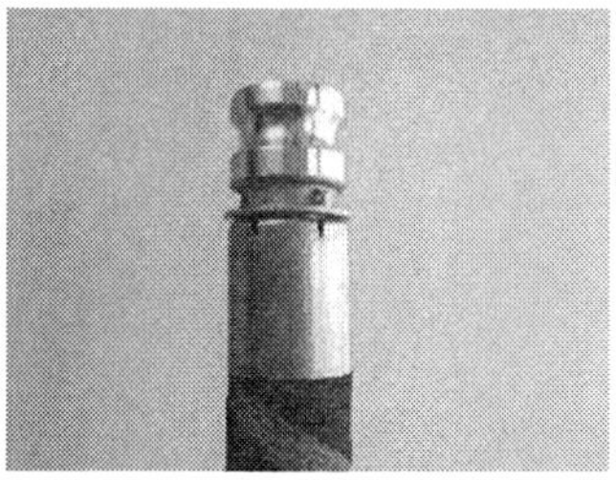

Good fitting - part of the hose is coming out of a 'hole' in the crimp. That tells you the hose has been crimped all the way up to the top of the fitting. If you can't see the hose then you don't know how far up the hose has been inserted and crimped.

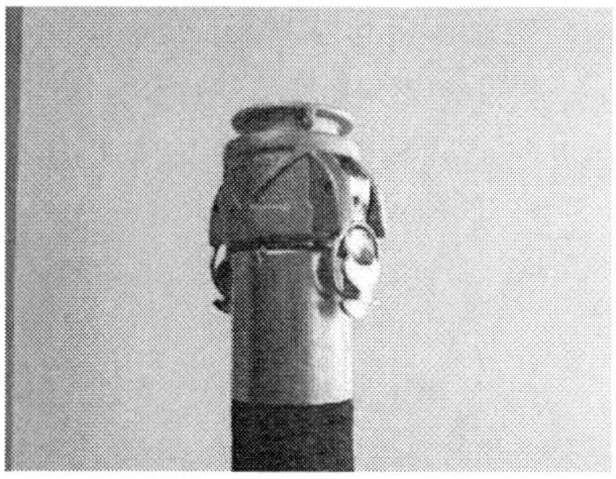

- ☐ Issues with Cam & Groove fittings

- ☐ Improperly banded hose: all of the buckles are in a line. Which may create a leak path and premature hose failure

- ☐ Other signs of significant deterioration such as

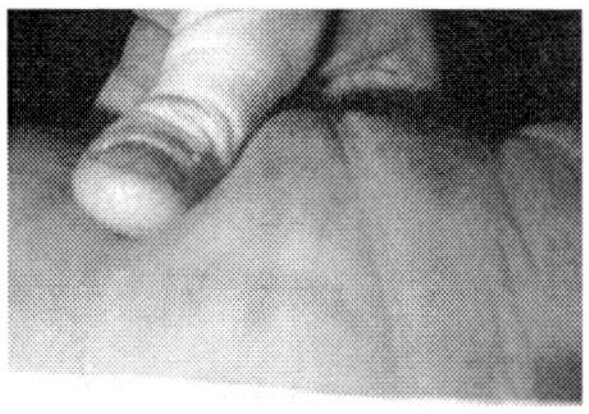

blisters, soft or bulging hose

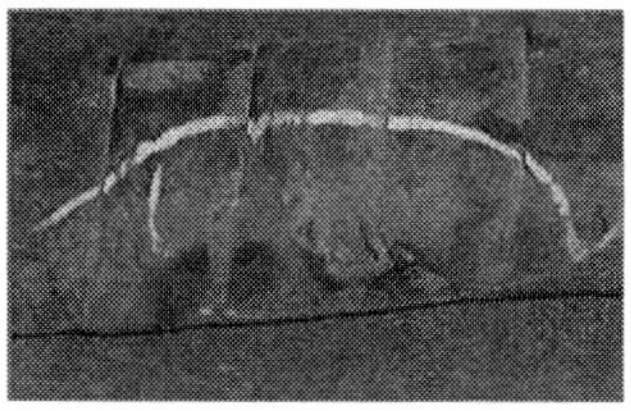

cover gouging

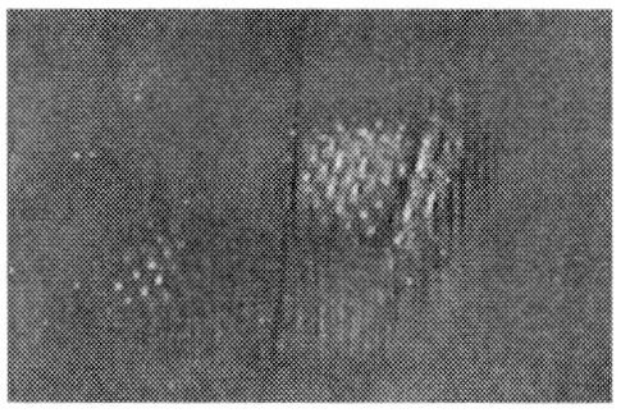

wear

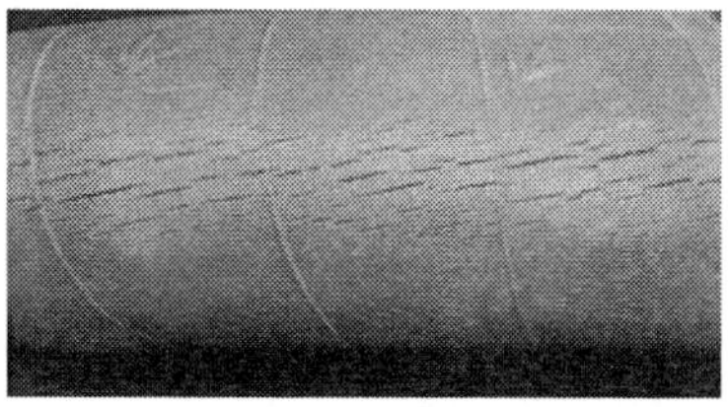

weathering/dry rot

☐ Compromised reinforcing wires.

☐ Dents, twists, or kinks

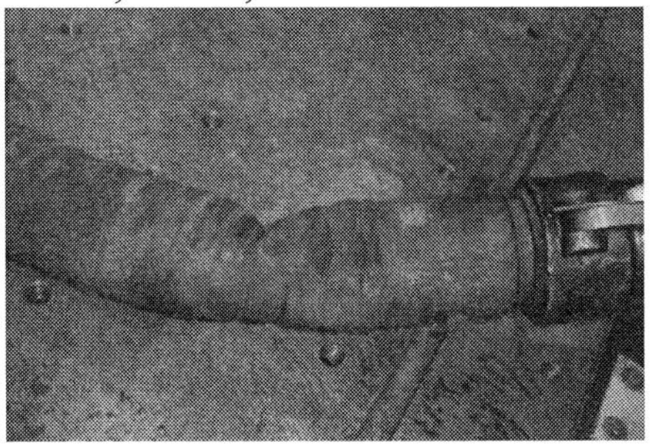

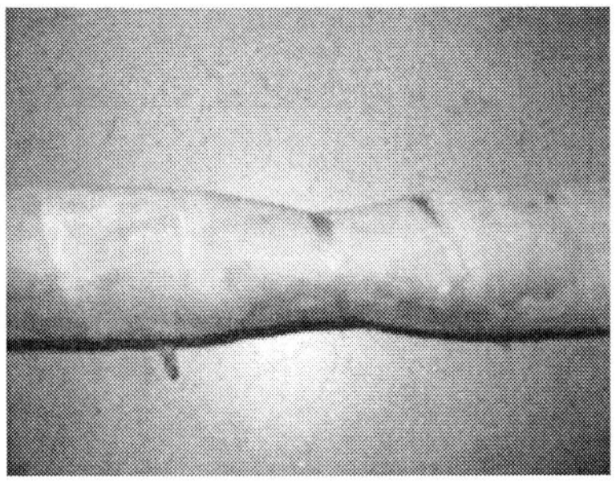

☐ Discoloration of color coded hose cover
☐ Verify test date and pressure is in conformity with requirements for the application.
☐ Check thread and seat connections

Internal Inspection: Small ID hoses are difficult to visually inspect without a special bore camera. As the IDs get larger an internal inspection is possible. 2" ID and larger hoses offer the ability to see several feet into the hose ID. What can be seen generally is representative of the rest of the hose. Look for any of the following:

- Loose hose tube (tube de-lamination) near the coupling

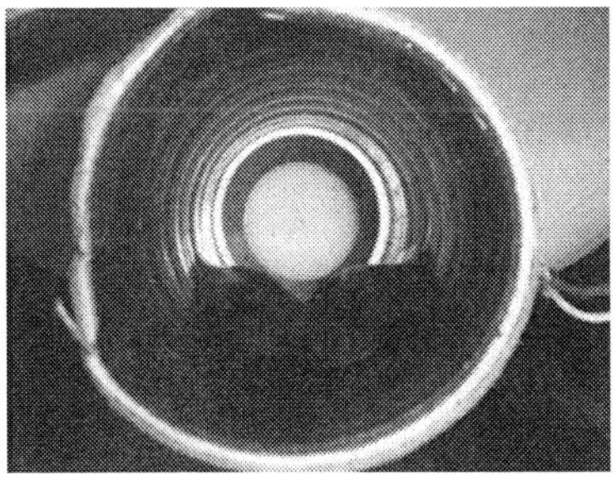

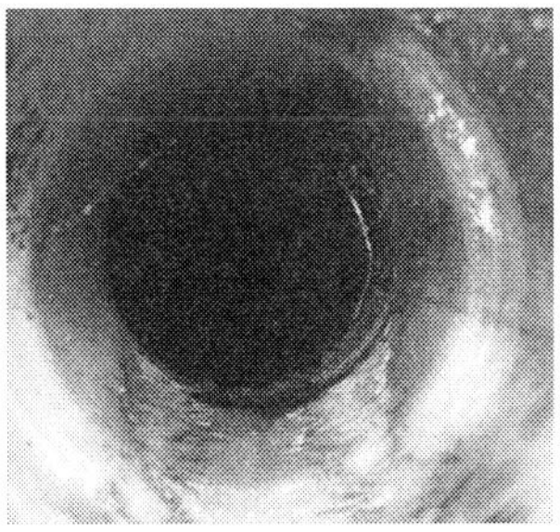

- Discolored hose tube due to chemical attack

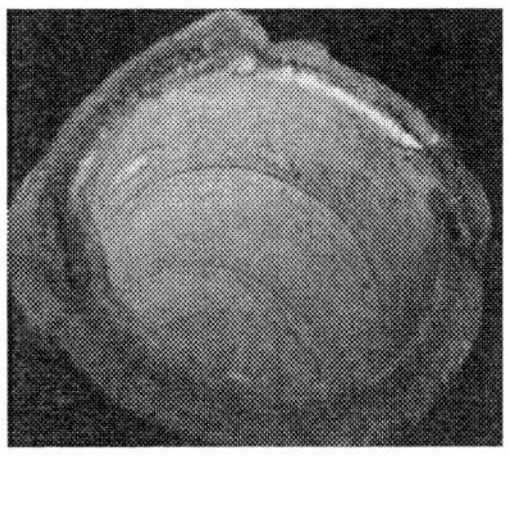

- Cracks or flaking in the hose tube

Cleaning - In many applications is best practice to clean each hose after each use. This prevents continued damage to the hose interior and couplings from the long term effects of potentially hazardous chemicals or the contamination of product when the hose is reused, and may improve hose service life.

The preferred process is typically accomplished by flushing the interior of the hose with water or a cleaning solution that is compatible with the hose tube. The temperatures used in this process should not exceed 212^0 F for longer than 20 minutes. The pressure used to circulate or flush the hose must never exceed the Working pressure of the hose or coupling.

Cleaning solutions should be able to dissolve or remove the residue material in the hose assembly and must be compatible with the hose tube and couplings. All material flushed from the hose must be processed in accordance with EPA requirements.

Inserting cleaning devices (wands, rods or brushes, etc.) into the hose ID is not recommended due to the potential of damaging the hose tube.

Steam cleaning is not the preferred method for cleaning any hose due to the possibility of overheating the hose and coupling. If steam cleaning is necessary the below procedure is recommended:

a. The hose should be placed in a straight line with one end higher in elevation to permit draining from the lower open end.
b. Never use superheated steam (steam temperature going into the hose should not exceed the max temperature of the hose or 212^0 F.)
c. Use a steam supply line not larger than ¾" ID into the adaptor to match the ID of the hose to be cleaned; the other end of the hose must be open and unrestricted.
d. Care must be taken to insure that the velocity of the steam injected in to the hose ID is minimal. A steam jet will damage the hose tube.
e. Steam cleaning duration not exceeding 15 minutes is recommended.
f. Hose should be flushed with cool clean water afterwards.

Processes used to clean hoses in service can dramatically impact both the service life of the hose and potential failures. Key considerations include:

- Excessive abrasion (both chemical and physical)
- Detergents/Degreasers/Solvents used in the cleaning process
- Temperatures of cleaning solutions or water (steam injector guns)
- Pressures of water used to clean the ID of industrial hoses
-

3. ACKNOWLEDGMENTS

Participating Companies:

GHX: Jim Reilly, Meg Adams, Jeff Davis

PSC: Rick Pitman, Bradley Coble
NAHAD: Debbie Mitchell
Groendyke Transport: Steve Niswander
Clean Harbors: Sean Benoit
HydroChem: David Thompson
Dixon Valve: Hazen Arnold
Campbell Fittings: Titus Jumper
Novaflex: Tom Wise
Marathon Petroleum – Roger Gautreau
Band-It: Kelly Kinney, Alton Williams

This whitepaper was developed under the auspices of the **Hose Safety Institute©**, operated by NAHAD, The Association for Hose and Accessories Distribution. The Institute provides a powerful forum for distributors, manufacturers, suppliers, end-users and industry organizations to support and promote hose assembly safety, quality and reliability, across all markets and industries.

The Institute's core deliverables are the **NAHAD Hose Assembly Guidelines;** performance standards for hose assembly specification, design, fabrication, handling and management as provided by qualified NAHAD Hose Safety Institute Members.

The Institute is managed by NAHAD's **Standards Committee** and supported by the end-users and industry experts serving on the **Hose Safety Institute Advisory Council.**

4. NOMENCLATURE/DEFINITIONS

API – American Petroleum Institute

Bonding – Bonding and grounding provide controls to help eliminate static electricity. Bonding provides electrical connections between isolated conductive parts of a system to equalize their electrical potential (voltage). Electrical resistance between two directly bonded connections should not exceed 10 ohms. Hose bonding refers to the continuous connection between the metal end of the hose, the helix wire inside the hose, and the connection to the liquid vacuum truck to equalize the static charge built up during the vacuum process and the process of maintaining a resistance level of <10ohms. Note: this applies to single lengths of hose only; multiple hoses strung together may need to be treated differently.

Grounding – Refers to the ability of the static charge to trace a path to a grounding point to "mother earth" for dissipation, and should be monitored to a level of <1000ohms. An electrical resistance <10 ohms should be achieved between the truck and the ground point.

Electrical Continuity Test – Electrical continuity testing determines if an electrical path can be established between two points; for hose assemblies, refers to testing the assembly to determine if there is a grounding path between end fittings.

Reference NFPA 77 and API 2219 for additional detail.

CFM – Cubic Feet per Minute

Coupling – also known as “fitting” – a device attached to the end of a hose to facilitate connection.

Cam and Groove couplings, also called cam lock fittings or couplings.
Cam and groove is a kind of quick connect coupling usually used in most industries as a hose connection. This kind of coupling is popular because it is considered to be a simple and economical way of re-connecting and disconnecting.
A Cam and groove fitting can be used in a system where rapid filling of chemical drums takes place. It can be used by factories that have needs of dye, paint and ink medium transfers. It is used where frequent changes of hoses are required to find the right mix. It also suitable for petroleum trucks, etc.

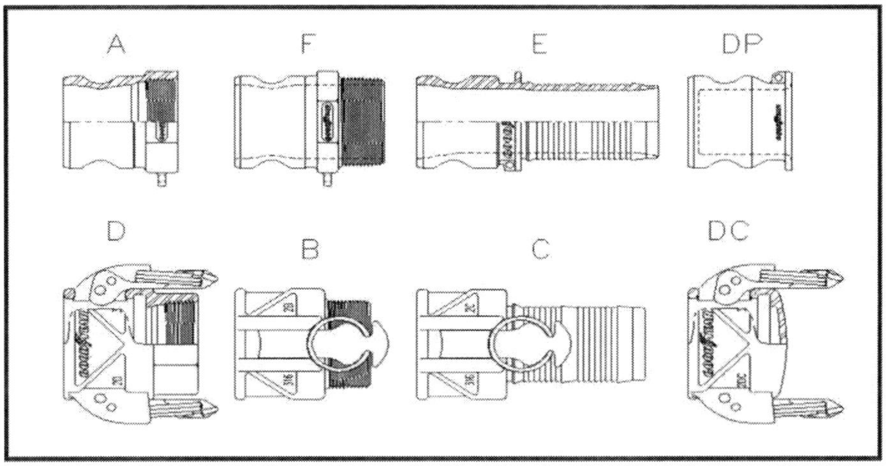

Part C Hose End “Hose Barb Coupler” (a coupler is always the female portion of a Cam and Groove fitting)
Part E Hose End “Hose Barb Adaptor” (an adaptor is always the male portion of a cam and Groove fitting)

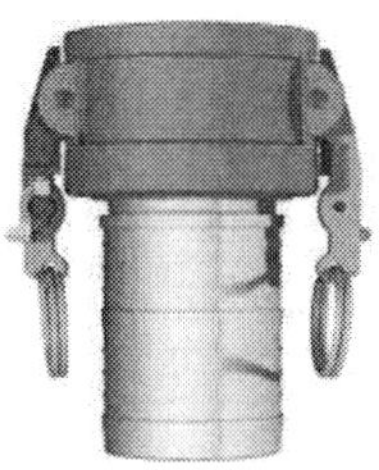

Female Cam & Groove (with Locking Arms)

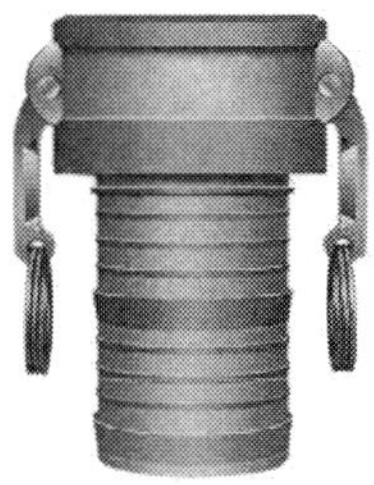

Female Cam & Groove (without Locking Arms)

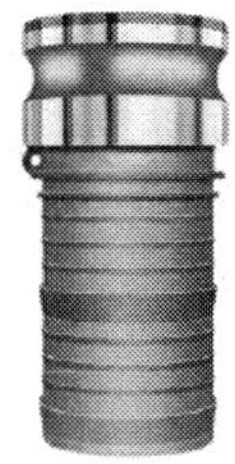

Male Cam & Groove

General Uses:

Cam and Groove female couplers and male adapters are used as a means of quickly coupling small to large hose assemblies. They connect hose to hose or hose to a pipe system for the purpose of transferring liquids. Connections are made by inserting the male into the female coupler and rotating ALL the cam arms into their locked positions.

Note: Cam and Groove couplings are not recommended for any type of compressed gas service, including steam or air.

Gaskets – the chemical compatibility of the media and the gasket material needs to be verified to optimize service life and the effectiveness of the seal.

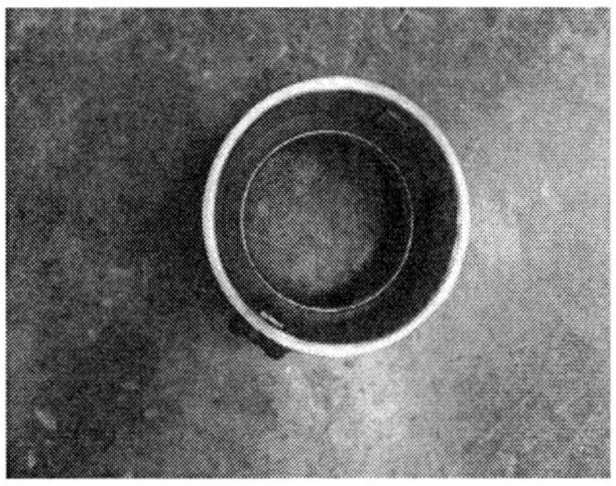

Hose - Vacuum hose, Liquid Vacuum Hose – Hoses used to transfer material to and from a liquid vacuum truck and a liquid source such as a sump, pond, tank, etc… The two main types of Liquid Vacuum hoses are **Petroleum Suction & Discharge Hose** and **Chemical Suction & Discharge Hose**.

Chemical Suction and Discharge Hose (plastic or rubber lined) – These hoses are typically used for the transfer of low or high pH liquids under pressure, gravity flow, or suction. Chemical rubber hoses manufactured from plastic liners such as Cross-Linked Polyethylene (XLPE) or Ultra-High Molecular Weight Polyethylene (UHMWPE), Nylon or Fluoropolymer (Teflon*) Hose. The hose and couplings, and gaskets or seals selected for use must be compatible with the chemical(s) to be conveyed under the stated service conditions. The application temperature range is -40°F (-40°C) to +350°F (+177°C) but that range may be shifted higher or lower depending on hose tube and construction.

Hose can be used in a vacuum application as a suction hose if constructed with a helical wire or reinforcement that will support a vacuum. Note: there are other considerations beyond a helix for creating a hose that will support a vacuum, but helical wire and/or reinforcements predominate – make sure the hose is specified for liquid vacuum use, constructed with appropriate ways to dissipate static buildup.

Petroleum Suction and Discharge Hose – These hoses are typically used for transfer of moderate pH liquids such as gasoline and other petroleum based products under pressure, gravity flow or suction. Typical pressures of up to 150 psi. Typical Constructions Include: Oil, Gasoline, Petroleum and Biodiesel Suction and Discharge. This hose is designed for petroleum based products, and is constructed of petroleum based materials such as oil resistant tubing. Consult manufacturer for compatibility of blended fuels, such as biodiesel and E85 fuels.

Hose can be used in a vacuum application as a suction hose if constructed with a helical wire and/or reinforcement that will support a vacuum. The normal application temperature range is 35°F to 200°F (2°C to 93°C) but can go higher depending on the hose tube and construction, or can go lower (e.g., arctic fuel suction and discharge hose) depending on hose tube and construction. -25 – 180F is suggested. Depending on the construction, usable temperature range may be shifted higher or lower.

Hydrostatic Proof Test - A hydrostatic test is a way in which the suction/discharge hose is tested for strength and leaks. The test involves filling the hose with a liquid, usually water, which may be dyed to aid in visual leak detection, and pressurization of the vessel to the specified manufacturer test pressure. Sometimes referred to as Proof testing or testing that "proves" the finished hose assembly meets the pressure rating required by the application for which it will be used, and that the end fittings have been correctly fitted and the assembly is leak free.

Life Cycle – The time frame in which an asset can be utilized before the asset should be destroyed and replaced.

Safety Data Sheet – A document containing information and instruction on hazardous materials present in the workplace; SDS's contain detail about the hazards and risks associated to the substance and the requirements for its safe handling as well as actions to be taken in the event of fire, spill, or overexposure.

Vacuum hose, Liquid Vacuum Hose – Hoses used to transfer material to and from a liquid vacuum truck and a liquid source such as a sump, pond, tank, etc… The two main types of Liquid Vacuum hoses are **Petroleum Suction & Discharge Hose** and **Chemical Suction & Discharge Hose**.

Vacuum Truck – A transportable vacuum system consisting of vacuum pump, vacuum cargo tank, and associated tools and accessory equipment mounted on a motor vehicle.

Wastewater Suction & Discharge Hose – Industrial and municipal. Not covered within this paper.

2015 WJTA-IMCA Conference and Expo
November 2-4 • New Orleans, Louisiana

Paper

SPECTRAL MEASUREMENT OF ELEMENT SOLUBILITY IN WATER UP TO 300 MPA

M. Bauer, O. Grünzel, D. Zaremba, H. J. Maier, T. Hassel
Institut für Werkstoffkunde (Materials Science), Leibniz Universität Hannover, Germany

ABSTRACT

High-pressure water jet technology is used progressively in industrial applications since 1971. The material behavior of pure water (H_2O without additions) has been studied for a wide temperature and pressure range. Usually, tap water is used as process medium for cutting applications. However, only little data is available for the behavior of tap water at elevated temperature and pressure changes. It is commonly known that the use of ion exchangers leads to a significant softening of the water, which has a negative effect on seal lifetime. Hard water has similar negative effects. It leads to calcification, which causes life reduction of high pressure pump components and also seals due to tribological effects.

In order to understand the relationship between water and its constituents at high pressure up to 300 MPa, a chamber for spectroscopic and photometric measurements was constructed and built. After loading the chamber with different water specimens, the pressure was increased in 25 MPa steps and the spectral light intensity was measured. In this study, deionized water was employed as a reference and compared to common tap water and pure water with added NaCl and $CaCl_2$. The results were evaluated statistically though a self-devised, "R"-based software. Significant alterations when applying given pressure could be observed both through photometric measurements and spectrometric investigations. By using deionized water as a reference, the pressure-dependent intrinsic behavior of the measurement setup was compensated.

With the described method, a variance in spectral intensities was shown when using different fluids and different pressure levels. Particularly, calcium chloride blended water solutions showed a behavior of the spectral intensities similar to tap water.

1. INTRODUCTION

Since its market launch in the 1970s, high-pressure water jet technology is used for industrial purposes. Due to high accuracy and quality of the cut, water jet technology offers many applications. The technique is used for example in the field of handicraft production, food industry and dismantling of nuclear facilities. These exemplary applications may be completely different, but have one thing in common: They need high pressure water. In classical mechanical engineering applications, various requirements exist for production, transport and supply of high-pressure water. The behavior of pure water (H_2O) is described in a wide temperature- and pressure range in several scientific papers [BAK75, HAR89, WAG08]. Pure water means water without any admixed components. For more complex systems such as tap water no extensive studies for the dependence of temperature or pressure exist.

Many studies have been carried out regarding the temperature behavior of water, because its temperature characteristics are very important for thermal power stations. In these studies, water is only considered as a pure substance and is thus idealized. In contrast to single observations of the temperature behavior, investigations at high pressure go along with an increased effort. For the investigation of its behavior at high pressure, only non-contact investigations with a defined examination volume are possible. State of art for non-contact investigations are spectrometric and photometric measurements.

Empirical data indicates that the water composition has a significant effect on the service life of high pressure components like pumps, valves or fittings. An excessive softening of water by ion exchangers results in a lifetime reduction, because this leads to increased corrosiveness. Likewise disadvantageous is hard water: In this case, deposits inside the machine can be detected. These are reported to lead to limited lifetime of seals and nozzles.

In this paper the implementation of spectrometric measurements at high pressure process-water up to 300 MPa is described. For the experiments, a suitable spectrometer and a high-pressure test chamber were constructed and built. Thereafter, specimens of tap water, pure H_2O and mixtures of H_2O with NaCl and $CaCl_2$ were investigated for their behavior under high pressure. The spectral intensity was measured, compared and analyzed. The change of spectral intensity is used to show a pressure depended compound solubility.

2. MATERIALS AND METHODS

2.1 MATERIALS

2.1.1 Used Fluids

For the experiments, it has to be exactly defined what is meant with the word "water". It is necessary to differentiate between the pure substance "water" and the complex medium water, such as "tap water". Tap water includes several dissolved, water-soluble elements (like salts), which differ in compilation and concentration dependent on where the sample was taken.. In Germany, the maximum concentration of certain constituents in tap water is regulated by law. The hereby available is however aimed at the hygienic harmlessness of drinking water, necessary specific values for industrial applications are still unknown. To achieve a certain water quality, water treatment can be carried out prior to the application to achieve a longer lifetime of valves or seals. To achieve this, there is also the possibility of applying additives to the water.

2.1.2 Spectrometry

In general, optical spectrometry can be differentiated between two functional modes of operation. These are "spectrometry by emission" on the one hand and "spectrometry by absorption" on the other.

"Spectrometry by emission" is based on the effect of a stimulated sample, which emits characteristic radiation. The radiation is directly diffracted and the hereby-generated energy levels are considered. This method only works with samples which emit radiation or can be stimulated to emit radiation, so the list of possible sample types is limited. Generally, spectrometry by emission only works with solids and gases [LOT06].

The main difference when regarding "spectrometry by absorption" is that the sample is loaded with radiation from an external beam source (see Figure 1). The output range can be determined as a reference value by measuring the emission spectrum of the radiation source without a sample in the beam path. Every sample absorbs the radiation energy level it was exercised with. The intensity of the corresponding energy levels decreases through irradiation through the sample, so a small specimen volume is advantageous [TIP04].

The radiation emanating from the specimen is spectrally diffracted by a prism or a diffraction grating. An optical sensor detects the hereby-generated diffraction pattern. This sensor detects the brightness values and the intensity of the individual spectral components. The detected emission spectrum quality is directly related to the intensity of the beam source. The resulting output spectrum is offset against the known input spectrum of the radiation source, hereby a difference spectrum can be deducted. The spectral energy levels of the difference spectrum reflect the alteration through the sample. Spectrometry by absorption is used for measurements of samples that cannot be stimulated to emit radiation itself. The only prerequisite is that the sample has to be transparent [LOT06, TIP04].

2.1.3 Spectrometer setup

The mechanical construction was designed in a 3D CAD system (Dassault Systems, SolidWorks). The spectrometer consists of two main optical components. Both are fixed in two different positions of the housing (compare Figure 2). On one side, the slit aperture is mounted through a carrier plate, while the other side houses the optical grating and the camera. The slit aperture and grating are adjustable in angle.

The slit aperture of the spectrometer is aligned to appear as a vertical line on the camera sensor. The alignment of the grid has a direct influence on the spectrum, which appears horizontally in the output image. For the calibration of the spectrometer, the housing can be closed, while the slot is illuminated and a picture is taken by the camera. The spectral image is analyzed by an image processing tool (GIMP). The imaging process and grid adjustment are iteratively repeated until the spectrum is mapped horizontally on the CCD sensor of the camera (see Figure 3). The used CCD sensor allows a resolution of 1280x720 pixels and can capture fixed images and videos.

2.1.4 Construction of the high-pressure cuvette

Cuvettes made for conventional spectrometric measurements are usually made of plastic or glass. This makes them solely usable for low pressure applications. Therefore, a high pressure cuvette was designed, which included two windows (see Figure 4). The windows (manufacturer SITEC) are intended for high pressure applications up to 400MPa and are sealed through a Bridgman seal. The cuvette is made of stainless steel (EN 1.4571, compare to AISI 316Ti). This is a common steal for high pressure components. The cuvette has a cylindrical shape with a diameter of 98 mm. Both windows have a leakage hole at the sealing position. The unit furthermore includes two high-pressure connectors (6 mm HP).

2.2 METHODS & EXPERIMENTAL SETUP

2.2.1 Implementation of experiments

For the experiments, the cuvette was integrated into the high-pressure piping system (compare Figure 5). Each experiment were started with a fluid-filled cuvette and closed valves V0, V1 and V2. After the opening of valve V0, the high-pressure pump applied the desired operating pressure. The pressure was iteratively increased in 25 MPa steps in order to apply static pressure to the test fluid. 2 minutes after the desired pressure was reached, the measurement was started. This waiting time allows the seals to settle and balancing processes to take place (due to changes in element solubility). Subsequently, the fluid-loaded cuvette is illuminated by the full-spectrum light source. The diffraction image is detected on the other side of the cuvette. Each experiment lasted 100 seconds with a picture recorded every 2 seconds.

2.2.2 Visualization

Results of the spectral measurements is a graph of the spectral intensity distributions for each pressure (compare Figure 6a). For a better evaluation of the occurring effects, a further processing of the measurement data was carried out (Figure 6c). In this graph, the blue colored curve was translated into a 3D bar chart. Every bar has an intensity interval range of 15 nm and represents the average intensity of this range. This way, 30 bars in the range of 400 and 850nm are generated. The bars are coded with color gradients from blue to yellow. Blue corresponds to a minimum light intensity (I = 0) and yellow to the maximum (I = 1). So the light intensity is characterized by bar height and color for one given pressure. To get information about pressure-dependent intensity variations, translation was done. In Figure 6b, the wavelength is now displayed on the y-axis while the wavelength is plotted on the x-axis. Now, a color-coded box represents the average light intensity of each measurement. This way, , the pressure-dependent changes of the wavelength range can easily be recognized as color change on the x-axis. This procedure is carried out for all the fluids. Finally, the results are normalized. For this purpose, the spectrum of pure water was subtracted from all graphs. An exemplary graph of the hereby-generated visualization of the normalized spectral measurement is displayed in Figure 6d.

3. RESULTS

H_2O The spectral intensity measurements of pure water show an influence of pressure on the spectral intensity distribution. The gradient alteration is displayed in Figure 7. In wavelength range between 475 nm and 520 nm, an increased spectral intensity can be recognized. Differences occur in the longer wavelengths range, for medium and short wavelengths no change can be detected.

Tap water Compared to H_2O, pressure increase shows a minor influence on the spectral intensity of tap water (see Figure 8a). In the low (< 460 nm) and high wavelength (> 800nm) ranges, no alteration of intensity detectable. In the middle wavelength range, intensity increases slightly. The normalized spectral intensity of tap water is displayed in Figure 8b. Differences to the reference H2O are displayed in green (higher intensity) and red (lower intensity). In the wavelength ranges between 520 nm and 610nm, color change from red to green can be recognized (Figure 8b). No differences in intensity compared to the reference H_2O are displayed in white.

NaCl-blended water The Sodium chloride blended water solution shows no alteration of spectral intensity changes on increased pressure (Figure 9a). However, four significant changes are visible regarding the complete wavelength range at the pressures 25MPa, 150MPa, and 280MPa. It can be assumed that these alterations occur through interference during measurement. A possible cause is a shift of the cuvette windows' seals. This process is reversible, a pressure reduction reverts the seal to its initial state.

In Figure 9b, clear differences can be seen compared to the behavior of H_2O. The spectral intensity of the sodium chloride solution is lower in the lower wavelength range. In the middle wavelength range, the behavior of the NaCl-blend approximately corresponds to H_2O. In the higher range, the spectral intensity of the NaCl-blended water is much higher than of H_2O. A clear intensity increase can be detected between 775 nm and 790nm. An increase of pressure also increases the light intensity.

$CaCl_2$-blended water The Calcium chloride blended water solution shows a spectral intensity shift at the pressure of 25MPa (compare Figure 10a). This shift can be attributed to a measuring fault. Within these two blocks, a consideration is possible. At the high (715 nm - 850nm) and low (400 nm - 475nm) wavelength ranges, a low light intensity becomes apparent. Considering the spectral intensity distribution, it can clearly be shown that the spectral intensity of the calcium chloride solution is pressure-dependent.

The biggest differences compared to H_2O are located in the higher and lower wavelength range (Figure 10b). These ranges show an almost constant behavior. Higher wavelength result in a higher, lower wavelengths in a lower intensity compared to H_2O. For the medium wavelength range, a shift in intensity from 0.5 can be 1.3 detected.

4. CONCLUSIONS & DISCUSSION

The presented spectral analysis of high-pressure process water differs a lot compared to methods currently available in literature [FOU77, SAW07]. By using the described experimental procedure, no samples have to be removed from the high-pressure volume. For this reason, even temporary pressure effects can be shown, which only occur on high pressure. In the analyzed fluids H_2O, tap water, NaCl-blended water and $CaCl_2$-blended water, a correlation between spectral intensity and pressure could be shown.

The behavior of water at increased high-pressure has not been subject to extensive investigations, so far. In 1977, the solubility of amorphous silicates was researched by Fournier [Fou77]. In this study, a solubility of silicates in water could be shown with increasing pressure. Sawamura et al. [Saw07] examined the solubility of sodium chloride in water. This work differs from the actual paper, because that Sawamura et al. [SAW07] took water samples and investigated them under atmospheric pressure. Therefore, reversible effects could not be shown using his method.

Another possible effect is the occurrence of oppositional effects at the measurements, which can affect rising and falling spectrometric intensity values. A possible reason are two competing effects which effect the permeability of the liquid inside the high-pressure volume and change the measurable intensity. If the solubility is low and crystals are settling on the bottom of the cuvette, spectrally measurable intensity rises. For the case that the omitted components form floating crystals, light dispersion can rise and the measurable intensity decreases.

The chosen rest period of 100s before starting the measurement does not seem to be sufficient. To adjust the effect of crystal formation, sufficient time must be maintained after pressurization. Additionally, the addition of a resistance measurement method could give more information about the crystal formation.

Furthermore, the current measurement set-up needs to be improved. The accuracy the measurements can be improved by substitution of the slit aperture with a pinhole. This way, it would be possible to map the spectrum across the diagonal of the image sensor. Thus, the definition of the analysis could be improved.

Finally, in this paper only tap water, pure H_2O and water-blends with chloride were investigated. Prospectively, H_2O fluids with sulfates ($MgSO_4$, K_2SO_4, Na_2SO_4) or silicates like sodium silicate should also be investigated.

5. ACKNOWLEDGEMENTS

The authors are members of the Waterjet Laboratory Hannover (WLH, Institut für Werkstoffkunde (Materials Science), Division Underwater Technology Center) and the German Working Group of Waterjet Technology (AWT).

6. REFERENCES

[BAK75] Bakanova, A. A., Zubarev, V. N., Sutulov, Y. N., Trunin, R. F.: Thermodynamic properties of water at high pressures and temperatures. Zh. Eksp. Teor. Fiz 68, No. 3 (1975): pp. 1099-1107.

[FOU77] Fournier, R. O., Rowe, J.J.: Solubility of amorphous silica in water at high temperatures and high pressure. Am. Mineral; 1977.

[HAR89] Harvey, A. H., Prausnitz, J. M.: Thermodynamics of High-Pressure Aqueous Systems Containing Gases and Salts. AIChE Journal 35, No. 4 (1. April 1989): 635–44. doi:10.1002/aic.690350413.

[LOT06] Lottspeich, F.: Bioanalytik. München; Heidelberg; Spektrum; 2006.

[NN15] N. N.: Physikalisch-chemische Beschaffenheit des von der Stadtwerke Hannover AG verteilten Trinkwassers. Data sheet, enercity Stadtwerke Hannover AG, https://www.enercity.de/infothek/downloads/broschueren/wasser/trinkwasseranalyse-enercity-wasser.pdf, 21.07.2015

[SAW07] Sawamura, S., Egoshi, N., Setoguchi, Y., Matsuo, H.: Solubility of sodium chloride in water under high pressure. Fluid Phase Equilibri; 254.1-2; 2007.

[TIP04] Tipler, P. A., Mosca, G.: Physik für Wissenschaftler und Ingenieure. 2. Aufl.; Heidelberg; Spektrum Akademische Verlag; 2004.

[WAG08] Wagner, W., Kretzschmar, H.-J.: International Steam Tables. Berlin, Heidelberg: Springer Berlin Heidelberg, 2008. http://link.springer.com/10.1007/978-3-540-74234-0.

7. TABLES

Table 1. Composition of the tested fluids

	H_2O in m.-%	NaCl in m.-%	$CaCl_2$ in m.-%	Rest m.-%
H_2O	100	0	0	0
Tap Water [NN15]	> 99.97	0.002	0.022	< 0.001
H_2O-NaCl	82.05	17,95	0	0
H_2O-$CaCl_2$	63	0	37.0	0

8. GRAPHICS

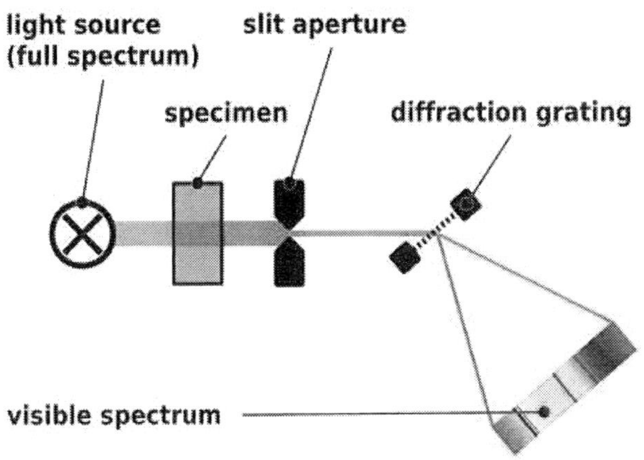

Figure 1. Functional principle of a grating spectrometer

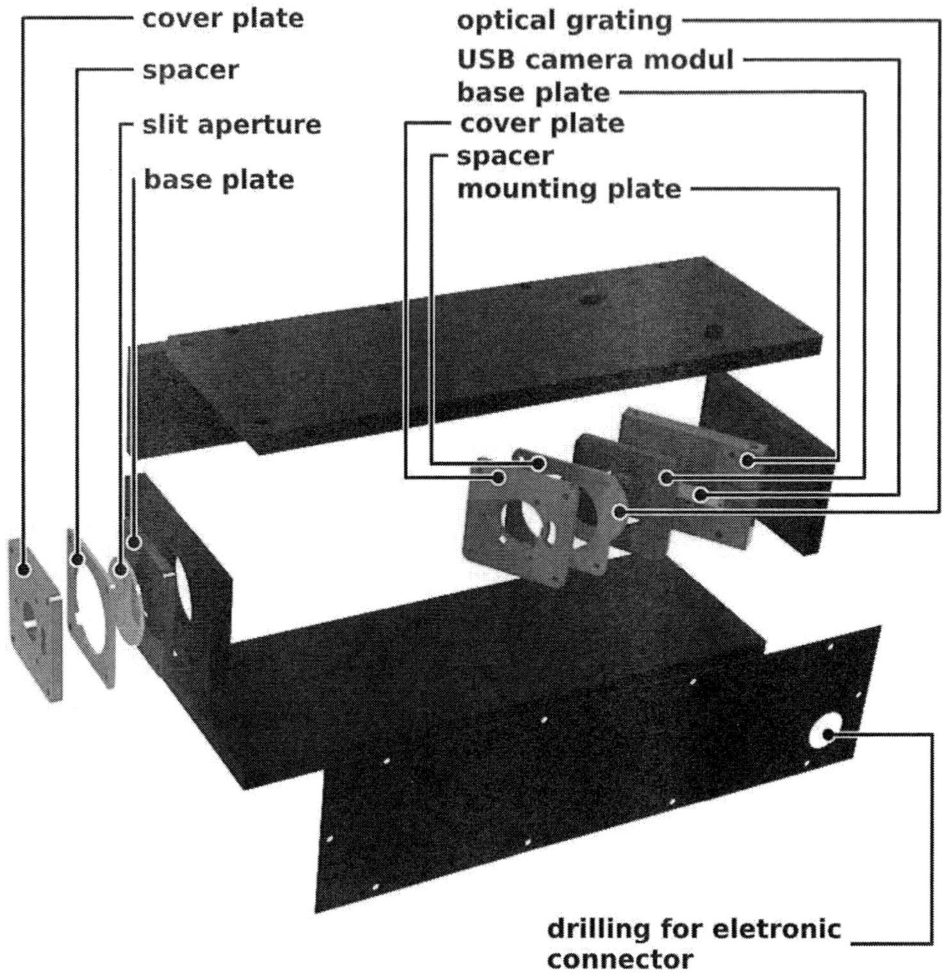

Figure 2. Exploded assembly drawing of the spectrometer setup

Figure 3. Calibration picture of the spectrometer

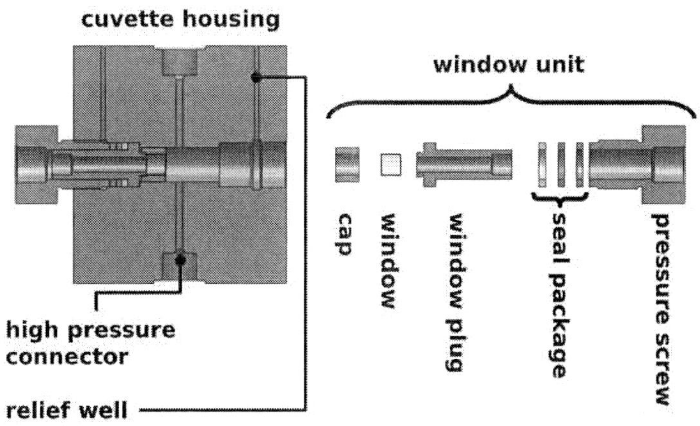

Figure 4. High-pressure cuvette

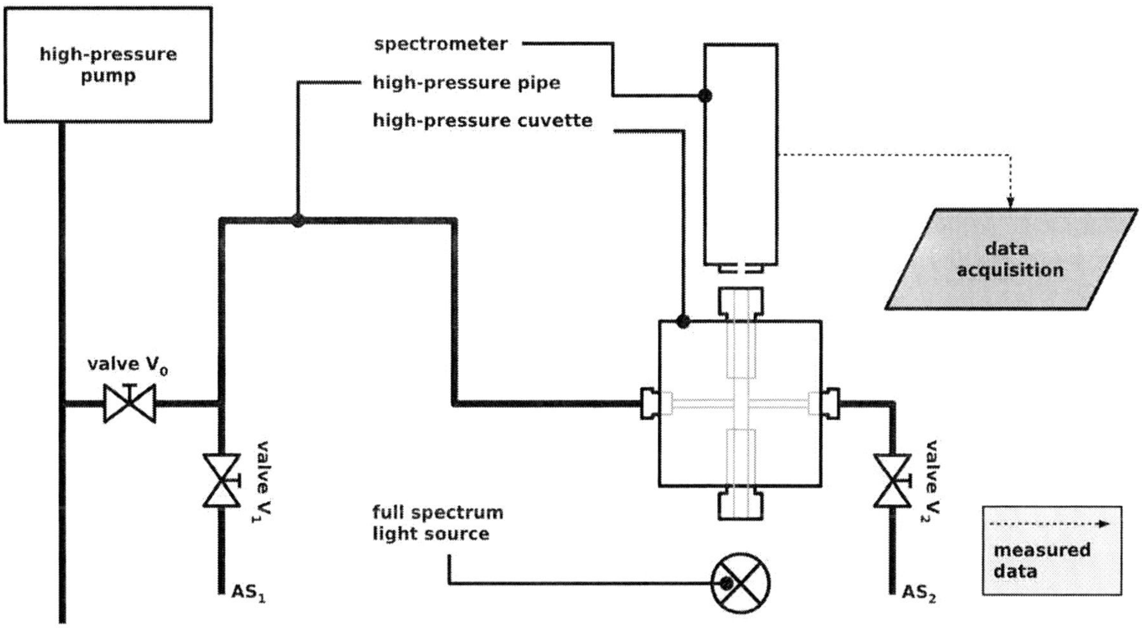

Figure 5. Experimental setup

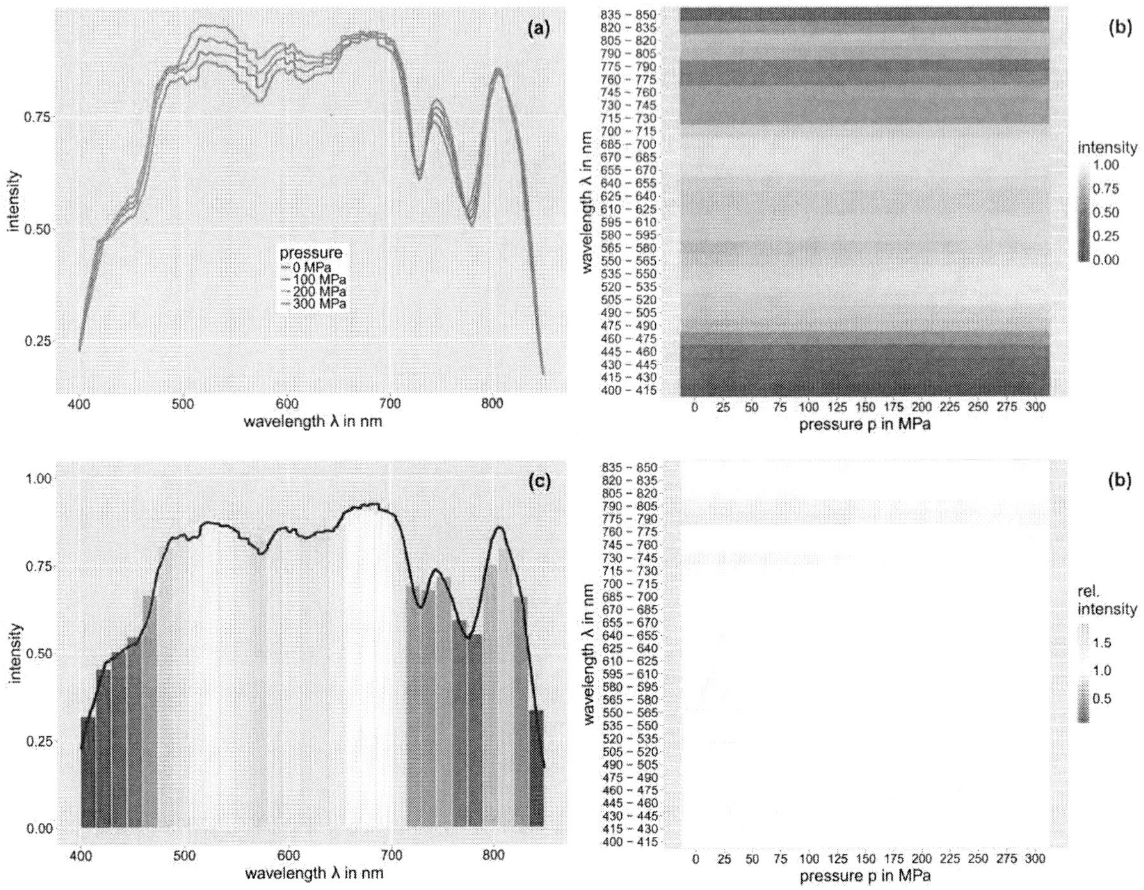

Figure 6. Graphical visualization methods: multiline plot (a), spectral intensity (b), combination of multiline plot and spectral intensity (c), relative intensity (d)

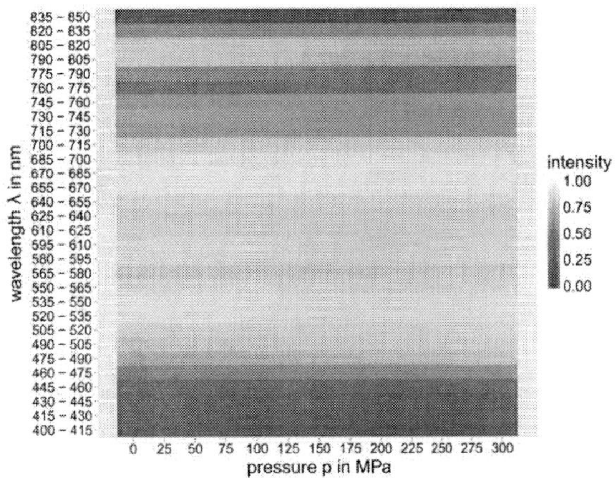

Figure 7. Spectral intensity of H_2O

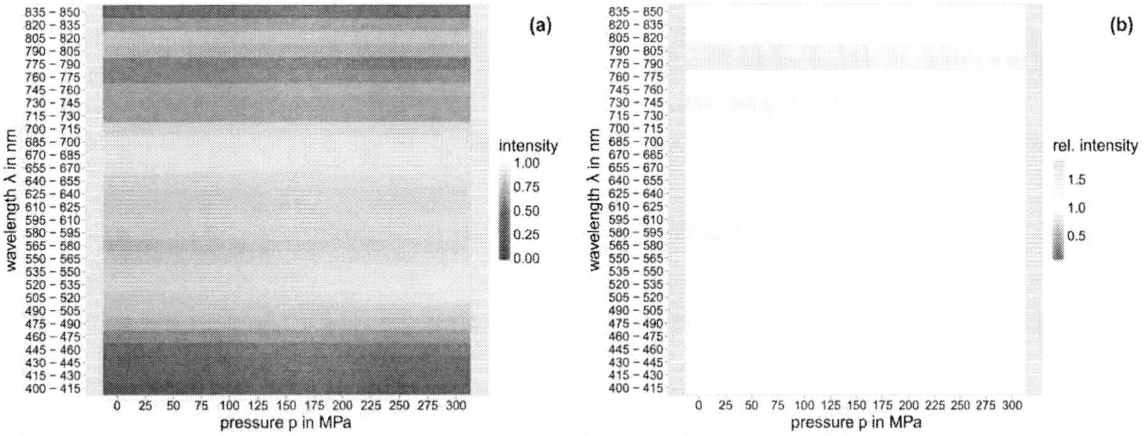

Figure 8. Spectral intensity (a) and normalized spectral intensity (b) of tap water

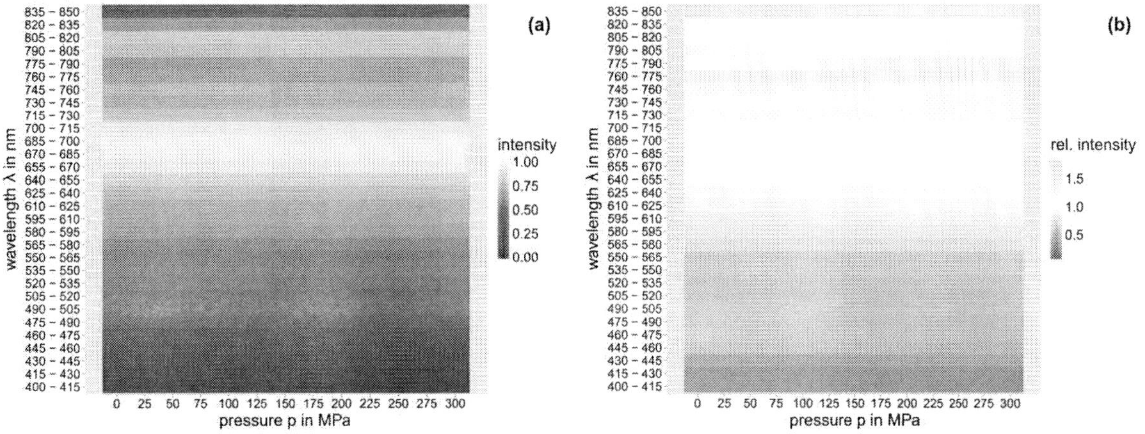

Figure 9. Spectral intensity (a) and normalized spectral intensity (b) of NaCl-blended water

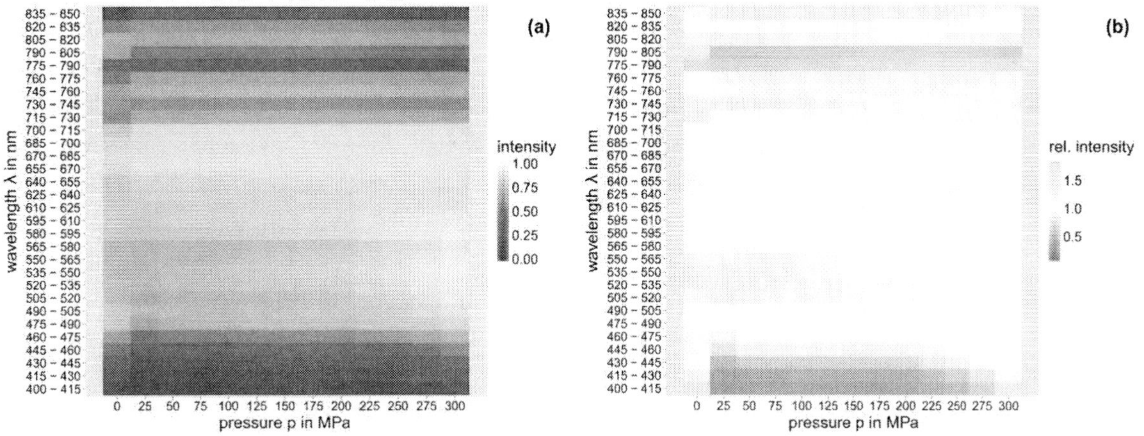

Figure 10. Spectral intensity (a) and normalized spectral intensity (b) of $CaCl_2$-blended water

2015 WJTA-IMCA Conference and Expo
November 2-4 ● New Orleans, Louisiana

Paper

A STUDY ABOUT THE EFFCT OF THE STRUCTURE PARAMETERS OF THE MULTI-ORIFICES NOZZLE ON NOZZLE DISCHARGE COEFFICIENT

Li Jingbin, Li Gensheng, Huang Zhongwei, Song Xianzhi
State Key Laboratory of Petroleum Resources and Prospecting, China U. of Petroleum
Beijing, Beijing 102249, China

ABSTRACT

As the "core" of radial jet drilling (RJD) technology, the multi-orifices nozzle must forcefully pulling the nozzle forward and effectively break the rock to form a radial hole. But nozzle discharge coefficient of the multi-orifices nozzle is always smaller than 0.6 in most cases that means only less than 36 percent hydraulic energy transfers to the fluid. It is extremely important to study the influence coefficients of the nozzle discharge coefficient. In this research, exact expression of the nozzle discharge coefficient was derived and effects of numbers of orifices and flow rate were studied by experiment. Results show that the nozzle discharge coefficient of multi-orifices nozzle depends on the discharge coefficient and weight ratio of each single orifice; the discharge coefficient of each orifice lies on its surface degree of finish, diameter, length, Reynolds number, and the complexity of the flow field; most of the energy loss of each orifice was caused by local resistance loss; the local resistance coefficient would slightly decrease with flow rate while greatly increase with the orifices due to the complicated flow. When a multi-orifices nozzle is designed, larger diameter and fewer number orifices are strongly recommended and optimal combination should be tested by experiment.

Organized and Sponsored by WJTA®-IMCA®

1 INTRODUCTION

Radial jet drilling (RJD) technology is typically used to provide an extended wellbore radius along multiple radial directions in a vertical wellbore (Dickinson et al. 1985, 1989). This technology has been applied or tested in the USA, Canada, China, Bolivia, Argentina, Egypt, and Russia, among other countries. The nozzle is the most critical component of RJD and it must pull the hose moving forward and break the rock in front of it to form a radial hole. For the limitation of flow rate and pressure in field, the exiting nozzle hardly meets the application requirements. It is great significant to present a highly effective nozzle.

The multi-orifice nozzle which consists of several forward orifices and backward orifices is a type of highly efficient nozzle applied in RJD applications (see Figure 1). Forward orifices form the forward jet to drill a big horizontal hole, while the backward orifices form the backward jet to generate self-propelled force to pull the nozzle into strata. Numerous researchers have studied these issues: P. Buset (2001) analyzed the mechanism of rock breaking and the self-propelled ability of multi-orifice nozzles; Guo Ruichang (2010) studied the internal and external flow field distribution and local flow characteristics of a jet bit by numerical simulations; Liao Hualin (2011) examined the effect of the number of forward orifices and hydraulic parameters on the rock-breaking properties; Ma Dongjun (2014) researched the mechanism of self-propelled force and the influence coefficients by experiments. However, the nozzle discharge coefficient of the multi-orifices nozzle is always smaller than 0.6 in most cases that means only less than 36 percent hydraulic energy transfers to the fluid. It is extremely important to study the influence coefficient of the nozzle discharge coefficient. In this research, exact expression of the nozzle discharge coefficient was derived and effects of numbers of orifices and flow rate were studied by experiment.

2 ENERGY CONVERSION EFFICIENCY OF THE MULTI-ORIFICES NOZZLE

According to Chen Tinggen's study, the pressure drop of the nozzle equation can be expressed as

$$\Delta p = \frac{1}{2}\frac{\rho Q^2}{C^2 A^2} \tag{1}$$

Where, Δp is the pressure drop of the nozzle, MPa; ρ is density, kg/m^3; Q is flow rate, m^3/s; A is the outlet area, m^2; C is the nozzle discharge coefficient, dimensionless.

According to the equation of continuity, average velocity can be calculated by:

$$\bar{v} = \frac{Q}{A} \tag{2}$$

Where, $\bar{v}$ is the average velocity of the nozzle, m/s.

By substituting Eq. 2 into Eq.1, the relationship of nozzle discharge coefficient and energy conversion efficiency is found to be

$$\frac{1}{2}\rho \bar{v}^2 = C^2 \Delta p \tag{3}$$

Eq.3 relates that only the square of the nozzle discharge coefficient times hydraulic pressure transfer to the kinetic energy. The nozzle discharge coefficient determines the energy conversion efficiency and is one of the most important performance indexes of the multi-orifices nozzle. It is really necessary to research the nozzle discharge coefficient to improve the utilization of hydraulic energy.

3 EQUATION OF THE NOZZLE DISCHARGE COEFFICIENT

3.1 Structure of the Multi-orifices Nozzle

A simplified physical model of a multi-orifice nozzle (Guo et al. 2009) is shown in Fig. 1. The nozzle consists of one center orifice, n_2 forward orifices, and n_3 backward orifices. The corresponding diameters, velocities and flow rates of the orifices are d_1, d_2, and d_3; v_1, v_2, and v_3; and Q_1, Q_2, and Q_3, respectively. The angles between the center axis of the forward and backward orifices and the nozzle are θ_2 and θ_3, respectively. The diameter of the nozzle is d_0. The velocity, flow rate, and pressure of the incoming flow are v_0, Q_0, and P_0, respectively. In addition, p_{out} is the outlet pressure of the orifices. Due to the high velocity which can break rock, the flow in the orifices is turbulent flow, and the velocity equals to the average velocity approximately. We identify different nozzles by the number of backward orifices added to the number of forward orifices and the number of center orifices, such as the $n_3 + n_2 + 1$ nozzle.

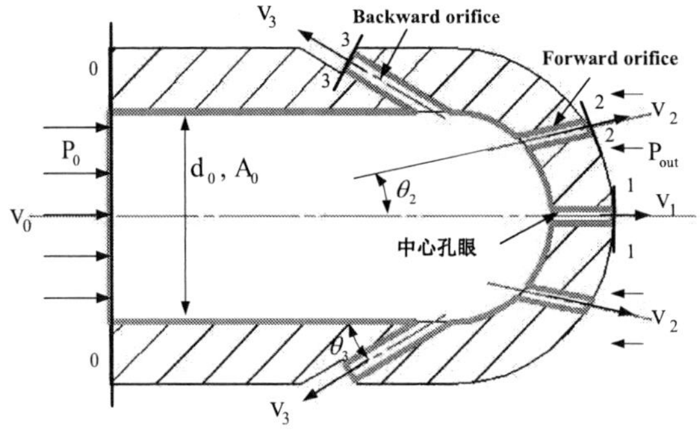

Figure 1. Structural Diagram of the Multi-orifices Nozzle

3.2 Energy Loss of the Multi-orifices Nozzle

In the study of the nozzle discharge of the conical nozzle, it was taken as a reducing pipe. Its main energy loss results from the frictional resistance loss and the contraction pressure loss (Liu et al. 2000; Fei et al. 1995, Yang et al. 2013).

Analogously, the multi-orifices nozzle is taken as the assembly of several sudden contraction tubes. Due to each orifice is cylindrical shape, there is no secondary shrinkage. The energy loss of the multi-orifices nozzle mainly results from the frictional resistance loss and local pressure loss.

a. Frictional resistance loss

All the flow in the orifices can be taken as the circular pipe flow. The frictional resistance loss can be calculated by the follow equation

$$h_f = \lambda \frac{L}{d} \frac{v^2}{2g} \tag{4}$$

Where, h_f is the frictional resistance loss, MPa; λ is the frictional resistant coefficient calculated by using the turbulent model in Yuan's study (1986); L is the length of the circular pipe, m; d is the diameter of the pipe, m; v is the average velocity, m/s; g is the gravitational acceleration, m/s^2.

b. Local Pressure Loss

Because there is no diversion section, rapid change in flow section and the flow direction leads to the local pressure loss. The multi-orifices nozzle is taken as the assembly of several sudden contraction tubes. The local resistance loss of each orifice can be obtained by

$$h_j = \zeta \frac{v^2}{2g} \tag{5}$$

Where, h_j is the local loss, MPa; ζ is the local resistance coefficient, dimensionless; v is the average velocity, m/s; g is the gravitational acceleration, m/s^2.

Due to the complex flow process of the sudden gravitational acceleration, there is only empirical formula for the calculation of the local resistance coefficient. The commonly used formula is as follow (Wu, 2007)

$$\zeta \approx 0.42\left(1 - \frac{A_2}{A_1}\right) \tag{6}$$

Where, A_1 is the cross-sectional area of the large tube; A_2 is the cross-sectional area of the small tube;

There are three types' orifices for the multi-orifices nozzle. Relevantly, there are three types' local resistance coefficients. They only can be obtained by experiment.

3.3 Derivation

As is shown in Fig.1, Bernoulli equation is set up respectively between 0-0 section and 1-1 section, 2-2 section, 3-3 section.

$$\frac{p_0}{\gamma}+\frac{\alpha_0 v_0^2}{2g}=\frac{p_{out}}{\gamma}+\frac{\alpha_i v_i^2}{2g}+h_{ji}+h_{fi}, \quad (i=1,\ 2,\ 3) \tag{7}$$

Where, P_0 is the inlet pressure, MPa; v_0, v_i is the velocity of inlet or orifices, m/s; α_0, α_i is the kinetic energy correction coefficient, if it is turbulent flow, then $\alpha_0=\alpha_i\approx 1$; γ is the product of density of water and the gravitational acceleration, N/m^3; P_{out} is the outlet pressure, if the atmosphere pressure is adopt, then $P_{out}=0$; h_f is the frictional resistance loss, MPa; hj is the local resistance loss, MPa.

Usually, the velocity in the orifices is far greater than that in the nozzle, which means $v_i > 10v_0$. Therefore, the inlet velocity (v_0) can be neglect. In addition, the experiment is conducted in the air, which means $P_{out}=0$. By substituting the relationships and Eq. 5 and Eq.6 in Eq.7, there results

$$\frac{\Delta p}{\gamma}=\frac{v_i^2}{2g}\left(1+\zeta_i+\lambda_i\frac{L_i}{d_i}\right) \tag{8}$$

According to the relationship of flow rate and velocity, Eq.8 can be expressed as

$$\frac{\Delta p}{\gamma}=\frac{1}{2g}\frac{Q_i^2}{A_i^2}\left(1+\zeta_i+\lambda_i\frac{L_i}{d_i}\right) \tag{9}$$

The across-section area can be expressed by

$$\frac{A_i}{A_{ne}}=\frac{d_i^2}{d_{ne}^2} \tag{10}$$

Where, A_{ne} is the equivalent across-section area, m^2; d_{ne} is the equivalent diameter, m. According to Eq. 9 and Eq. 10, the flow rate of each orifice can be written as

$$Q_i=\sqrt{\frac{2\Delta p}{\rho}}*\frac{d_i^2}{d_{ne}^2}*A_{ne}\sqrt{1\Big/\left(1+\zeta_i+\lambda_i\frac{L_i}{d_i}\right)} \tag{11}$$

According to the continuity, the relationship of the flow rate can be obtained

$$Q_0=Q_1+n_2Q_2+n_3Q_3 \tag{12}$$

By substitute Eq.11 in Eq.12, the total flow rate is found to be

$$Q_0 = \sqrt{\frac{2\Delta p}{\rho}} * A_{ne}\left(\frac{d_1^2}{d_{ne}^2}\sqrt{1\Big/\left(1+\zeta_1+\lambda_1\frac{L_1}{d_1}\right)} + n_2 * \frac{d_2^2}{d_{ne}^2}\sqrt{1\Big/\left(1+\zeta_2+\lambda_2\frac{L_2}{d_2}\right)} + n_3 * \frac{d_3^2}{d_{ne}^2}\sqrt{1\Big/\left(1+\zeta_3+\lambda_3\frac{L_3}{d_3}\right)}\right) \quad (13)$$

But if it is the ideal fluid, the total flow can be expressed as

$$Q_0^{'} = \sqrt{\frac{2\Delta p}{\rho}} * A_{ne} \quad (14)$$

According to its definition, the nozzle discharge is the ratio of the flow rate which flows through the nozzle with the ideal fluid and the real fluid under the same pressure. The nozzle discharge coefficient of the multi-orifices nozzle can be got

$$C = \frac{Q_0}{Q_0^{'}} = \frac{d_1^2}{d_{ne}^2}\sqrt{1\Big/\left(1+\zeta_1+\lambda_1\frac{L_1}{d_1}\right)} + n_2 * \frac{d_2^2}{d_{ne}^2}\sqrt{1\Big/\left(1+\zeta_2+\lambda_2\frac{L_2}{d_2}\right)} + n_3 * \frac{d_3^2}{d_{ne}^2}\sqrt{1\Big/\left(1+\zeta_3+\lambda_3\frac{L_3}{d_3}\right)} \quad (15)$$

In most cases, orifices of the multi-orifices nozzle have the same diameter. Eq.15 can be reduced to

$$C = \frac{1}{1+n_2+n_3}\sqrt{1\Big/\left(1+\zeta_1+\lambda_1\frac{L_1}{d_1}\right)} + \frac{n_2}{1+n_2+n_3}\sqrt{1\Big/\left(1+\zeta_2+\lambda_2\frac{L_2}{d_2}\right)} + \frac{n_3}{1+n_2+n_3}\sqrt{1\Big/\left(1+\zeta_3+\lambda_3\frac{L_3}{d_3}\right)} \quad (16)$$

Eq.16 relates that the nozzle discharge coefficient of the multi-orifices nozzle depends on that of each orifice and the weight of each type orifice. When the number of orifices is set, the more the orifice with large nozzle discharge coefficient, the larger nozzle discharge coefficient can be got. The nozzle discharge coefficient of each orifice depends on the frictional resistance coefficient and the local resistance coefficient. The friction resistance coefficient of each orifice lies on its surface degree, diameter, length, Reynolds number, and so on. But the length of orifice is generally short. It can be inferred that the friction resistance coefficient would be small. So the study should pay great attention to the local resistance coefficient which results from rapid change in flow section and the flow direction. Eq.16 only can be used to analyze the effect of the structure parameters on the nozzle discharge coefficient qualitatively for the uncertainty of local resistance coefficient which only can be measured by experiment.

4 EXPERIMENTS

4.1 Facilities

A high-pressure plunger pump is used as a power source at a rated pressure of 60 MPa and a certified capacity of 100 L/min. A hydraulic sensor with a measuring range of 30 MPa, an output current of 4~20 mA, and an accuracy of 0.1 % F*S is used to measure the jetting pressure. The multi-orifice nozzle named as the 6+3+1 nozzle has one center orifice with $d_1 =$ 0.7 mm, three forward orifices with $d_2 =$ 0.7 mm, and six backward orifices with $d_3 =$ 1.0 mm. The diameter of the inlet is $d_0 =$ 10 mm, and the angles are $\theta_2 = 12$ °and $\theta_3 = 40$ °.

Figure 2. Picture of the Multi-orifices Nozzle

4.2 Experiment Scheme

According to the reliability definition, the nozzle discharge coefficient can be expressed as

$$C = \frac{\sqrt{\rho Q^2/(2pA^2)}}{1000} \tag{17}$$

Where, ρ is the density of water, kg/m^3; Q is the flow rate, m^3/s; p is the jetting pressure, MPa; A is the equivalent outlet area, m^2.

For a particular nozzle, the jetting pressure data and relevant flow rate data are needed to calculate the nozzle discharge coefficient. The 6+2+1, 6+1+1, 6+0+1, 6+0+0 nozzles are created by blocking the forward and center orifices of the 6+3+1 nozzle. To ensure the accuracy of the experiment, the nozzle discharge coefficient with different flow rates are measured.

5 RESULTS AND DISCUSSION

5.1 Flow Rate-Jet Pressure Relationship

The flow rate-jet pressure curves of different nozzles are given in Figure 3. As is shown, there is an approximate power relationship between the flow rate and jetting pressure. But it is not exactly in accordance with the pressure drop equation of the nozzle. It can be inferred that the nozzle discharge coefficient is not a constant for a multi-orifices nozzle which makes it different from the conical nozzle obviously and is needed to be researched; meanwhile, it can tell that the jet pressure increases slowly with the flow rate when there are more orifices which means bigger equivalent diameter, while it increases rapidly with the flow rate with fewer orifices. Therefore, six tests are conducted for the nozzle with more orifices, while four/five tests for the nozzle with fewer orifices. In other words, little equivalent diameter allows very limit flow rate which is very hard to meet the requirement of rock breaking energy for the safety pressure of the RJD system.

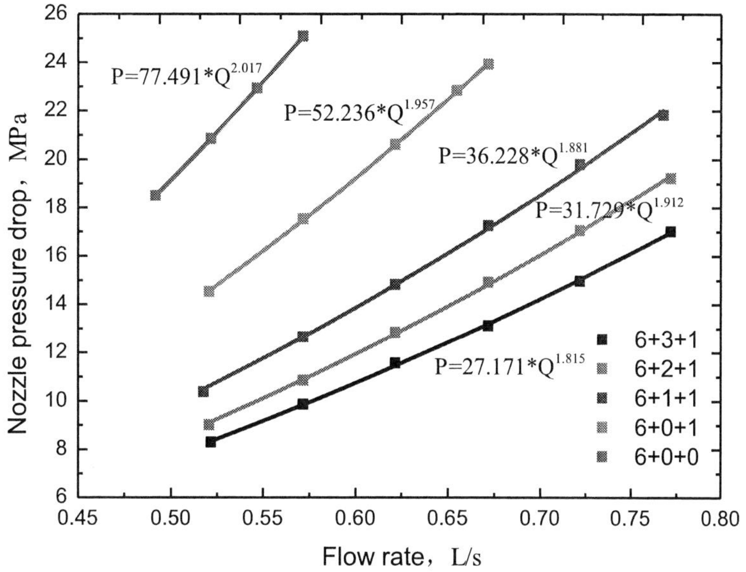

Figure 3. Flow Rate-Jet Pressure Curves of Different Nozzles

5.2 Effect of Flow Rate on Nozzle Discharge Coefficient

Because there is no diversion section in the multi-orifices nozzle, it is taken as the assembly of several sudden contraction tubes, and the rapid change in the flow cross-section leads to a lot of energy loss. According to Yuan Enxi's study (1986), the local resistance coefficient will increase with flow rate. The effect of flow rate on the nozzle discharge coefficient is studied in this research. According to the data in Figure 3, the nozzle discharges coefficient with different flow rates are obtained by Eq. 17. As is shown in Figure 4, all nozzle discharges coefficients are less than 0.6 which means only less than 36 percent hydraulic energy transfers to kinetic energy. It is really necessary to improve the hydraulic performance of the multi-orifices nozzle. Meanwhile, except for the 6+0+0 nozzle, the nozzle discharge coefficient of the nozzle increases with the flow rate slightly; and the nozzle with more orifices are much greater affected, for example, the increase of the nozzle discharge coefficient of the 6+3+1 nozzle up to 3.3 percent with the flow rate increase from 0.522 L/s to 0.772 L/s. It can be inferred that lager flow rate can decrease the local resistance coefficient and improve the hydraulic performance of the nozzle. But because the improvement is small, the hydraulic performance of multi-orifices nozzle must be improved by changing the structure of the nozzle.

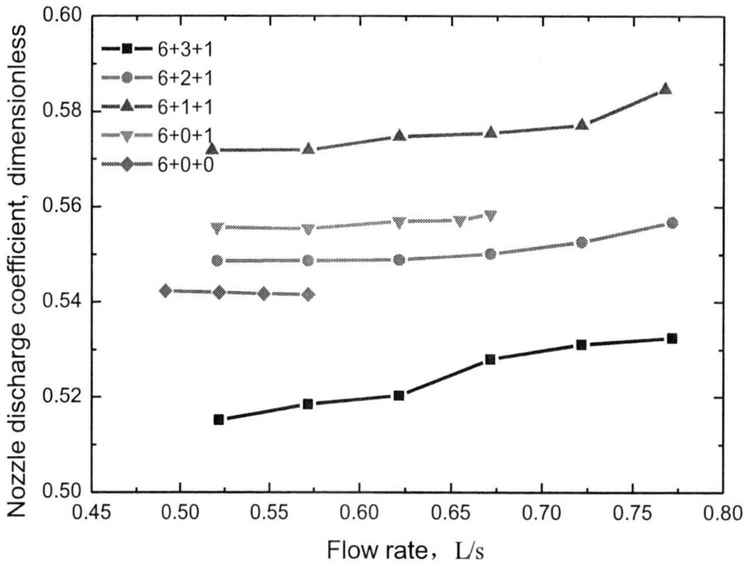

Figure 4. Nozzle Discharge Coefficient-Flow Rate Curves of Different Nozzles

5.3 Effect of the Number of Orifices on Nozzle Discharge Coefficient

It is easy to understand that the nozzle discharge coefficient of the forward orifice is bigger than that of the backward orifice. The nozzle discharge coefficient of the multi-orifices nozzle should increase with the number of forward orifices linearly, but it turns out complex in experiment. According to the above analysis, the average of the nozzle discharge coefficient is used in this part. The changing law of the nozzle discharge coefficient with forward orifices is given in Figure 5. As is shown, the nozzle discharge coefficient increases first and then decreases with forward orifices. According to Eq. 16, with bigger nozzle discharge coefficient the forward orifice improves the hydraulic performance of the nozzle; but with the increase of the forward orifice the flow in the nozzle becomes more complex which leads to the increase of the local energy loss. For example, the nozzle discharge coefficient of the 6+3+1 nozzle is even lower than that of the 6+0+0 nozzle. The same law should be found in the backward orifices.

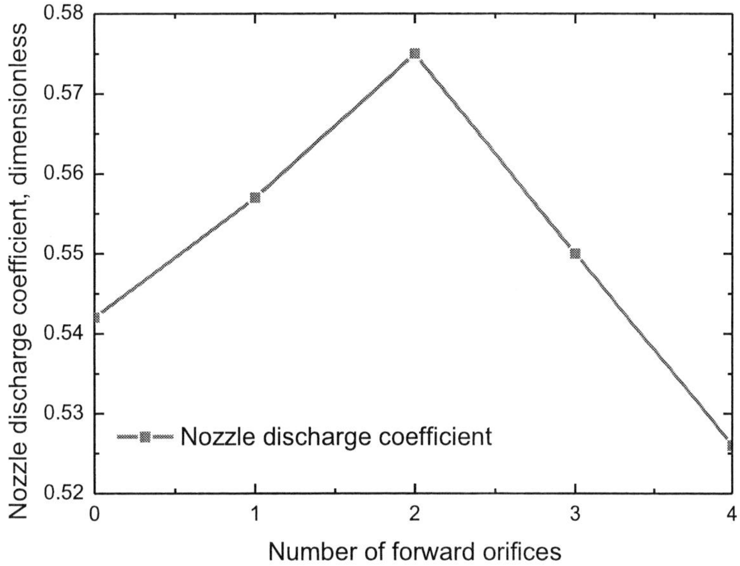

Figure 5. Nozzle Discharge Coefficient-Number of Orifices Curve

5.4 Effect of the Number of Orifices on the Local Resistance coefficient

To deep understand the changing law of the nozzle discharge coefficient, the effect of orifices on the local resistance coefficient is analyzed. The wall thickness of the nozzle is 3mm. The lengths of orifices can be obtained by multiplying the wall thickness with the reciprocal of the cosine the angles, which turns out L_1=3mm, L_2=3.5mm, L_3=4.5mm. The orifice is similar with the clean seamless steel tube. The absolute roughness can be set as Δ=0.01mm. According to Yuan's (1986) study, the frictional resistant coefficient of the center orifice can be got, $\lambda=0.038$. The local resistance coefficient can be calculated by Eq. 6, $\zeta_1=0.415$. Due to different orifices only have different angles, the local resistance coefficients of different orifices are correlative, which depends on the angle. The relationship can be expressed as

$$\zeta_2=\beta_2\zeta_1,\quad \zeta_3=\beta_3\zeta_1,\quad \beta_3>\beta_2>1 \tag{18}$$

Where, β_2 and β_3 are dimensionless constants which depend on the angle.

When there is only the backward orifice, the local resistance coefficient can be obtained by substituting the nozzle discharge coefficient of the 6+0+0 nozzle into Eq. 16 which turns out $\zeta_3=2.229$. According to Eq. 18, the dimensionless constant β_3 turns out 5.371. Because it is hard to figure out the exact relation of the resistance coefficient, the linear relation is used in this study. The dimensionless constant β_2 turns out 1.350. According to the average nozzle discharge coefficient and Eq. 16, the nozzle discharge coefficients of different nozzles are obtained and listed in Table 1. As is shown, for the 6+0+1 nozzle, the local resistance coefficient is $\zeta_1=0.477$, while the frictional part is only $\lambda_1 \frac{L_1}{d_1}=0.114$. It can be inferred that the main energy loss is caused by local resistance coefficient.

Table 1 Local Resistance Coefficient of Each Orifice of Different Nozzles

Nozzle type	λ	C	ζ_1	ζ_2	ζ_3
6+0+0	0.038	0.542	0.415	0.560	2.229
6+0+1	0.038	0.557	0.477	0.644	2.561
6+1+1	0.038	0.575	0.55	0.743	2.954
6+2+1	0.038	0.550	0.645	0.871	3.464
6+3+1	0.038	0.526	0.822	1.110	4.414

For the linear relationship of different local resistance coefficients, only that of the center orifice is analyzed. The curve of the local resistance coefficient and the orifices is shown in Figure 6. As is shown, the local resistance coefficient increases quickly with the orifices, which is caused the complicated flow. It explains that the more forward orifices may lead to lower the nozzle discharge coefficient. Fewer orifices are strongly recommended when a multi-orifices nozzle is designed. The best combination of the backward orifices and forward orifices should be tested to make sure the rock breaking capacity and self-propelled ability.

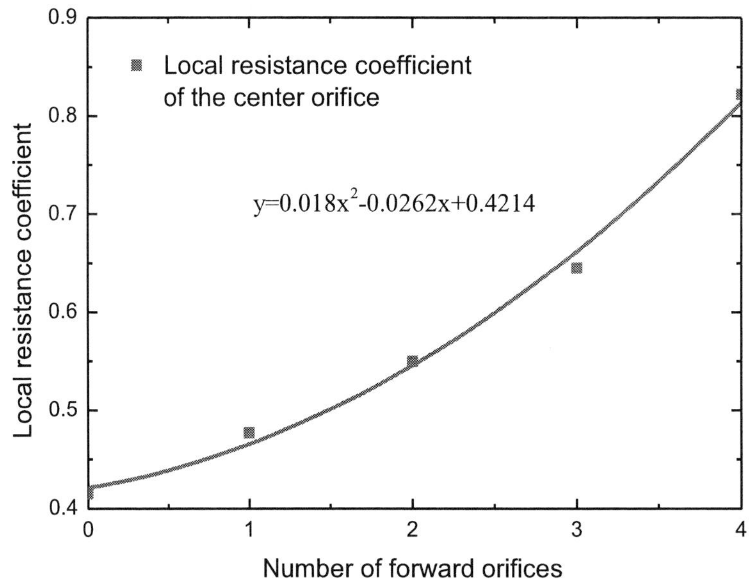

Figure 6 The Curve of the Nozzle Discharge Coefficient and Orifices

6 CONCLUSIONS

The energy conversion efficiency of the multi-orifices nozzle was found only 36 percent which seriously limits its hydraulic performance and is needed to be improved urgently. The nozzle discharge coefficient calculation equation was derived, and the effects of flow rate, number of orifices on it were researched. The following conclusion can be achieved:

(1) The nozzle discharge coefficient of the multi-orifices depends on that of each orifice and the weight of each type orifice. When the number of orifices is set, the more the orifice with large nozzle discharge coefficient, the larger nozzle discharge coefficient

can be got.

(2) The main energy loss of the multi-orifices nozzle was caused by the local resistance loss which would slightly decrease with flow rate and would great increase with the number of orifices.

(3) When a multi-orifices nozzle was designed, the large diameter and fewer orifices are strongly recommended. The best combination of the backward orifices and forward orifices should be tested to make sure the rock breaking capacity and self-propelled ability.

ACKNOWLEDGMENTS

This study was funded by the National Science and Technology Major Project (No. 2011ZX05036002) and Science Foundation of China University of Petroleum, Beijing (No. 2462013BJRC002). This support is gratefully acknowledged by the authors, who are also grateful to the reviewers of this paper for their detailed comments.

REFERENCES

Chen Tinggen and Guan Zhichuan. 2006. Drilling Engineering Theory and Technology. Dongying : China U. of Petroleum Press, 145-146. Print.

Fei Jianguo. 1995. Flow Coefficient's Expression and its Relation with Energy Conversion Efficiency of High Pressure Water Jet Nozzle[J]. J. Changchun Inst. OPT. & Fine Mech, 1995, 18(01): 32-36. Print.

Guo Ruichang, Li Gensheng, Huang Zhongwei, et al. 2010. Numerical Simulation Study on Flow Field of Multi-hole Jet Bit[J]. Fluid Machinery, 38(4): 13-17. Print.

Liao Hualin, Niu Jilei, Cheng Yuxiong, et al. 2011. Experiment study on water jet breaking rock by multi-orifice nozzle[J]. Journal of China Coal Society, 36(11): 1858-1862. Print.

Liu Chengwen, Li Zhaomin. 2000. A theoretical calculation method for flow rate coefficient and hydraulic parameters of conical nozzle[J]. Drilling & Production Technology, 23(05): 1-3. Print.

Ma Dongjun, Li Gensheng, Huang Zhongwei, et al. 2014. Mechanism of a self-propelled multi-hole jet bit and influencing rules on its self-propelled force[J]. Periodical of Gas, 34(04): 99-104. Print.

P. Buset, M. Riiber, and A. Eek, 2001. Jet Drilling Tool: Cost-Effective Lateral Drilling Technology for Enhanced Oil Recovery[C]. The SPE/ICoTA Coiled Tubing Roundtable, Houston, Texas,7-8 March. SPE-68504-MS. Print.

Shen Zhonghou. 1998. Theory and Technology of Water Jet[M]. Dongying: China U. of Petroleum Press, 269-270. Print.

W. Dickinson and R.W. Dickinson, 1985. Horizontal Radial Drilling System[C]. The SPE California Regional Meeting, Bakersfield, California, 27-29 March. SPE-13949-MS. Print.

W. Dickinson, R.R. Anderson, and R.W. Dickinson.1989. The Ultrashort-Radius Radial System[J]. SPE Drilling Engineering, 4(03): 247-254. SPE-14804-PA. Print.

Wu Chigong. 2007. Fluid Mechanics: Volume one [M]. Beijing: Higher Education Press, 129-155. Print.

Yang Yousheng, Zhang Jianping, Nie Songlin. 2013. Energy Loss of Nozzles in Water Jet System[J]. Journal of Mechanical Engineering, 49(02): 139-145. Print.

Yuan Enxi. 1986. Engineering Fluid Mechanics[M]. Beijing: Petroleum Industry Press, 121-131. Print.

NOMENCLATURE

A_1 Cross-sectional area of the large tube, m^2
A_2 Cross-sectional area of the small tube, m^2
A_{ne} Equivalent across-section area, m^2
C Nozzle discharge coefficient, dimensionless
d_{ne} Equivalent diameter of the multi-orifices nozzle, mm
d_i Diameter of different orifices (i=1,2,33), mm
g Gravitational acceleration, m/s^2
h_f Frictional resistance loss, MPa
h_j Local resistance loss, MPa
L_i Length of the orifice (i=1,2,33), mm
v_0 Velocity of the incoming flow, m/s
$\bar{v}$ Average velocity of the nozzle, m/s
v_i Average velocity of different orifices(i=1,2,33), m/s
P_0 Pressure of the incoming flow, MPa
P_{out} Outlet pressure, if the atmosphere pressure is adopt, then P_{out}=0, MPa
Δp Pressure drop of the nozzle, MPa
Q_0 Flow rate of the incoming flow, L/s
Q_i Flow rate (i=0, 1, 2, 3), L/s
ρ Density of water, kg/m^3
α_0, α_i Kinetic energy correction coefficient, if it is turbulent flow, then $\alpha_0=\alpha_i\approx 1$
β_2,β_3 Dimensionless constants which depend on the angle, dimensionless
θ_i Angles between the center axis of the forward and backward orifices and the nozzle (i=2, 3)
λ Frictional resistant coefficient, dimensionless
ζ Local resistance coefficient, dimensionless
γ Product of density of water and the gravitational acceleration, N/m^3

2015 WJTA-IMCA Conference and Expo
November 2-4 • New Orleans, Louisiana

Paper

NUMERICAL STUDY OF THE CAVITATING JET AND ITS EROSIVITY

Jin-Keun Choi, Chao-Tsung Hsiao, and Georges L. Chahine
DYNAFLOW, INC.
Jessup, Maryland, USA

ABSTRACT

Cavitation enhances waterjets effectiveness for cutting, cleaning, and drilling applications. Understanding the mechanisms involved with the increased erosivity is crucial for improving and controlling jet performance. Here, the cavitating jet flow field is investigated numerically solving the Navier-Stokes equations coupled with bubble dynamics. The jet shear layer forms large structures, which trap the bubble nuclei and enable them to grow explosively. Near the target bubble deformation and reentrant jet formation are found to be very important in generating concentrated shear and high impulsive loads on material layers to be removed.

Organized and Sponsored by WJTA®-IMCA®

1 INTRODUCTION

Cavitation enhances the erosivity of submerged waterjets and the efficiency or the removal rate for cutting, cleaning, and drilling applications (Johnson Jr et al.; G L Chahine, Conn, et al.; G L Chahine, Conn, et al.; G L Chahine, Choi, et al.). Cavitating jets are also used for material erosion testing, surface treatment, peening, etc. (Singh, Choi, and Chahine; March; G L Chahine, Choi, et al.; Choi, Jayaprakash, and Chahine; Odhiambo), and understanding the mechanisms involved with the increased erosivity is crucial for designing and improving jet performance.

Microbubble nuclei present in water grow as the water exits the nozzle orifice into the host liquid of lower pressure. The expanded bubbles then collapse energetically on the target surface producing microjets and inducing high impulsive pressures, which result in cleaning or erosion. In order to understand the mechanisms at play, we have investigated the cavitating jet flow field numerically coupling Navier-Stokes and bubble dynamics solutions.

It has been known for more than a century (Besant; Lord Rayleigh) that volume implosion of a bubble can generate very high pressure and shock waves. Many studies have also shown, experimentally as well as analytically, that the collapse of these bubbles near rigid boundaries results in high-speed reentrant liquid jets, which penetrate the highly deformed bubbles to strike a nearby rigid boundary generating water hammer like impact pressures (Knapp, Daily, and Hammitt; Hammitt). The resulting reentrant jet has been found a key element in pit formation on the material surface (C.-T. Hsiao et al.; G. L. Chahine; C-T. Hsiao and Chahine).

Bubble collapse near boundaries has been extensively investigated numerically by assuming flow incompressibility and using the boundary element method (BEM) (Blake; Jayaprakash, Chahine, and Hsiao; Zhang, Duncan, and Chahine; Plesset and Chapman; Chahine, Duraiswami, and Kalumuck). The BEM enables accurate description of the reentrant jet and provides the jet geometric and dynamic characteristics as functions of time and space since the liquid velocities (including the reentrant jet) are small relative to the speed of sound in water. On the other hand, during the bubble explosive growth, rebound, and at jet impact, compressible effects need to be included (A. B. Wardlaw and Luton; A. Wardlaw and Luton). For accurate simulations, we use here a hybrid numerical procedure involving time domain decomposition into incompressible and compressible stages in order to capture the full bubble dynamics including the shock phase occurring during reentrant jet impact, bubble collapse, and bubble rebound.

2 NUMERICAL APPROACH

In order to account for viscous effects in the submerged jet simulations we consider the flow field as a two-phase continuum with a gas volume fraction, α. The density, ρ_m, and kinematic viscosity, μ_m, are related to properties of the gas, ρ_g and μ_g, and those of the liquid, ρ_l and μ_l, through:

$$\rho_m = (1-\alpha)\rho_l + \alpha\rho_g, \qquad \mu_m = (1-\alpha)\mu_l + \alpha\mu_g. \tag{1}$$

The continuity and momentum conservation equations for the mixture are as follows:

$$\frac{\partial \rho_m}{\partial t} + \nabla(\rho_m \mathbf{u}_m) = 0, \qquad \frac{\partial \rho_m \mathbf{u}_m}{\partial t} + \nabla \rho_m \mathbf{u}_m \mathbf{u}_m = -\nabla p + \nabla \boldsymbol{\tau}, \tag{2}$$

where t is time, $\boldsymbol{u}_m$ the velocity, p the pressure, and $\boldsymbol{\tau}$ the stress tensor. These equations are solved using 3DYNAFS-VIS© (Hsiao, Jain, and Chahine; C.-T. Hsiao et al.; Raju et al.) and a finite volume formulation. 3DYNAFS-VIS has been extensively validated and used to study a range of two-phase problems and its results have compared favorably with available experimental data (G L Chahine).

2.1 Modeling of Bubble Dynamics and Motion

The bubbles are tracked individually and are treated as point sources and dipoles to represent the time variations of their volumes and translation speeds. Volume variations are obtained with a bubble dynamics equation using bubble surface averaged pressures and velocities, to account for flow non-uniformities, and an additional pressure term to account for bubble motion relative to the liquid (G. L. Chahine; Hsiao, Chahine, and Liu). The equivalent bubble radius, R(t), is obtained using a modified Rayleigh-Keller-Herring equation (Prosperetti and Lezzi), which accounts for the compressibility of the surrounding bubbly medium and for flow field non-uniformities:

$$\left(1 - \frac{\dot{R}}{c_m}\right) R\ddot{R} + \frac{3}{2}\left(1 - \frac{\dot{R}}{3c_m}\right)\dot{R}^2 = \frac{\left(\mathbf{u}_{enc} - \mathbf{u}_b\right)^2}{4} + \frac{1}{\rho_m}\left(1 + \frac{\dot{R}}{c_m} + \frac{R}{c_m}\frac{d}{dt}\right)\left[p_v + p_{g_0}\left(\frac{R_0}{R}\right)^{3k} - p_{enc} - \frac{2\sigma}{R} - 4\mu_m \frac{\dot{R}}{R}\right], \tag{3}$$

where c_m is the local sound speed in the two-phase medium and $\mathbf{u}_b$ is the bubble translation velocity. p_{enc} and $\mathbf{u}_{enc}$, the "encountered" pressure and velocity respectively, are the pressures and velocities in the two-phase medium averaged over the bubble surface. The introduction of $\mathbf{u}_{enc}$ and p_{enc} in Equation (3) is to account for non-uniform pressures and velocities around the bubble and for slip velocity between the bubble and the host medium. Using these average values instead of the conventionally used values at the bubble center, results in a major improvement over classical models (G. L. Chahine; Hsiao, Chahine, and Liu).

The bubble trajectory is obtained using the following motion equation

$$\frac{d\mathbf{u}_b}{dt} = -\frac{3}{\rho_m}\nabla p - 2g + \frac{3}{4}\frac{C_D}{R}\left(\mathbf{u}_{enc} - \mathbf{u}_b\right)\left|\mathbf{u}_{enc} - \mathbf{u}_b\right| + \frac{3}{2\pi}\frac{C_L}{R}\sqrt{\frac{\mu_m}{\rho_m}}\frac{\left(\mathbf{u}_{enc} - \mathbf{u}_b\right)\times\boldsymbol{\omega}}{\sqrt{|\boldsymbol{\omega}|}} + \frac{3}{R}\left(\mathbf{u}_{enc} - \mathbf{u}_b\right)\dot{R},$$

$$C_D = \frac{24}{R_{eb}}\left(1 + 0.197 R_{eb}^{0.63} + 2.6\times10^{-4} R_{eb}^{1.38}\right), \qquad R_{eb} = \frac{2\rho_m R\left|\mathbf{u}_{enc} - \mathbf{u}_b\right|}{\mu_m}. \tag{4}$$

where $\boldsymbol{\omega}$ is the local vorticity, g is the acceleration of gravity, C_L is the lift coefficient, and C_D is the drag coefficient given by (Haberman and Morton). The added mass term is indirectly represented by the first term of the right hand side where $d\mathbf{u}_{enc}/dt$ has been replaced by the

$\nabla p / \rho_m$ term. The last term is a force due to coupling between bubble volume oscillations and bubble motion (Johnson and Hsieh).

2.2 Bubble-Flow Motion Coupling

The two-way coupling between the medium and the bubbles has as key the deduction of the void fraction distribution from the instantaneous bubble sizes and locations. This is realized as follows:

- The volume change and the motion of the individual bubbles in the flow field are controlled by the two-phase medium local properties and flow field gradients.
- The local properties of the mixture (void fractions and local densities) are determined by the spatial distribution of the bubble and by their volumes.
- The mixture flow field with evolving mixture density distribution satisfies mass and momentum conservation.

2.3 Bubble Collapse Modeling

Our numerical approach to model bubble collapse and material response uses a hybrid approach able to simulate fluid structure interaction (FSI) problems involving shock and bubble pressure pulses (C.-T. Hsiao and Chahine; Chahine, Raju, and Hsiao). During growth of a cavitation bubble an incompressible Boundary Element Method (BEM) code 3DYNAFS-BEM© (Chahine & Perdue 1989, Chahine & Kalumuck 1998, Choi & Chahine 2003) is used to simulate most of the bubble period until the end of the bubble collapse where, due to high liquid speeds or to the bubble reentrant jet impacting on the liquid or on the structure, compressible flow effects prevail again. The BEM method has been shown to provide reentrant jet parameters and speed accurately (Chahine and Perdue; Zhang, Duncan, and Chahine; J.-K. Choi and Chahine; Jayaprakash, Singh, and Chahine). The solution of the BEM code is then passed to a fully compressible code capable of shock capturing to simulate reentrant jet impact and bubble ring collapse. The compressible flow solver 3DYNAFS-COMP© (Kapahi, Hsiao, and Chahine) is used and enables modeling shock propagation, liquid-liquid, and liquid-solid impacts.

3 NUMERICAL RESUTS

3.1 Cavitating Jet Flow Field

A nozzle with hemispherical contraction upstream of the orifice is considered. The nozzle has an orifice with diameter, dj = 0.064", and the length of the orifice is 0.064". We considered a flow of 5.5 gpm at a pressure drop of 4,200 psi across the orifice. Figure 1 shows the computational mesh to model this nozzle and the flow field between the nozzle and a target wall to be cleaned or eroded.

As shown in Figure 2, the jet exiting the nozzle has a hat velocity profile with the maximum speed on the axis. Viscous friction with the surrounding liquid diffuses the jet, widens it, and generates vortices, which become structured and modify significantly the flow. The jet speed is seen in Figure 2 to form regular organized structures close to the nozzle, while the jet speed on the jet edge decay significantly. Similarly, the jet speed decays as the liquid moves away from the nozzle.

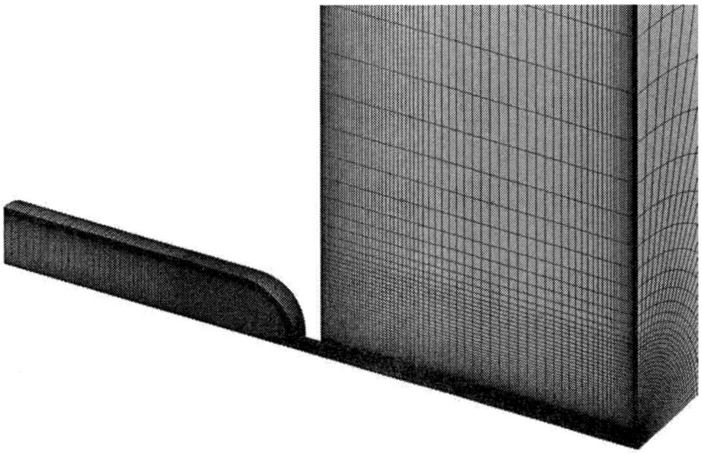

Figure 1. Computational mesh for the jet flow modeling.

The thickness of the boundary layer at the wall depends on the local liquid speed. This has strong implications on cleaning as microscopic particles at the wall are often embedded in the very low speed boundary layer and the shear exerted on them is unable to dislodge them. This strong vortex generation can then be purposely sought and enhanced through jet nozzle design for strong acoustic resonance (G L Chahine, Conn, et al.; Kalumuck et al.). At the wall, as the incoming jet velocity fluctuates, the shear flow has also to adapt and as seen in Figure 3, and ends up exhibiting a very intensely varying shear flow configuration in both time and space.

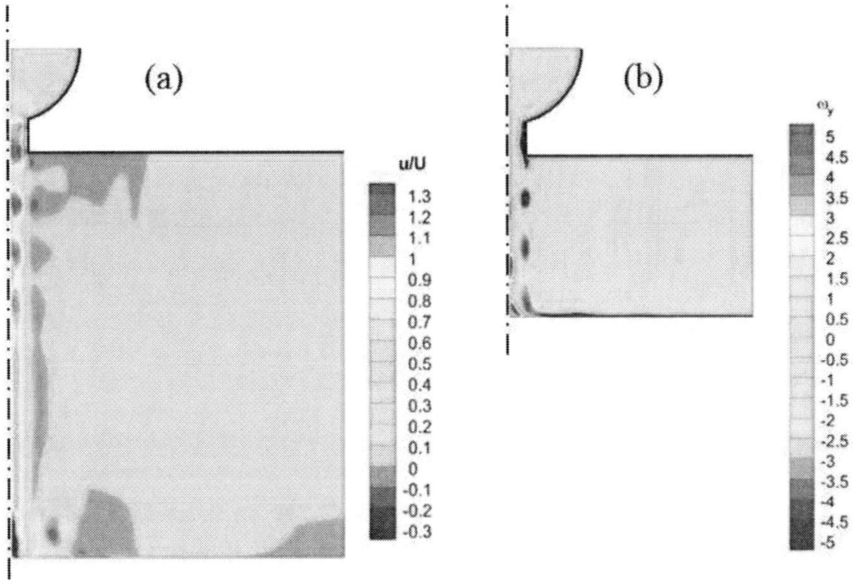

Figure 2. (a) Contours of non-cavitating axial velocity normalized by the mean jet velocity. Wall distance = 13 orifice diameters. (b) Vorticity contours when the wall is distance is 5 orifice diameters.

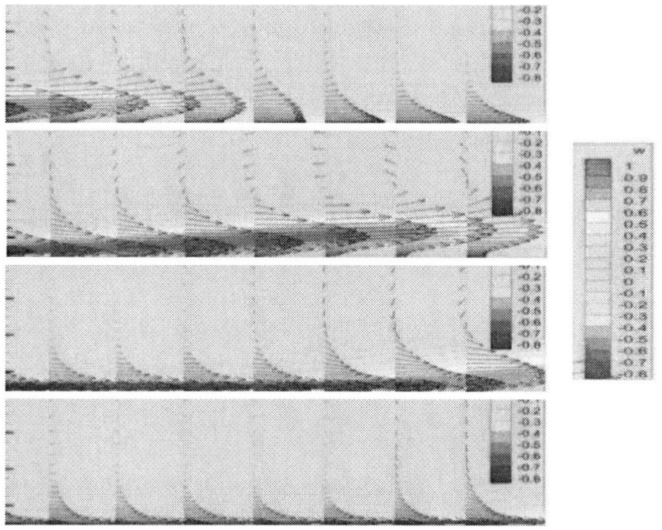

Figure 3. Time sequence of radial velocity contours near the wall at a distance from the jet axis of two jet diameters. Time between images is 1.44 μs. Wall at 13 orifice diameters from the orifice.

In order to account for bubble dynamics, bubble nuclei are randomly dispersed in the liquid and tracked in the simulations. These microscopic or nanoscopic nuclei are omnipresent in any liquids (G L Chahine; Mørch) unless extreme precautions are taken to produce ultra-pure liquids, in which case there still is a chance to have undetected nanoscopic bubble nuclei. As described earlier, a Lagrangian procedure is used to track these nuclei and determine their position and volume in space and time. This is fed back to the mixture flow and provides the void fraction at each location necessary to advance the solution. Knowledge of the bubble dynamics and the fluid dynamics then enables one to determine fluctuating shear and pressures at the cleaned surface.

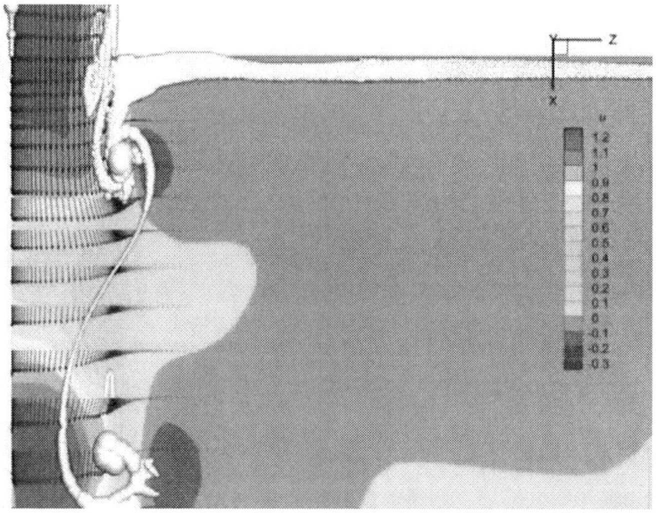

Figure 4. Flow field of a cavitating jet showing bubble capture and growth of ring and longitudinal vortices. White contours are larger cavities represented by a level set scheme.

In this study, two nuclei sources are considered. The first is surface nucleation where nuclei from the rigid surfaces are emitted when the pressure decreases below the vapor pressure. The second source is free field nuclei: here for illustration 20 μm bubbles with a void fraction of about 1%.

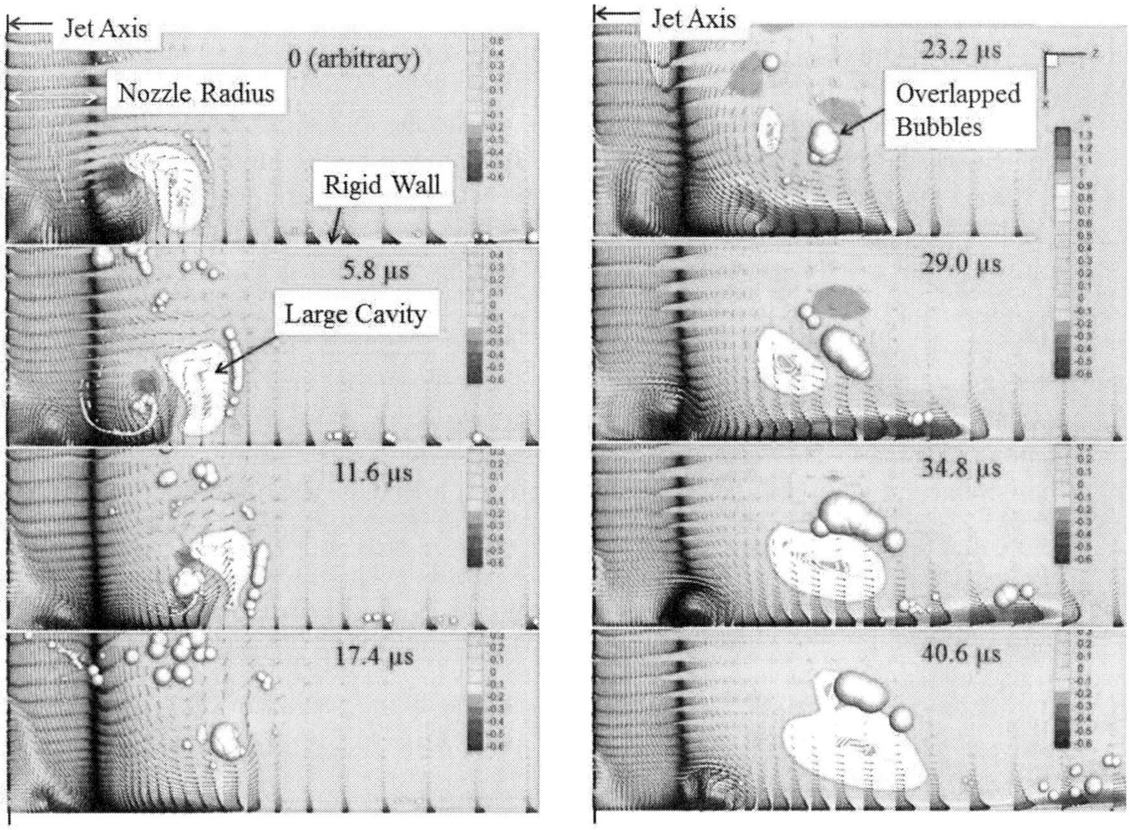

Figure 5. Velocity vectors, cavitation bubbles, and contours of radial velocity (parallel to the wall) for a cavitating jet impinging on a wall. The time sequence is left top to bottom, then right top to bottom. Time between images is 5.8 μs. Wall at 13 jet diameters from the orifice. The radial velocities are normalized by the mean jet velocity.

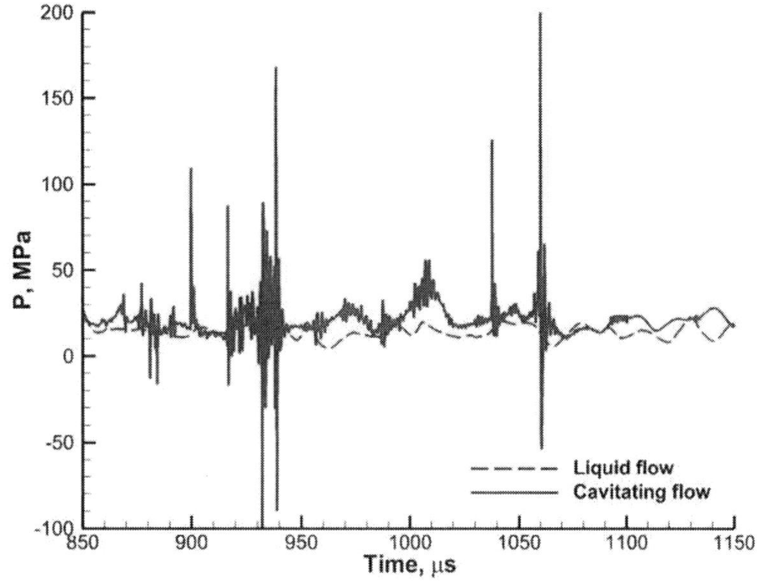

Figure 6. Time history of pressure at wall at the jet impingement center. The red curve is obtained for the jet flow in absence of nuclei seeding, while the green curve includes nuclei seeding and cavitation.

Figure 4 shows an illustrative example for the two-way coupled (bubble-liquid and liquid-bubble) simulations of a cavitating jet. The flow field near the nozzle orifice is nicely illustrated by the velocity vectors and their evolution as the distance from the nozzle increases. A more impressive illustration of the flow field is seen through the nuclei captured from the face of the nozzle into the shear layer. These nuclei grow and are captured by the large vortical structure at the edge of the submerged jet. These structures form two modes: an azimuthal mode resulting in vortex rings and a longitudinal mode that results in elongated vortices which join two vortical ring structures. This is very well visualized by the bubbles captured in the vortices and clearly seen in the figure. Later on as the jet advances to hit the wall (Figure 5), the bubbles grow into larger clouds, which collapse at or near the wall. The passage of the structures and the bubbles further induces fluctuations in the wall pressures and shear. A very large difference in the generated pressure field between cavitating and non-cavitating jet is clearly seen in Figure 6. Cavitation induces orders of magnitude larger pressures relative to the non-cavitating jet (red curve) in both number and amplitude of impulsive pressures due to the collapsing bubbles (green curve).

The vortices generated in the shear layer of the jet forms vortex rings first, and then the vortex rings tends to break into longitudinal vortices. In Figure 7 left, iso-vorticity surfaces are shown where the longitudinal component develops as the jet flows away from the orifice. Small bubble nuclei tend to grow when they are entrained by the vortices. Figure 7 right shows an overview of the cavitating jet. Numerous bubbles flow with the jet, some of them grow as they are entrained into the vortex structures of the jet. Due to the low pressure in the vortex core, the cavitation bubbles grow and then collapse at the wall.

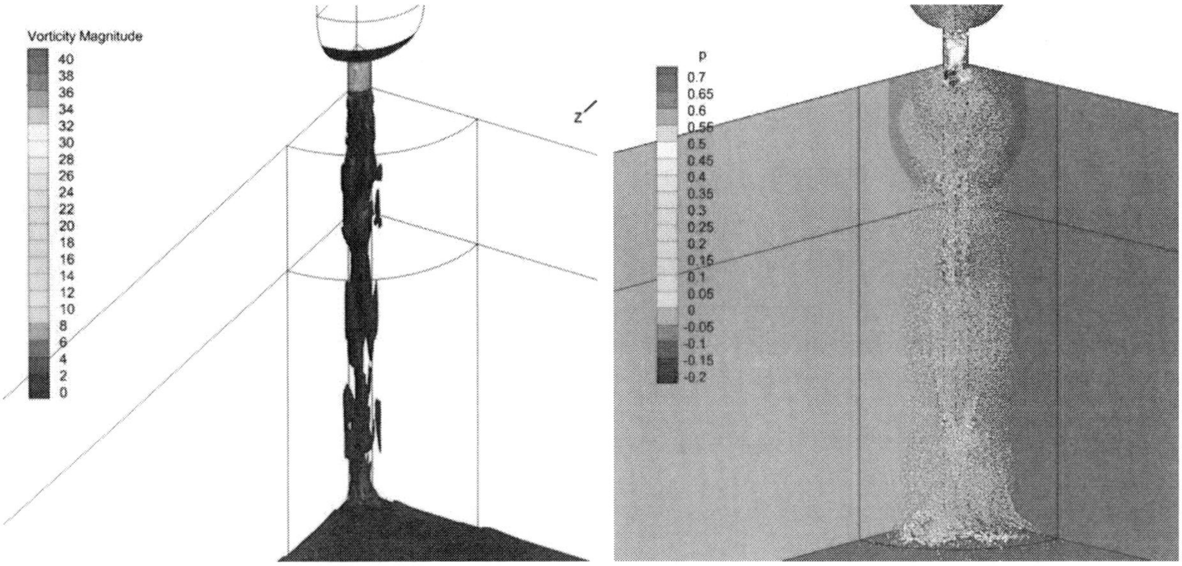

Figure 7. Vorticity contours and iso-vorticity surface sowing the dimensionless vorticity, $|\omega| = 2$ (left), and cavitation bubbles which tend to concentrate along the vortex structures in the jet (right).

3.2 Cavitation Bubble Collapse near Wall

In this section, we examine the pressure loading generate by a collapsing bubble. This requires studying the dynamics of single bubbles near the wall. We consider an initially spherical bubble of radius 50 μm, located at a distance of $X = 1.5\text{mm}$ from a flat material surface and subject it to a time-varying pressure field as represented in Figure 8 and expressed as follows:

$$p(t) = \begin{cases} 10^5\,\text{Pa}; & t < 0, \\ 10^3\,\text{Pa}; & 0 \le t \le 2.415\text{ms}, \\ 10^7\,\text{Pa}; & t > 2.415\text{ms}. \end{cases} \tag{5}$$

Figure 8b shows the time history of the bubble radius in response to the imposed pressure field obtained by solving the Rayleigh-Plesset equation ignoring for now the presence of the wall. It is seen that the bubble starts an explosive grow as soon as the pressure suddenly drops to a low pressure below the critical pressure. The bubble continues to grow until the pressure rises again to a high value. This time-varying pressure field is set up to mimic the encountered pressure for a traveling cavitation bubble in a cavitating jet and arriving at the stagnation region near the wall.

To simulate the bubble dynamics near the wall, the incompressible BEM solver is first applied with a total of 400 nodes and 800 panels used to discretize the bubble surface. This corresponds to a grid density, which provides grid independent solution (Chahine, Duraiswami, and Kalumuck). Figure 9 shows the variations of the bubble outer contours as time advances. As the bubble grows between $t = 0$ and $t \sim 2.4$ms, it behaves almost spherically on its portion away from the wall, while the side close to the material flattens and expands in the direction parallel to the wall actually never touching the wall as a layer of liquid remains between the bubble and the wall. Such a behavior has been confirmed experimentally by (Chahine, Duraiswami, and Kalumuck).

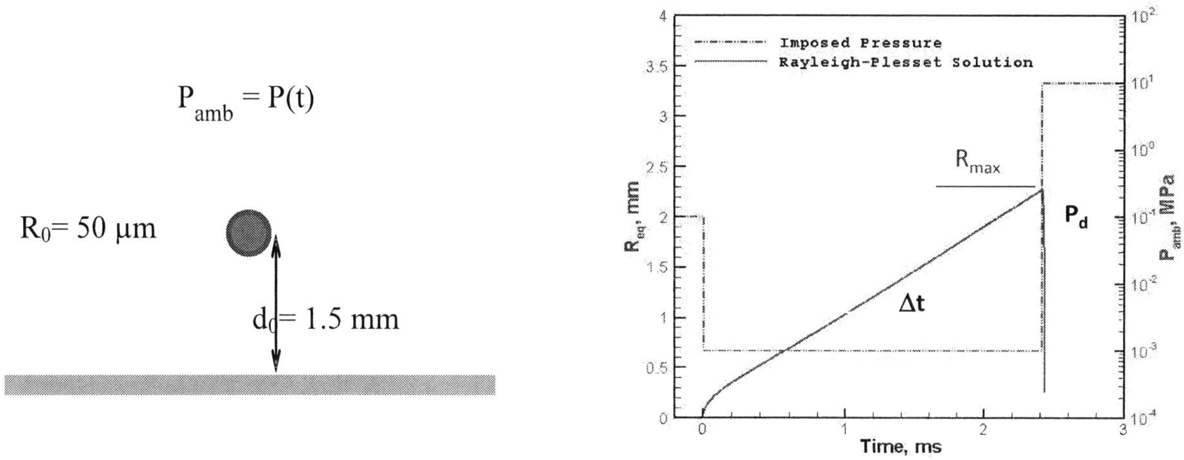

Figure 8. (a) Illustration of the problem of a bubble growth and collapse near a wall. Initial spherical bubble radius 50 μm at 1.5 mm distance from the wall. (b) Imposed time-varying pressure field and resulting time history of the bubble radius obtained by solving the Rayleigh-Plesset equation.

The bubble continues expanding following pressure reversal due to the inertia of the outward flow of the liquid. Due to the asymmetry of the flow, the pressures at the bubble interface on the side far from the wall are much higher than those near the material; thus the collapse proceeds with the far side moving towards the material wall. The resulting acceleration of the liquid flow perpendicular to the bubble free surface develops into a reentrant jet that penetrates the bubble and moves much faster than the rest of the bubble surface to impact the opposite side of the bubble and the material boundary.

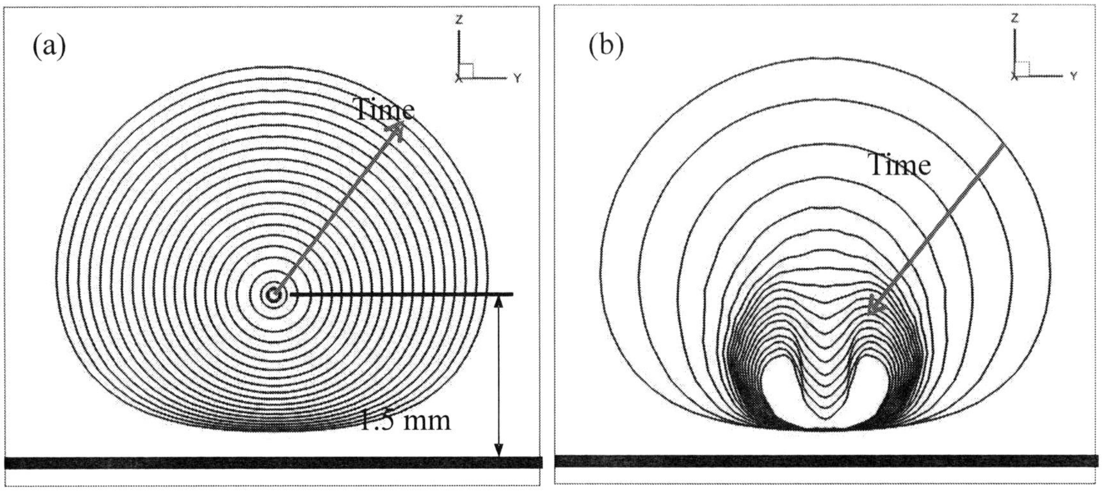

Figure 9. Bubble shape outlines at different times showing (a) bubble growth ($0 < t < 2.415$ ms) and (b) collapse (2.415 ms $< t <$ 2.435ms). Results obtained by the 3DYNAFS-BEM© simulations for $R_0 = 50$ µm, $P_d = 10$ MPa, $\bar{X} = 0.75$, and p(t) described by Equation (5).

3.3 Reentrant Jet Impact and Bubble Ring Collapse

The simulation of the bubble dynamics is then switched from the BEM to the compressible flow solver right before the jet touches the opposite side of the bubble (dubbed touchdown). The compressible flow solver is used to continue the simulation. An axisymmetric 1 m × 1 m domain with a total of 220 × 1,470 grid points with stretched grids concentrated in the immediate region surrounding the bubble is used. The grids are distributed such that there is a uniform fine mesh with a size of 10 µm in the area of interest where the interaction between bubble and plate is important as shown in Figure 10. A reflection boundary condition is imposed on the axis of symmetry, i.e. all physical variables such as density, pressure, velocities and energy are reflected from the axis, while transmission non-reflective boundary conditions (i.e. the flow variables are extrapolated along the characteristic wave direction) are imposed at the far field boundaries.

Figure 11 shows the bubble shapes and the corresponding pressure contours computed at six time instances after jet impact. It is seen that after 0.05 µs the jet has completely penetrated the bubble and touched the opposite side. The liquid-liquid impact event generates a localized high pressure region which then expands quasi spherically to reach the material liquid interface after 0.2 µs. The volume of the bubble ring remaining after the jet touchdown shrinks and reaches a minimum at $t - t_{link} = 0.7$ µs. The collapse of the bubble ring generates another high pressure wave, which then propagates toward the axis of the cylindrical domain and reaches the wall at $t - t_{link} = 0.9$ µs.

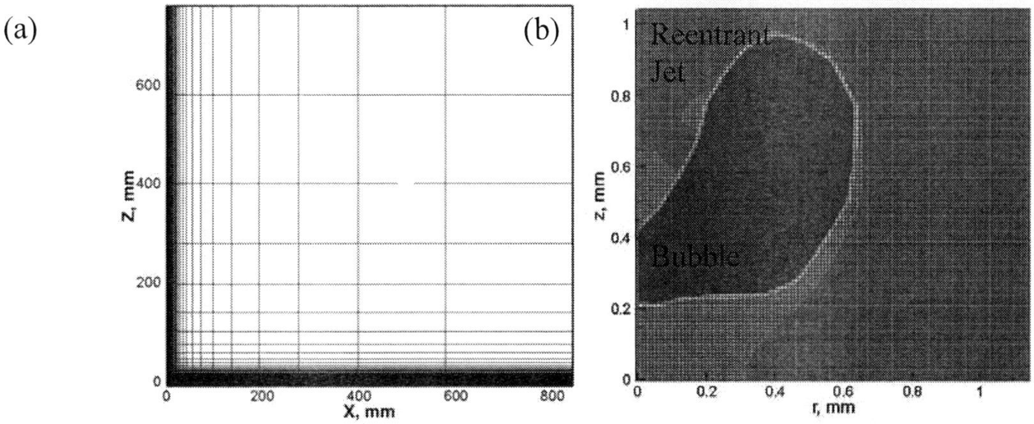

Figure 10. Axisymmetric computational domain used for the computation of the bubble dynamics by the compressible flow solver: (a) full domain, (b) zoom on the bubble/wall region. The blue region is the inside of the bubble after it formed a reentrant jet on the axis of symmetry, which is the Z-axis here.

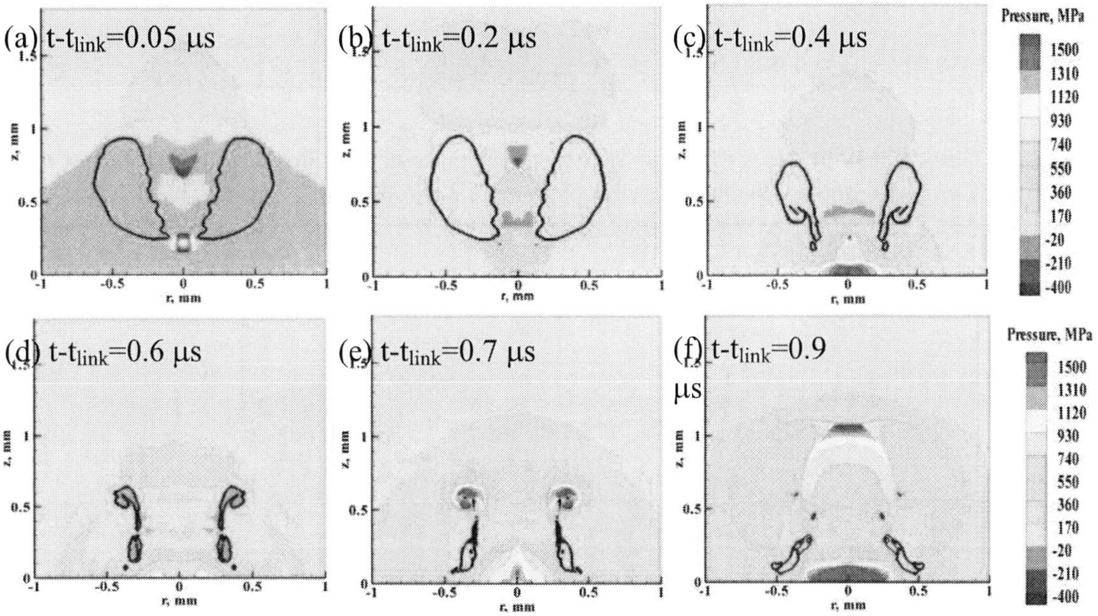

Figure 11. Pressure contours and bubble outlines at different instances following reentrant jet touchdown. Initial bubble radius R_0 = 50 μm, R_{max} = 2 mm, d_0=1.5mm, the initial standoff $\bar{X}$ = 0.75, and collapse driving pressure P_d = 10 MPa.

The pressures generated are seen to exceed 1.5 GPa and explain the strong erosive effect of cavitating jets. Material response for Al7075 was obtained by using an FEM code, fully coupled with the fluid simulation. The time histories of the liquid pressure and the vertical displacement of the material surface at the center of the Al7075 plate / liquid interface are shown together in Figure 12. The material starts to get compressed as the high pressure loading due to the reentrant jet impact reaches it, and the plate surface center point starts to move into the material direction at t - t_{link} = 0.45 μs. The maximum deformation occurs when the highest pressure loading peak due to the bubble ring collapse reaches the center of the plate at time t - t_{link} = 1.15 μs. Once the pressure loading due to the full bubble dynamics has virtually vanished at t - t_{link} = 4 μs, the surface elevation continues to oscillate due to stress waves propagating back and forth through the metal alloy thickness and lack of damping in the model. Finally, a permanent deformation in the form of a pit remains as a result of the high pressure loading causing local stresses that exceed the Al7075 elastic limit. The vertical displacement of the monitored location eventually converges to a non-zero value of about 9 μm.

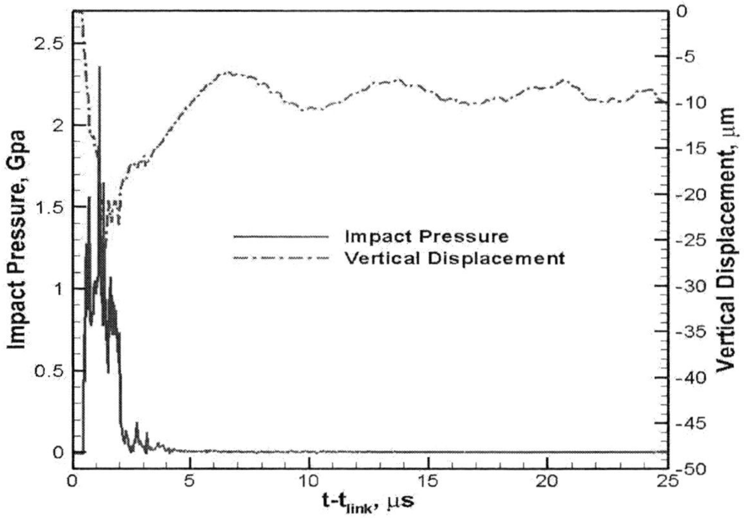

Figure 12. Time history of the pressure and vertical displacement at the top surface center of an Al7075 plate following the collapse of a cavitation bubble. $R_0 = 50\ \mu m$, $R_{max} = 2\ mm$, $P_d = 10\ MPa$, and $\overline{X} = 0.75$.

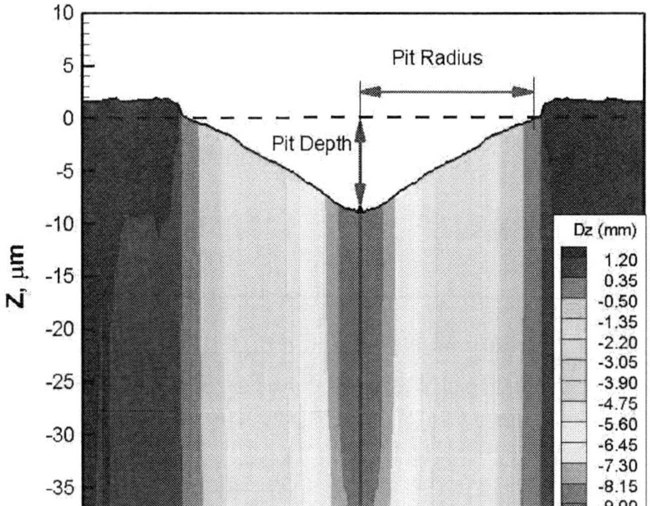

Figure 13. Profile of the permanent deformation of an Al7075 plate following the collapse of a cavitation bubble. $R_0 = 50\ \mu m$, $R_{max} = 2\ mm$, $P_d = 10\ MPa$, and $\overline{X} = 0.75$.

Finally, the impacted region bears a permanent deformation as shown in Figure 13. The profile of the permanent deformation generated on the plate surface reproduces that of the observed cavitation generated pit shapes (Philipp and Lauterborn; Kim et al.). Further details can be found in (Chahine and Hsiao; Chao-Tsung Hsiao et al.; Choi et al.).

4 CONCLUSIONS

Cavitating jet flow field was modeled using a coupled viscous flow and bubble dynamics approach, which handles the bubbly mixture flow and tracking and volume change of the bubble nuclei in the flow. The two schemes were fully coupled, where the mixture flow fields provides pressure and velocity for bubble simulations, whereas the results of bubble dynamics modifies the void fraction distribution in the mixture flow field. For individual bubbles, a Boundary Element Method was used first to follow the reentrant jet development, and then a compressible flow solver was used during the shock wave propagation and reentrant jet impact on the wall.

The modeled cavitating jet flow shows the tendency of the jet to form large structures, which trap the nuclei and enable them to grow into relatively large bubbles. Bubble deformation and reentrant jet formation are seen to be very important on generating concentrated shear and high impulsive loads on the material surface to be cleaned. The pressure loading on the material surface during the bubble collapse is shown to be due to the reentrant jet impact and to the collapse of the remaining bubble ring.

5 ACKNOWLEDGEMENTS

This work was conducted under support from DYNAFLOW, INC. internal IR&D and partial support from the Office of Naval Research under Contract N00014-12-M-0238, monitored by Dr. Ki-Han Kim. We would like to thank Dr. Kim for his support.

6 REFERENCES

Besant, W. A Treatise on Hydrostatics and Hydrodynamics. Deighton: N.p., 1859. Print.

Blake, J. "Cavitation Bubbles Near Boundaries." Annual Review of Fluid Mechanics 1987: 99–123. Print.

Chahine, G L, A F Conn, et al. "Cleaning and Cutting with Self-Resonating Pulsed Water Jets." 2nd U.S. Water Jet Conference. Rolla, Missouri: N.p., 1983. 167–176. Print.

2Chahine, G L, J-K. Choi, et al. "Development of a Diver-Operated Propeller Cleaning System Using the DYNAJETS® Enhanced Cavitation Jets ." Dynaflow, Inc. Technical Report 2M13001-1-ONR (2014): n. pag. Print.

Chahine, G L. "Nuclei Effects on Cavitation Inception and Noise." 25th Symposium on Naval Hydrodynamics. St. John's, NL, Canada: N.p., 2004. Print.

---. "Numerical Simulation of Bubble Flow Interactions." Journal of Hydrodynamics. Vol. 21. Warwick University, UK: N.p., 2008. 316–332. Print.

Chahine, G L, A F Conn, et al. "Passively Interrupted Impulsive Water Jets." 6th International Conference on Erosion by Liquid and Solid Impact. Cambridge England: N.p., 1983. Print.

Chahine, G L, J-K. Choi, et al. Study of Cavitation Erosion Prediction: Testing, Modeling, and Scaling. N.p., 2009. Print.

Chahine, G L, and K Kalumuck. "BEM Software for Free Surface Flow Simulation Including Fluid Structure Interaction Effects." International Journal of Computer Applications for Technology 11.3-5 (1998): 177–198. Print.

Chahine, G L, and T O Perdue. "Simulation of the Three-Dimensional Behavior of an Unsteady Large Bubble Near a Structure." 3rd International Colloquium on Drops and Bubbles. Ed. T G Wang. Monterey, CA: American Institue of Physics, 1988. 188–199. Print.

Chahine, G L, R Raju, and C-T. Hsiao. "Simulation of the Bubble Collapse Near a Wall: HandBack from 3DynaFS (c) to Gemini." Dynaflow, Inc. Technical Report 2M9007-NSWCIH-1 (2010): n. pag. Print.

Chahine, G. L. "Modeling of Cavitation Dynamics and Interaction with Material." Advanced Experimental and Numerical Techniques for Cavitation Erosion Prediction. Ed. A. Kim, K-H, Chahine, G.L., Franc, J-P., Karimi. Berlin: Springer, 2014. 123–173. Print.

Chahine, G. L., R. Duraiswami, and K. M. Kalumuck. "Boundary Element Method for Calculating 2-D and 3-D Underwater Explosion Bubble Behavior in Free Water and Near Structures." Naval Surface Warfare Center, Weapons Research and Technology Department Report NSWCDD/TR-93/46 (1996): n. pag. Print.

Chahine, G. L., and C.-T. Hsiao. "Modeling Cavitation Erosion Using Fluid-Material Interaction Simulations." Royal Society Meeting Amazing (Cavitation) Bubbles: Great Potentials and Challenges. Buckinghamshire, UK: N.p., 2014. Print.

Choi, J. K., and G. L. Chahine. "Non-Spherical Bubble Behavior in Vortex Flow Fields." Computational Mechanics. Vol. 32. Austin, TX: N.p., 2003. 281–290. Web.

Choi, J.-K., and G L Chahine. "A Numerical Study on the Bubble Noise and the Tip Vortex Cavitation Inception." 8th International Conference on Numerical Ship Hydrodynamics. Vol. 7. Busan, Korea: N.p., 2003. 13–33. Print.

Choi, J-K Jin-Keun et al. "Relationship between Space and Time Characteristics of Cavitation Impact Pressures and Resulting Pits in Materials." Journal of Materials Science 49.8 (2014): 3034–3051. Web.

Choi, J-K., A Jayaprakash, and G L Chahine. "Scaling of Cavitation Erosion Progression with Cavitatin Intensity and Cavitation Source." Wear 278-279 (2012): 53–61. Print.

Haberman, William L., and Rose K. Morton. "An Experimental Study of Bubbles Moving in Liquids." Transactions of the American Society of Civil Engineers 121.1 (1956): 227–250. Web. 26 Aug. 2015.

Hammitt, F. G. Cavitation and Multiphases Flow Phenomena. McGraw-Hill, 1980. Print.

Hsiao, C.-T. et al. "Numerical and Experimental Study of Bubble Entrainment Due to a Horizontal Plunging Jet." International Shipbuilding Progress 60.1 (2013): 435–469. Print.

Hsiao, C.-T., and G. L. Chahine. "Development of Compressible-Incompressible Link to Efficiently Model Bubble Dynamics near Floating Body." Advances in Boundary Element & Meshless Techniques XIV. Paliseau, France: N.p., 2013. Print.

Hsiao, C.-T., A Jain, and and G L Chahine. "Effect of Gas Diffusion on Bubble Entrainment and Dynamics around a Propeller." 26th Symposium on Naval Hydrodynamics. Rome, Italy: N.p., 2006. Print.

Hsiao, Chao-Tsung et al. "Modelling of Material Pitting from Cavitation Bubble Collapse." Journal of Fluid Mechanics 755 (2014): 142–175. Web.

Hsiao, C-T., and G L Chahine. "Dynamic Response of a Composite Propeller Blade Subjected to Shock and Bubble Pressure Loading." 83rd Shock and Vibration Symposium. New Orleans, LA: N.p., 2012. Print.

Hsiao, C-T., G L Chahine, and H Liu. "Scaling Effects on Prediction of Cavitation Inception in a Line Vortex Flow." ASME Journal of Fluids Engineering 125.1 (2003): 53–60. Print.

Jayaprakash, A., G. L. Chahine, and C.-T. Hsiao. "Numerical and Experimental Study of the Interaction of a Spark-Generated Bubble and a Vertical Wall." ASME Journal of Fluids Engineering 134.3 (2012): 31301–31312. Print.

Jayaprakash, A., S. Singh, and G. L. Chahine. "Experimental and Numerical Investigation of Single Bubble Dynamics in a Two-Phase Bubbly Medium." ASME Journal of Fluids Engineering 133.12 (2011): 121305–121309. Print.

Johnson Jr, V et al. "The Development of Structured Cavitating Jets For Deep-Hole Bits." 57th Annual Fall Technical Conf & Exhibition of the Society of Petroleum Engineers of ASME. New Orleans, Louisiana: N.p., 1982. Print.

Johnson, V E, and T Hsieh. "The Influence of the Trajectories of Gas Nuclei on Cavitation Inception." Sixth Symp. Naval Hydrodynamics (1966): 163–179. Web. 24 Feb. 2015.

Kalumuck, K M et al. "Development of a DynaJet Cavitating Water Jet Cleaning Tool for Underwater Marine Fouling Removal." Proceedings of the 9th American Water Jet Conference. Vol. 2. Dearborn, MI: N.p., 1997. 541–554. Print.

Kapahi, A., C.-T. Hsiao, and G. L. Chahine. "A Multi-Material Flow Solver for High Speed Compressible Flow Applications." Computers in Fluids (2015): DOI: 10.1016/j.compfluid.2015.03.016. Web.

Kim, K-H. et al. Advanced Experimental and Numerical Techniques for Cavitation Erosion Prediction. Ed. Ki-Han Kim et al. Vol. 106. Springer Netherlands, 2014. Print. Fluid Mechanics and Its Applications.

Knapp, R T, J W Daily, and F G Hammitt. Cavitation. London: McGraw-Hill, 1970. Print.

Lord Rayleigh. "On the Pressure Developed in a Liquid during the Collapse of a Spherical Cavity." Philosophical Magazine Series 6 34.200 (1917): 94–98. Print.

March, P A. "Evaluating the Relative Resistance of Materials to Cavitation Erosion: A Comparison of Cavitating Jet Results and Vibratory Results." Cavitation and Multiphase Flow Forum. Cincinnati, OH: American Society of Mechanical Engineers, 1987. Print.

Mørch, K.A. KA. "Cavitation Nuclei: Experiments and Theory. Journal of Hydrodynamics." Journal of Hydrodynamics, Ser. B 21.2 (2009): 176–189. Print.

Odhiambo, D. "Cavitation Shotless Peening for Improvement of Fatigue Strength of Carbonized Steel." International Journal of Fatigue 25.9-11 (2003): 1217–1222. Web. 28 July 2015.

Philipp, A., and W. Lauterborn. "Cavitation Erosion by Single Laser-Produced Bubbles." Journal of Fluid Mechanics 1998: 75–116. Print.

Plesset, Milton S., and Richard B. Chapman. "Collapse of an Initially Spherical Vapour Cavity in the Neighbourhood of a Solid Boundary." Journal of Fluid Mechanics 1971: 283. Print.

Prosperetti, A., and A. Lezzi. "Bubble Dynamics in a Compressible Liquid. Part 1. First-Order Theory." Journal of Fluid Mechanics 1986: 457. Print.

Raju, Reni et al. "Study of Pressure Wave Propagation in a Two-Phase Bubbly Mixture." ASME Journal of Fluids Engineering 133.12 (2011): 121302–121312. Web.

Singh, S., J-K. Choi, and G. L. Chahine. "Characterization of Cavitation Fields From Measured Pressure Signals of Cavitating Jets and Ultrasonic Horns." Journal of Fluids Engineering 135.9 (2013): 9111–91302. Web.

Wardlaw, A, and J A Luton. "The Gemini Euler Solver for the Coupled Simulation of Underwater Explosions." NSWCIH - Technicak Report 2500 (2003): n. pag. Print.

Wardlaw, Andrew B, and J Alan Luton. "Fluid-Structure Interaction Mechanisms for Close-in Explosions." Shock and Vibrations 7 (2000): 265–275. Print.

Zhang, S, J Duncan, and G L Chahine. "The Final Stage of the Collapse of a Cavitation Bubble Near a Rigid Wall." Journal of Fluid Mechanics 257 (1993): 147–181. Print.

2015 WJTA-IMCA Conference and Expo
November 2-4 • New Orleans, Louisiana

Paper

WHAT'S IN THE WATER?

P.J. Engelbert, CIH, CSP, Grad IOSH, RPIH, CHST, CET
Job Safety Associates, LLC
Hammond, Indiana, USA

ABSTRACT

For decades, Ultra High Pressure lances have been used to break up waste material and clean. Often used in coal fired boilers in electric generating plants to cut and remove coal slag and refractory. While using multi-gas meters in these confined spaces with moisture traps, condensate can accumulate at about 44 ml in as little as 3 hours with 6 or more lances working in boiler cyclones. This condensate can include as much as 5 mm deep X 31.75 mm diameter of solid particulate. Sizes of particulate were as large as 1.5 mm in diameter. These are potential vectors for infectious material. Further potential for harm can be found in various forms of Legionella. Legionella is common in many environments, including soil and aquatic systems, with at least 50 species and 70 serogroups identified. Between 1995 and 2001 over 50% of crews became ill with flu symptoms during hydro work. This is far more than group cross-infection. (For over 5 years, numerous outages at coal fired power stations throughout Illinois, USA included from 80 to 120 employees per shift, round the clock for 2 to 4 weeks.) One type of Legionella is called Pontiac Fever, caused by any of 4 types of Legionella, which presents with the same symptoms of flu but without the pneumonia of Legionnaire's Disease. Because it resolves itself just like common flu, it often goes undiagnosed. These species have been traced to a variety of water sources including those often used for water jetting and water cleaning, even to natural ponds and potting soils. With these omnipresent species, have we been seeing the result of their presence but not recognizing their hazard. Research into the causes of employee illness and risks of respiratory exposure need to be done. What's in the water?

Organized and Sponsored by WJTA®-IMCA®

1. INTRODUCTION

This paper is intended to raise awareness and start discussion of Legionella bacteria exposure to the water jetting industry. Field observations are all that currently exist. However, the extent of the presence of Legionella in water logically leads one to believe we have in the past and continue to channel contaminated water through our equipment.

The water jetting industry has not been accustomed to investigating respiratory issues. Physical hazards of the cutting action of water have dominated. But respiratory issues may have been present without us recognizing them. In the late 1990's, Midwest coal fired generating plants often used water jetting to remove coal slag from inside boilers, especially cyclones. These were confined spaces and multi-gas monitors were used to monitor the atmosphere. Because of high humidity and mist, water traps were installed to protect the monitors. Water quickly filled the traps. Condensate accumulated to about 44 ml in as little as 3 hours with the sample tube in only one cyclone with one 68.9 MPa (10K) lance operating. This condensate included as much as 5 mm deep X 31.75 mm diameter of solid particulate. Sizes of particulate were as large as 1.5 mm in diameter. These are potential vectors for infectious material.

For years, we have been concerned with bacteria as a complication only if one were to get cut by water streams. But we have not seriously considered respiratory routes of entry. If such large and numerous solids can be suspended in mist generated at 10K psi, what else is contained within them? One potential bacteria could be Legionella. While Legionnaire's Disease is widely known, it is only one of 50 species, 60 serogroups and hundreds of strains of Legionella. During the above referenced work in boilers, 80 to 120 employees worked 12 hour shifts for up to three weeks at a time. Many would fall to a flu, up to half the shift and on both shifts. This was the norm. One of the forms of Legionella can cause a condition called Pontiac Flu. This presents as common flu. Were these employees infected by Legionella? Like so many outbreaks in the past, no connection was made to Legionella, so no testing was done. Should we be screening water sources for Legionella now? What is the likelihood that Legionella is present in the water used in the water jetting industry?

2. DISCUSSION OF LEGIONELLA

Legionella occurs naturally in water sources. It has been found in all 50 states, but more common along the Eastern states and Midwest. It is highest in New York, Maryland and Illinois. It has been found in fresh, brackish and salt water. Legionellosis (infection by one of the Legionella types) infects an estimated 75,000 persons a year in the USA. It is not transmitted from one person to another. We will address its particulars by following the common causation model. Viable Bacteria, Amplification, Dissemination, Transmission and Route of Exposure.

Viable bacteria are found under normal conditions. Its reservoir is water, surface or underground. Legionella is resistant to normal levels of water treatment for potable water. It can lay dormant below 68 degrees F and tolerate pH from 4.5 to 10.5. It can survive over 135 days in distilled water and up to a year in tap water. Rubber and some plastics can facilitate growth.

Amplification can occur when temperatures reach 77 degrees F to 120 degrees F. Amplification takes place in a biofilm which can form in stagnant or slow moving water and resembles a slime. Amplification sites include just about every type of engineered system for holding water, including CPAP machines, vegetable sprayers, body misters, even potting soil and mines.

Dissemination occurs in the water courses and aerosolized droplets. Bacteria do not survive long in suspended mist, but can survive (under ideal conditions) up to one mile from cooling towers. Any activity that can create a mist of contaminated water can carry Legionella to a host. Common sources are waterfalls, fountains and spraying of water in industrial applications such as metal cooling fluids. Certainly, water jetting, even pressure washing pressures can aerosolize the Legionella. Bacteria cannot survive drying out. Other limiting factors are low humidity, UV light, temperature and smog.

Routes of exposure include contact with open wounds from contaminated water directly or through soils. Inhalation and ingestion are the primary routes of entry. It is common for healthy persons to have Legionella antibodies, showing past exposure. There is no consensus on infectious levels, so any detectable amount may cause infection in susceptible hosts. Susceptible hosts have been identified as those with suppressed immune response by disease or by medication, elderly, smokers, premature infants, and persons with COPD or diabetes.

The attack rate (those infected out of the possible susceptible hosts exposed) varies between types. Legionnaire's Disease has only a 5% attack rate with up to 80% mortality. For survivors, long term health suffers from brain function to joint pain. Other forms, which are less life threatening, have higher attack rates. Pontiac Fever, which presents as common flu, has a 95% attack rate with a 15% mortality. Non-pneumonia Legionellosis infects heart values and joint replacements.

3. CURRENT STATUS OF HAZARD RECOGNITION AND RESPONSE

Some facility owners have recognized that Legionella is a risk and have started monitoring programs. However, results are often turned over to contractors that will do maintenance cleaning without further assistance nor guidance. Knowing the presence of Legionella does not in itself solve the problem. One needs to find sources and engineer or limit the infestation before PPE can truly be effective. While knowing the areas to be cleaned have Legionella, one must be sure the water used in water jetting does not also contain Legionella. Even if the area to be cleaned tests negative for Legionella, one might be shooting contaminated water through the water jetting equipment.

Testing can be lengthy. The Centers for Disease Control in the USA recommends a culture test for Legionella. This is the only accepted test for Legionella in the USA. The culture can take up to two weeks for a final result. In the United Kingdom, the Health and Safety Executive (the UK version of OSHA) uses dip slides. These are similar to our pH paper. They are dipped into the water, then inserted into a tube and allowed to grow at elevated temperatures and can give a reading within 45 minutes. However, their results cannot be verified and are not accepted by the CDC. DNA testing has advanced for detecting Legionella, but is not yet proven itself. To date, culture test method for Legionella is the only CDC recognized method.

Even using the CDC/ISO culture method only shows a snapshot of what is going on in a given system. Since no infectious level of Legionella is known, testing can only tell one about the presence of Legionella. Periodic monitoring should be done to determine if bacteria counts are increasing or decreasing in response to maintenance of water storage systems. OSHA has recommendations in this regard. The relevant section from OSHA's Technical Manual is shown below.

The levels requiring action vary for the source of exposure based on the assumption that some routes or exposure result in a greater dose to the lung. For this reason, humidifiers and similar devices such as misters and evaporative condensers which produce an aerosol mist that can be directly inhaled should be controlled to lower levels. Remember that these numbers are only guidelines, and the goal is zero detectable Legionella in a water source. Levels of Legionella equal to or greater than the values in the table constitute a need for action, as described below.

Action 1: Prompt cleaning and/or biocide treatment of the system.
Action 2: Immediate cleaning and/or biocide treatment. Take prompt steps to prevent employee exposure.

Table III: 7-1. Colony Forming Units (CFU) of Legionella Per Milliliter

Action	Colling Tower	Domestic Water	Humidifier
1	100	10	1
2	1,000	100	10

Figure 1. Text and graphic taken from OSHA's Technical Manual

These are only guidelines, no regulations yet exist specific to Legionella. They are however, consistent with UK and Australian levels.

Only advance planning prior to water jetting work would allow accurate testing of water sources. In regular maintenance outages with water jetting, the use of fire hydrants to supply water is common. Water sources from fire hydrants could be tested when scheduling occurs and give enough time for results and remediation to be done. This of course of action would need to be done in cooperation of both the facility owner and contractor. For mobile crews that supply their own water and transport it to work sites, testing can be accomplished well in advance of mobilization. Of particular note is that Legionella does not amplify (grow) outside of biofilm. As long as biofilm does not form, Legionella cannot increase in count. Maintaining water storage systems free from biofilm prevents spreading Legionella. Regular monitoring can make one aware of biofilm forming within the water system.

4. CONCLUSION

The water jetting industry may have already undergone multiple outbreaks of Legionella in the form of Pontiac Fever and other Non-pneumonia Legionella forms. The reliance of 3 micron filters to ensure water cleanliness has no bearing on Legionella. The prevalent use of fire hydrant water, a stagnant and untreated water storage system, can harbor biofilm and a reservoir for Legionella. The nature of the work often results in weaker immune systems from long hours, shortened rest and reduced food intake on long outages. Water jetting equipment produces large quantities of mist. These factors make any water source with Legionella a hazard.

While general water jetting PPE augmented with a minimum of half mask respirator using HEPA filters with R or P 100 pre-filters changed each shift is certainly called for. Any presence of slime or biofilm around pumps or hoses should be collected by swab for testing. Sampling should be carried out by a Certified Industrial Hygienist or persons trained by them. Without routine monitoring of water used in water jetting equipment, we don't know what's in the water.

5. ACKNOWLEDGEMENTS

Diane Miskowski, MPH of EMSL Analytical of Cinnimonson, New Jersey for her substantial work in this field and generous sharing of information through written material and seminars.

Frank Messana, DO, MRO, CIME of Comprehensive Care in Gary, Indiana for his support and reference in this paper.

6. REFERENCES

EMSL Analytical, Legionella Testing, downloaded 11 March 2015, Web. <http://www.legionellatesting.com/>

Health and Safety Executive (UK), Legionella, downloaded 7 August 2015, Web. <http://www.hse.gov.uk/metalworking/legionella.htm>

Miskowski, D., Advanced Legionella Webinar, 10 March 2015 and sources listed therein

New South Wales (AU) Department of Health, NSW Code of Practice for the Control of Legionnaire's Disease, 2nd, June 2004

Occupational Safety and Health Administration, OSHA Technical Manual, Section III, Chapter 7, Legionnaire's Disease, downloaded 5 August 2015

7. NOMENCLATURE

Attack Rate – Those infected out of all susceptible host that were exposed

Amplification – Growing in numbers

Dissemination – Dispersed or spread out

Routes of Exposure – 4 primary routes of exposure are inhalation (breathing something into the lungs), ingestion (swallowing something), skin absorption (passing through the skin without damage and getting into the cells or blood stream) and injection (tearing a whole in the skin and leaving material inside the body)

Vector – The vehicle that transmits the disease-causing organism from the reservoir to the host

2015 WJTA-IMCA Conference and Expo
November 2-4 • New Orleans, Louisiana

Paper

WHY AUTOMATION

Randy Decker

HydroChem LLC

Deer Park, Texas, U.S.A.

ABSTRACT

Safety is what drives our automation innovations. There is an inherent safety hazard associated with personnel handling pressurized lances during the cleaning process. Getting employees out of the line of fire, and on the other side of the safety barricade were the objectives behind the creation of automated tooling. With automation, we are able to reduce crew size, effectively eliminating exposure with reduced man hours. A secondary benefit of automation centers on the adherence to established cleaning standards and procedures. With automation, the ability to set specific hose speed, feed rates and flow settings gives you continuous cleaning at a controlled rate improving quality, value and efficiency.

Refinery and petrochemical owners receive these benefits from automation and at the same time see an overall reduction in their total cost of ownership (TCO). The use of automated tooling in place of manual hydroblasting tooling is not a short term trend. Instead, a complete shift in the perspective, practice and expectations of industrial cleaning. Automation may completely eliminate manual hydroblasting methods in the near future because of the tremendous value received from the advancements in hands-free industrial cleaning.

Organized and Sponsored by WJTA®-IMCA®

1.0 INTRODUCTION

The days of having three crews shotgunning on the pad is becoming a thing of the past. Over the last several years, it has been proven time and time again that using automated tooling is safer and more efficient than technicians working a shotgun or holding a lance. In some cases, the available automation is a perfect fit for the application, like cleaning a 20' heat exchanger on the pad. More often than not, significant planning and engineering is required to properly apply automation to the customer's needs. In the past, it was the norm for refinery and petrochemical customers to favor the contractor applying more physical muscle than brains to the project. Recently, we have begun to see significant shifts in customer attitude. Now, several very large national petrochemical companies have taken more of a leadership role by requiring the industrial cleaning contractor(s) to use automation in all projects 100% of the time within their facilities. Sometimes this will even require the facility owner to perform physical modifications to unit structures to facilitate the use of automated equipment. We believe that this recent customer interest in automation is because of the resulting improvement in the cleaning contractor's safety performance, service quality, productivity and reduced cost. In this paper, we summarize the value elements associated with industrial cleaning automation to explain why the use of automation benefits all involved.

2.0 BACKGROUND

In 2009 HydroChem funded a full time, dedicated Technology Center with the goal of eliminating water cuts. Since then, we have invested over 25 million dollars designing, testing and implementing new technologies to support our 40 branches and hundreds of customers in the US and abroad. Our Technology Center allows us to modify tooling in a safe and timely manner to fit site specific needs. We judge our success by the value we provide our customers. Sometimes an application requires our engineers to develop a tool from scratch, other times off the shelf tooling from a supplier makes the most sense. Whatever tooling option will add the most value, then that's what is used. Now, in 2015, we are fine tuning our automation program to guarantee everything we do is adding value and saving our customers time and money.

2.1 MANUAL CLEANING TECHNIQUES VERSUS AUTOMATION

Automated tooling has been a part of industrial cleaning for many years. Some of the early automated tools included roto-moles for line cleaning, ATLs for tube cleaning and 3D nozzles for tank or column cleanings. Lower invoice cost, as well as faster setup/rig-down times continued to give manual cleaning options the edge over automation through the 1990's and into the early 2000's.

2.2 AUTOMATION ADVANCES CLOSE THE GAP

At HydroChem, we have focused on decreasing the total job time and employee count. In most cases using automation decreases job time and number of employees required when compared to manual cleaning. Automated tooling also provides a safer environment, better quality, greater value and maximum efficiency.

2.2.1 SAFETY

Stopping water cuts and getting employees out of the line of fire, and on the other side of the safety barricade was the objective behind the creation of automated tooling. Tangled lances in work areas is another safety issue. It causes both a safety hazard as well as increased down time. The FlexReel^{TM} system was designed to eliminate safety concerns caused by tripping hazards, entangled flex lances, kinking, and lance damage. The ability of having crew set up outside the barricade and eliminating the slicker suit is one more safety opportunity. Overheating in the summer heat is consistently problematic.

Figure 1. FlexReel^{TM}

2.2.2 QUALITY

The specific hose speed, feed rate and flow settings on automated tooling assure you receive a consistent clean every time. The angles and turning access without additional entry points that using automation provides, is another advantage to using technology. The Axys^{TM} system automates tube-to-tube indexing. Job quality improves and exchanger cleaning times can be significantly reduced. This new generation of indexing technology makes the set-up a breeze and indexing between tubes is faster and more consistent. Using our Axys^{TM} indexer systems allows you to move from tube to tube with greater precision, speed and safety. Axys^{TM} reduces the chance of missed tubes and ensures a quality job. Automated tooling can dramatically improve heat transfer and flow rates leaving you with ideal operating conditions.

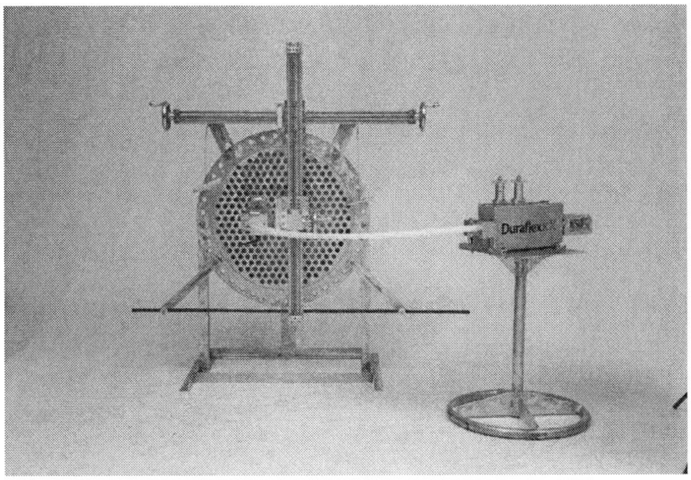

Figure 2. The Axys™ Indexing System

2.2.3 VALUE

Each job now requires fewer and better trained crew. For example, one person operating our PadBlaster™ takes the place of three operators on the pad. The main reason for this are the flow, trigger time, and smaller crew size. Employees not wearing a slicker suit are more comfortable, but also means they require fewer, less lengthy breaks. Using technology, you have effectively turned "heavy" work into "light" work.

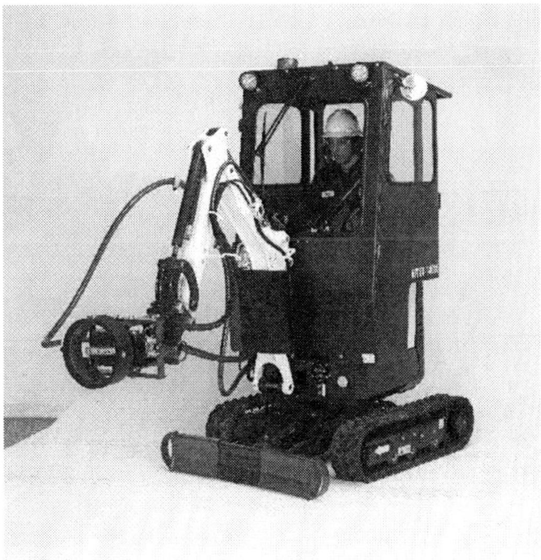

Figure 3. PadBlaster™

2.2.4 EFFICIENCY

Our AutoBlaster™ can handle 3 times the flow of a manual shotgun, allowing you to maximize the horse power of the pump you are using. Trigger time triples over the use of a manual gun. The extra flow makes cleaning tough deposits much easier and thus increases the productivity.

With automation, you receive maximum efficiency with consistent cleaning the first time and every time.

Figure 4. AutoBlaster™

2.2.5 MORE CUSTOMIZED SOLUTIONS

Our dedicated Technology Center delivers a deep bench of talent. Every day we take on challenging projects and design custom tooling solutions to fit each customer's unique requirements. In many cases, our Custom Engineered Solutions group has given our customers the confidence to set the goal of a 100% automated facility.

3.0 THE EDUCATED CUSTOMER

Today our customers are very well educated on the value automation brings. As a result, several of the largest petrochemical companies in the world have decided to require 100% automation for hydroblasting services in all of their facilities. There are many key performance indicators (KPI's) that justify the use of automation over manual cleaning techniques including the ones that matter the most, safety and total cost of ownership. Service providers have improved their ability to measure and report value through metrics. In the sections to come, we will outline the improved performance and added value the customer can expect to receive when considering the critical KPI categories.

4.0 HOW AUTOMATION POSITIVELY IMPACTS KEY PERFORMANCE INDICATORS (KPI)

When generating a metric set for a customer, several primary areas are targeted for measurement and performance review. These metric sets include the categories of safety, quality, productivity and value. Really, the first three categories are sub-categories to overall value. For customer sites that have moved toward 100 percent automation, significant improvements in each of the critical KPI categories have been achieved.

4.1 SAFETY

Safety was the initial driver of automaton innovations. Most of the available automated tooling today allows the workers to operate outside of the barricade, once the system is energized. In addition, sophisticated tooling like indexers that also serve as anti-withdraw devices and hose reels that keep the work area free of tangled lances have helped to eliminate many of the hazards historically associated with hydroblasting activities. It is no coincidence that the OSHA rate at HydroChem has decreased by over 70% in the years following 2008, while the percentage of automation used in conjunction with the hydroblasting work has increased by approximately the same percentage.

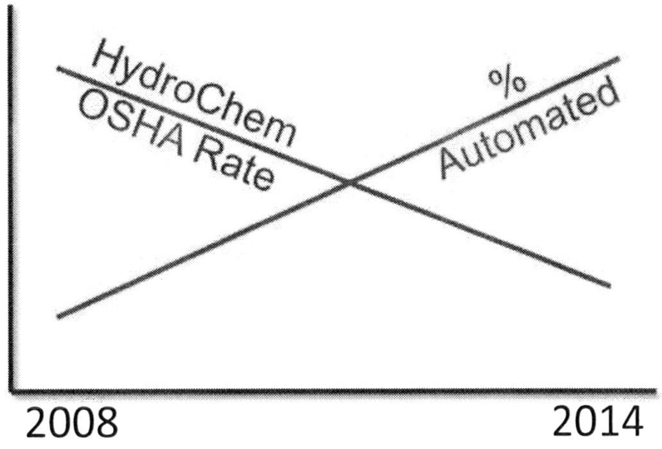

Figure 5. HydroChem OSHA rate versus % automation

Also, it is worthy to note that incident severity is much lower. From a safety perspective, the OSHA incident rate reduction is only one benefit realized from the implementation of automation. Insurance rates and Workers Compensation cases are also decreasing. Savings we are able to pass on to our customers. We briefly touch on additional safety benefits realized by automation below.

4.1.1 PERSONAL PROTECTIVE EQUIPMENT (PPE)

Typically, the suite of PPE required at a worksite is a slicker suit, hard hat, rubber gloves, face shield, FRC, safety glasses, ear plugs and rubber boots. When using automation, the worker is often able to stand outside of the barricade while operating the equipment. PPE will no longer become contaminated with product from the customer's equipment. This reduces the overall operating cost to perform the service and can lower the invoice cost to the customer as well.

4.1.2 FATIGUE AND HEAT STRESS

Automation eliminates almost all of the physical demands often associated with hydroblasting. Many times during a root cause incident investigation, fatigue and heat stress are identified as the primary or secondary cause of the incident. Therefore, automation helps to eliminate factors associated with most hydroblasting related incidents.

4.1.3 CHEMICAL EXPOSURE

Hydroblasting workers wear slicker suits in order to create a barrier between their bodies and the blast water or effluent exiting the customer's equipment. Often the effluent will contain toxic chemicals contained within the scales that are being removed from the plant equipment. Exposure to these chemicals can cause both acute and chronic health problems. Many times, the slicker suit tears or rips, exposing the worker to these toxic chemicals. The only way to eliminate this type of chemical exposure is to change the job setup to include automated tooling and move the worker outside of the barricade.

4.1.4 STAFFING

Average crew size is typically a metric the customer will request to review during the monthly and quarterly review meetings. When a manual hydroblasting project is performed, the crew consists of a crew leader and two technicians for a crew size of three. For manual hydroblasting work, at least two technicians are required in order to adequately address fatigue and heat stress. Today, using automation, fatigue and heat stress are significantly reduced, therefore, an automation crew can be reduced to a crew leader and one technician in some circumstances. The lower average crew size reduces the customer's costs for labor and with reduced crew size, exposure to safety threats are lessened as well.

4.2 QUALITY

The primary purpose of a job procedure is to define the series of steps that provide an expected result that meets the required specifications. For hydroblasting, the primary variables that must be defined within a job procedure to achieve repeatable results include pressure, flow and hose feed rate. Hose feed rate is defined as the time for the lance to travel the length of the heat exchanger. Typically, the pressure is set at the pump and flow is driven by the tip selection. However, the only way to guarantee a consistent hose feed rate is through automation. For most modern automated tooling the hose feed rate can easily be set, thus guaranteeing that the feed rate defined in the job procedure is applied to each and every tube in the heat exchanger. This same level of precision is not achievable using manual cleaning techniques.

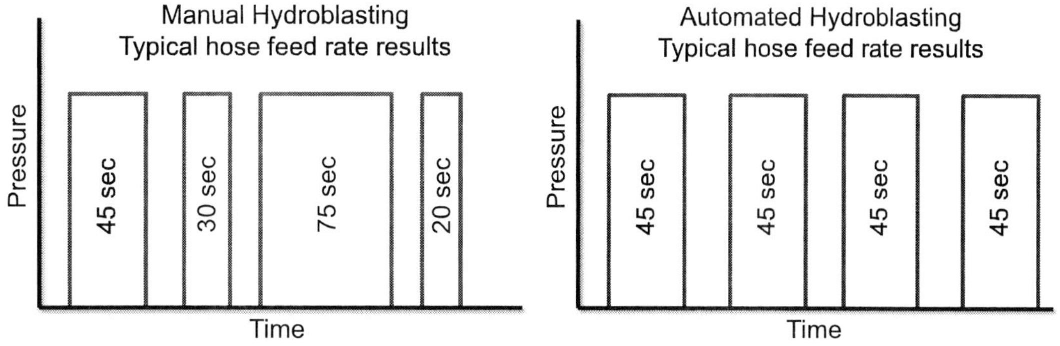

Figure 6. Example of hose feed rates for manual versus automated hydroblasting

Additionally, today's indexing tools have reduced the chances of skipped tubes or missed rows. When tubes are not properly cleaned, it can cause the customer's equipment to fail sooner than normal, costing thousands of dollars in lost production opportunity and additional cleaning cost. Often, borescope images are part of the customer's final inspection process before the job is considered complete. At HydroChem, we see 0% rework when a job procedure has been established and automation is used. This is compared to historic percentage of 5 to 10 percent rework using manual cleaning methods.

Edwards Deming, considered by many to be the grandfather of quality systems once said "A bad system will beat a good person every time." In the industrial cleaning industry, the bad system has been the manual cleaning methods. Often wet, hot and dirty work associated with manual cleaning has defeated and driven away promising talent from the industrial cleaning industry. With the new age of automation, we are discovering that talent is staying longer, thus driving down turnover rates. Our industry is now attracting and keeping new talent coming in from colleges and technical schools. They seem to match up well with the automation approach that includes challenging upfront procedural design, followed by the use of handheld control systems resembling that of an Xbox. Both the contractor and the owner wins in this environment. The contractor is reducing hiring and training costs, while the customer is getting a higher quality worker in their facility.

4.3 PRODUCTIVITY

Unquestionably, automation has changed the game when it comes to productivity. There are so many different positive impacts automation brings to cleaning productivity that we will focus only on what we consider to be the top areas of impact.

4.3.1 INCREASED PRODUCTION RATES

An amazing characteristic of automation is the ability to double, triple, or quadruple productivity. This can easily be demonstrated with heat exchanger cleanings. With manual flex lancing, a single tube is cleaned with each pass of the lance. With automation, tools have been developed that can easily clean up to 5 tubes with a single pass. Obviously, the significant reduction in cleaning times can lower the customer's invoice cost and increase production unit uptime.

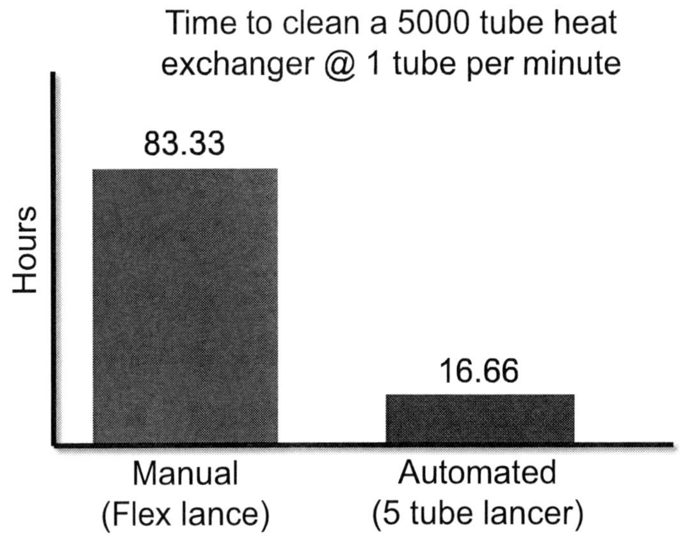

Figure 7. Example of the automation impact on cleaning times

4.3.2 DECREASED INDEXING TIME

The term indexing is defined as the time to move from tube to tube when cleaning a heat exchanger. We will once again use the 5000 tube heat exchanger example to calculate the improvement automation brings to the activity of indexing. We have measured the average index time for a typical heat exchanger located on a cleaning pad to be 15 seconds for a manual cleaning operation. This takes into account all typical fatigue and environmental delays the workers will experience during the course of the project. Using HydroChem's PowerIndexer™ technology, we can achieve average index rates of 8 seconds. Assuming that both the manual and the automated technology is setup to clean 1 tube per cleaning pass, the PowerIndexer™ will reduce the job time by 11 hours. Again, this will lower the customer's invoice cost and potentially increase production unit uptime.

4.3.3 SETUP TIME

For now, generally speaking, manual cleaning methods currently win out over automation when it comes to job setup times. Each iteration of automated tooling design brings to market smaller, lighter, more agile features. Today, our employees receive better training, more experience, and use improved tooling with fewer pieces and parts. The gap between manual and automated job setup time is continuing to shrink and most likely will tip in the favor of automation in the near future.

4.3 CASE STUDY

Refractory Removal in 70 inch Riser

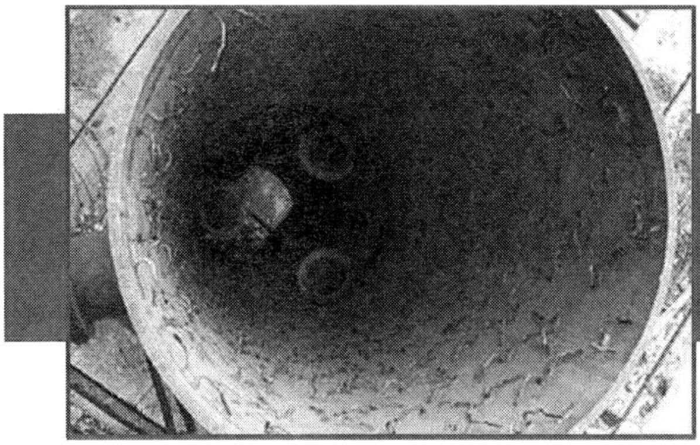

Figure 8. FCC reactor riser

Figure 9. 2250 HP Pump

Challenge

An FCC reactor riser line was scheduled for relining during the turnaround and a safer and faster solution to manual chipping was necessary. There was need to eliminate the risks associated with confined space entry as well as falling debris caused by stacking workers in this vertical riser line. A significant production improvement was also needed to achieve the refractory removal and reline in a turnaround time frame.

Solution

A specially trained team of operators mobilized our MoarJet™ high volume, high pressure and high horsepower refractory removal system to the refinery. Using this highly aggressive and technical system, we successfully broke up 17EC carbon impregnated refractory at a rate of over 20 sq. ft. per hour without entering the riser. Historically, these riser units are replaced during the turnaround which is not only capital intensive, but creates many logistical challenges and negatively affect other areas of the turnaround, thereby stretching the critical path. By using our MoarJet™ system, refractory can be removed, risers can be relined and put back in service at a small fraction of the cost of replacement. Our advanced technology MoarJet™ system is changing expectations of safety in the workplace and providing a more cost effective solution for risers and similar equipment.

Value Received

- Greatly Improved safety and work conditions
- Eliminated vessel entry
- Reduced job time by 60%
- Improved turnaround schedule

4.4 TOTAL VALUE – PUTTING IT ALL TOGETHER

Automated tooling is the secret to driving down cost, while maintaining the highest level of safety, service, and execution. Below, we provide a visual representation of the automation value model, when taking into account the elements of safety, quality and productivity.

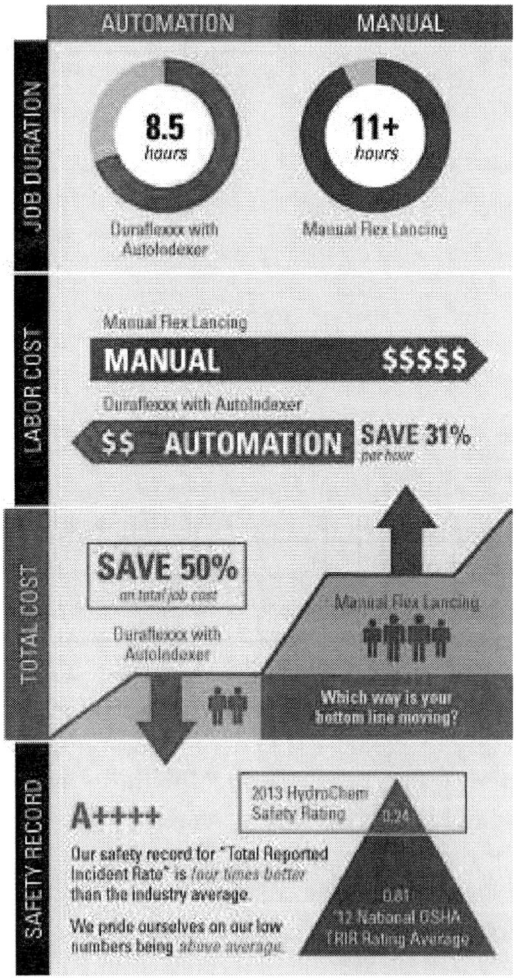

*Results based on typical 20ft 800 tube exchanger clean with a 2 man crew

Figure 10. Automation Value Model

This paper has focused on why refinery and petrochemical customers will continue to support and often drive the conversion from manual to automation cleaning techniques in our industry. We need to briefly point out that this conversion is also in the best interest of the industrial cleaning company.

- Safer working conditions
- Lower turnover and higher skilled workforce
- Scheduling accuracy by accurately predicting job times

Automation is the proverbial win-win approach.

5.0 CONCLUSION

Safety was the initial driver of automation. However, refinery and petrochemical customers are realizing tremendous value in the areas of quality and productivity, which results in a lower TCO. Since automation also brings value to the service provider's company, we will continue to see automation replace manual cleaning tools and techniques. The likely outcome will be the extinction of manual hydroblasting altogether.

6.0 ACKNOWLEDGEMENTS

Gregory Rice, COO and President, HydroChem LLC

Will Frederking, VP, General Counsel and Secretary, HydroChem LLC

Frank Moll, VP of Technology, HydroChem LLC

Brook Hull, Executive Assistant, HydroChem LLC

Leanne Trainer, Executive Assistant, HydroChem LLC

Erica Eckel, Marketing Coordinator, HydroChem LLC

Daniel Mahon, President, Mahon Studios

2015 WJTA-IMCA Conference and Expo
November 2-4 • New Orleans, Louisiana

Paper

Mechanical Tube Cleaning: The Emerging Trend in Maintaining Heat Exchanger Efficiency

Christopher K. Van Name
Conco Services Corporation
Verona, Pennsylvania, U.S.A.

Jason R. Wilburn
Conco Services Corporation
Verona, Pennsylvania, U.S.A.

ABSTRACT

Industry today places a high importance on safeguarding the health and wellbeing of both its employees and the environment. Managers in chemical plants and refineries have been tasked to find better, safer ways to clean their heat exchangers while minimizing the impact of wastewater. An emerging trend in maintaining the cleanliness of heat exchangers has been the use of mechanical tube cleaners. These mechanical systems utilize a variety of shooting, brushing and drilling methods combined with low-pressure water (under 700 PSI) to safely and quickly remove even the most tenacious deposits, restoring heat transfer efficiency. As an alternative to current practices, these mechanical tube cleaning systems offer a much smaller footprint than traditional high-pressure water methods, and allow the equipment to be placed much closer to the heat exchanger further reducing congestion during the overhaul. By utilizing lower water pressure, these mechanical tube cleaning systems are much safer to use and generate far less wastewater than high-pressure water methods. The mechanical cleaning technology has proven to be safe, effective and environmentally sound. This paper will address five case studies where positive results were achieved with this emerging technology.

Organized and Sponsored by WJTA®-IMCA®

SECTION 1 – INTRODUCTION

No matter what the type of heat exchangers nor which industries they are being employed in, eventually all of these units will foul. As fouling builds up on the interior walls of a heat exchanger's tubes, it forms an insulating barrier that interferes with heat transfer. While the type of deposits, their tenacity to adhere to the tube walls and their insulating value depend on the fluids or gases flowing within the tubes, the end result of their presence is the same, a loss in productivity. As the heat transfer rate drops, so does the heat exchanger's efficiency. This loss of efficiency will not only negatively impact the unit's production capacity, but will also drive up the cost of the final product. To reduce this impact and help keep production flowing at peak efficiency, heat exchanger performance must be monitored and when fouling is identified, the deposits removed. While high-pressure water cleaning (hydroblasting) has historically been the method of choice for fouling mitigation, in industrial markets the use of mechanical tube cleaners with low-pressure water has been gaining popularity. Early identification of fouling characteristics and a fundamental knowledge of cleaning system capabilities are essential in determining the most effective cleaning technology to employ.

Safety

Now more then ever, safety on the jobsite has become the number one priority of plant management. With this focus on safety, conventional approaches to heat exchanger cleaning have been revised, and new and innovative cleaning methods have been developed. Recent strides in cleaning technologies using mechanical tube cleaners and low-pressure water provide for very safe cleaning applications with unprecedented results. Low-pressure water means less pressure and a smaller safety zone. Figure 1. Low-Pressure Water Footprint, shows a typical safety zone for a low-pressure water application, while Figure 2. High-Pressure Water Footprint, shows the larger safety zone required for high-pressure water cleaning of the same heat exchangers. High-pressure water applications often use from 10,000 to 40,000 or more PSI to remove tube deposits while low-pressure water mechanical methods use under 700 PSI to provide a thorough cleaning. In addition, low-pressure water mechanical cleaning requires significantly smaller crew sizes than high-pressure water methods. As an added benefit, low-pressure water mechanical cleaning is done in place; it is no longer necessary to dismantle the heat exchanger and move it to a separate work zone for cleaning. This reduces the risk of accident, injury and component damage.

Environmental Impact

When considering the various options available for cleaning heat exchangers, the environmental impact of the various cleaning methods must weigh into the selection process. The amount of secondary waste (wastewater) generated by each cleaning method and the ease and cost of disposal of this waste is often a concern. One great environmental advantage of the low-pressure water mechanical cleaning methods is water conservation. Less water use means less water to clean up or to reclaim in the post-cleaning phase.

Speed

"Time is money," and for plants in the processing industries that saying holds true. When a unit is down for maintenance, that unit is not making money. Giving consideration to the environmental impact of how a heat exchanger is cleaned is important. However, the reality is that for many managers, budget constraints will drive the decision-making process on how to clean. Because new and innovative low-pressure water mechanical cleaning applications require far less unit downtime than the high-pressure water approach, it is the cleaning method that is most economically advantageous for industry. The time required for low-pressure water mechanical cleaning can be 70% less than high-pressure water methods. This can mean a unit down time of hours cleaning with mechanical cleaners versus days with hydroblasters.

SECTION 2 – HYDROBLASTING VERSUS MECHANICAL TUBE CLEANING

For many industries, hydroblasting with up to 40,000 or more PSI of water pressure has always been the preferred method for cleaning heat exchangers. However, there are significant trade-offs associated with this approach. When compared to the footprint required for low-pressure water mechanical cleaning methods, the hydroblasting footprint is sizable. Hydroblasting requires multiple water trucks and cleaning apparatus/pumps on site, and a large number of technicians must be present to not only operate the equipment but also to ensure that the large safety zone is maintained. In addition, the duration of cleaning is lengthy compared to mechanical tube cleaning because water alone is not the best cleaning agent. Furthermore, the environmental impact of using thousands of gallons of water to clean one heat exchanger is a scenario drawing greater scrutiny than ever. In drought-stricken parts of the world where industry and agriculture are bound by strict water usage laws that can come with stiff penalties, the water usage requirements of high pressure water cleaning have made this approach unviable.

Newer cleaning systems utilizing mechanical tube cleaners operate at far lower water pressures than hydroblasting. A typical water pressure for a low-pressure application is under 700 PSI compared to the average range of 10,000 to 40,000 PSI used in hydroblasting. Also, the cleaning components of these mechanical systems are smaller, more specialized and require fewer technicians than high-pressure water methods. Fewer technicians means less unit congestion, lowering the safety risk for the labor force and the equipment being cleaned. Additionally, the environmental impact of mechanical cleaning with low-pressure water is far less than with high-pressure water methods. Mechanical cleaning uses on average 90% less water than hydroblasting. That dramatic reduction in water consumption equates to 90% less contaminated wastewater to be contained and treated, saving money. It also means less exposure to contaminated wastewater for personnel and nearby aquifers. Figure 3. HPW vs. LPW Charts, compares the crew size, safety footprint, wastewater generated and water pressure requirements for hydroblasting versus mechanical tube cleaning of 21 heat exchangers.

In Figure 3. HPW vs. LPW Charts, cleaning the 21 heat exchangers using hydroblasters at 20,000 PSI generated 48,000 gallons of wastewater while the mechanical tube cleaning method at 500 PSI generated only 5,000 gallons of wastewater, see Figure 4. HPW vs. LPW Wastewater. This large disparity in water usage despite the superior cleaning results of the low-pressure method is in part explained by the differing mechanics of the two methods. The Hydroblast method pumps water 70% of the time the system is operating, dramatically driving up water consumption and creating vast quantities of wastewater. The mechanical tube cleaner shooting method pumps water for only the three seconds it takes for the mechanical cleaner to be propelled through the tube. The flow is then stopped, conserving water.

There are three basic types of low-pressure water mechanical tube cleaning systems on the market today: tube shooting, brushing and drilling. Depending on the type of deposit and the extent of the fouling, these systems are often used in combination to provide superior cleaning to hydroblasting, see Figure 5. HPW vs. LPW Cleaning Results. Low-pressure water mechanical tube cleaning is effective at removing the most tenacious deposits including:

- particulate and biological fouling
- calcium-carbonate
- asphalt
- baked-on hard deposits
- acrylic
- high-density polyethylene
- iron oxide and others

Tubes cleaned with these low-pressure water systems are ready for Eddy Current or other non-destructive testing and require no additional cleaning or preparation. Once brought back online, these units will show an immediate recovery of production capacity and heat transfer. Through low-pressure mechanical cleaning, these results are achieved safely, quickly and efficiently.

Tube shooting methods, such as with Conco's TruFit™ system, utilize a mechanical tube cleaner (pig) propelled through a tube using low-pressure water at under 700 PSI, see Figure 6. Tube Shooting. Mechanical tube cleaners are available in a variety of sizes, materials and configurations, see Figure 7. Mechanical Tube Cleaners. The best mechanical tube cleaners to use are ones custom engineered to match the interior diameter of the tubes to be cleaned, the tube materials and the types of deposits. Typical configurations include mechanical tube cleaners with spring tension metal blades and stainless steel wire brushes. Special application mechanical tube cleaners are also available such as the Cal-Buster™, see Figure 7d. Calcium Cutter, equipped with "glass-cutter wheels" to score and break apart Calcium-Carbonate deposits or U-Tube Cleaners, see Figure 7b. U-Tube Cleaner, designed to navigate and thoroughly clean the bends of U-Tube units. While low-pressure water is used to propel the mechanical tube cleaner down the length of the tube, it also serves to flush out the deposits as they are loosened.

When dealing with thick, hard, baked on deposits, brushing or drilling may become necessary. For tube brushing applications only, a flexible shaft system like the Excaliber™ can be used. This system uses a brush mounted to the tip of a flexible shaft rotating at up to 2,500 RPM and a low-pressure water to remove deposits. As the flexible shaft rotates, the unit pumps water at 30 to 125 PSI through the shaft casing to the brush, pushing out deposits as they are loosened, see Figure 8. Tube Brushing. Because of their flexible shafts, these type of systems are ideal for cleaning heat exchangers in tight locations, and with an output water flow of only 3 to 7 GPM, generate far less wastewater then hydroblasting operations.

For those situations where the tube is completely blocked or the deposit is too tenacious for a flexible shaft system to handle, then a rigid shaft system or HydroDrill™ is recommended. This system uses a drill bit mounted to the tip of a rigid shaft. As the shaft rotates, the unit pumps water at 250 PSI through the shaft and orifices in the drill bit, pushing out deposits as they are loosened, see Figure 9. Tube Drilling. The HydroDrill has a very low output water flow at only 2 to 3 GPM, keeping wastewater to an absolute minimum. The HydroDrill is designed for maximum cleaning effectiveness. Drill bits are sized to be 0.005" below the minimum tube I.D. They feature long shanks to ensure that the axis of the bit and the axis of the tube are in complete alignment. In addition, the bits are designed with carbide tips on the leading edge only and rounded corners to ensure no sharp edges directly impact the tube wall. The drill also rides on a thin layer of water for lubricating bearing surfaces between the bit and tube.

SECTION 3 – FIVE HEAT EXCHANGERS AND THE LOW-PRESSURE WATER TECHNOLOGIES USED TO CLEAN THEM

Crude Bundle Shooting

A Midwest refinery historically used hydroblasting to clean U-tube style crude bundles. However, the plant's management found that this high-pressure water method was not getting the job done to their satisfaction. Since hydroblasting equipment is unable to pass through the U-bend of the exchanger tubes, the heat transfer surface areas in the bends were not being cleaned. They believed the deposits left in the bends were interfering with the heat transfer, preventing the heat exchangers from performing at maximum efficiency. To mitigate this problem and see if there would be a measurable difference in heat transfer efficiency recovery by cleaning all the way through the U-bends, management decided to try the mechanical cleaning shooting method using U-Tube cleaners custom sized to the Tube I.D.s, see Figure 7b. U-Tube Cleaner, and low-pressure water to remove the deposits. After shooting all the tubes in two sequential cleanings, spaced five months apart, the heat exchangers showed a 20% increase of the Heat Transfer Coefficient upon initial start-up. Management was pleased with the results. The refinery is now utilizing this low-pressure water mechanical cleaning method on their other U-tube exchangers and experiencing similar results.

Polyethylene Unit Maintenance

A major Gulf Coast chemical manufacturer routinely requires cleaning of the U-tube heat exchangers in its polyethylene unit to remove the build-up of product in the tubes. The cleaning is required not only to recover the heat transfer efficiency, but also to restore production rates. The plants management knew from experience that the rigid lances used by hydroblasters were incapable of navigating through and completely removing the deposits in the bend of the tubes. They also knew that any deposits left in the tubes would negatively impact both the unit's heat transfer efficiency and production rates. Leaving deposits in the tubes was unacceptable. To remove the build-up of product throughout the tubes, management turned to mechanical tube cleaners and low-pressure water to get the job done. The tubes were first loaded with specialized mechanical cleaners, see Figure 7b. U-Tube Cleaner, custom sized to the tube I.D.s and designed to navigate through the bend of the tubes. The cleaners were then shot through the tubes using a hand held water gun shooting a stream of low-pressure water. The water not only served to propel each cleaner down the length of its tube, but also pushed out the product build-up as it was loosened, leaving the tube thoroughly cleaned. This mechanical method not only allowed the entire tube lengths to be cleaned, but by positively displace the fouling material, management found this cleaning process to be faster than hydroblasting. They now recognize mechanical tube cleaning with low-pressure water as the most effective cleaning method for these exchangers. The plant has since developed a written standard procedure based on using specialized U-tube mechanical cleaners for these exchangers.

Hydrocracker Fin-Fan Tube I.D. Cleaning

For years, a West Coast refinery had been hydroblasting the tube interiors of a set of eight fin-fan coolers in its Hydrocracker Unit with 20,000 PSI of water. The hydroblast work typically took several weeks to complete and was not always done to test ready standards. Often, the cleaning results from hydroblasting were not sufficient to obtain good test data. In addition, environmental regulations required that the plant capture and treat the vast amounts of wastewater generated by this high-pressure water cleaning method. Furthermore, set-up and tear-down of the wastewater capture system was a major factor in the duration of the hydroblasting operation, increasing the time it took to bring these units back online. In searching for a better, faster way to clean these tube I.D.s, plant management found a technology that combined low-pressure water mechanical tube cleaning with a unique manifold bridging system, see Figure 10. Tube Bridge. This bridge allowed the tube cleaners to safely traverse the gap in the manifolds at the ends of the tubesheets, making low-pressure water mechanical tube cleaning viable for cleaning fin-fan units. Management decided to give this new technology a try. Requiring a much smaller safety area/footprint and generating far less wastewater than hydroblasting operations, the low-pressure water mechanical tube cleaning system, tube bridge and wastewater capture device were quickly set up on the first fin-fan unit. The tube bridge was then loaded with standard metal tube cleaners, see Figure 7a. Metal Tube Cleaner, custom sized to the tube I.D.s. The cleaners were then shot with low-pressure water through the tubes, exiting the tube bridge on the far side of the unit. Utilizing this method, cleaning the set of eight fin-fan coolers took only ten 12-hour shifts. This far less than the time required by hydroblasting, allowing the units to get back online much sooner. It also produced 90% less wastewater that had to be captured and treated. Post-project reports from the plant also indicated that the tube cleanliness was better using mechanical tube cleaners with low-pressure water than any other cleaning method. The tubes were test-ready clean, allowing the plant to obtain much better test results.

Tantalum Tube Brushing

A chemical manufacturer with a heat exchanger containing 151, 6 ft. long Tantalum tubes experienced a process upset, the result of which was a hard baked on “varnish” in the tubes. While Tantalum has great corrosion resistance, the wall thickness of these ¾” tubes is only 0.020”. With such thin walls, plant management felt that hydroblasting and even “shooting” cleaners was not an option. Working with its contractor, the management determined that the best way to restore the heat exchanger to acceptable operating cleanliness, while at the same time safeguarding the integrity of the thin tube walls, was through a combination of mechanical brushing and low-pressure water flow to remove the deposits. To mitigate the “varnish” problem, the cleaning crew used a pneumatic rotary tube cleaner equipped with a flexible shaft and brush, custom sized to the tube I.D.s. While the spinning brush was fed down the length of each tube by the cleaning crew, water was pumped through the shaft casing to the brush, pushing out deposits as they were loosened. Cleaning was completed in a single 12-hour shift, with no tube damage. As an added benefit, with an output water pressure under 125 PSI and only 3 to 7 GPM of flow, the crew required a very small safety area/footprint, minimizing disruption to other plant operations.

Calcium Hydroxide Drilling

A chemical manufacturer has a heat exchanger where Calcium Hydroxide deposit build-up will completely plug the tubes. As this build-up occurs it greatly reducing the flow rate and of the unit. Past attempts to mitigate the problem and restore the heat exchanger to maximum efficiency included hydroblasting, as well as chemical cleaning with Inhibited Hydrochloric acid. However, no matter how long the cleaning crews hydroblasted or chemically cleaned, they could not get all the tubes 100% clean. During a recent outage the plant's management had the heat exchanger acidized for five days. Upon completion of the acidizing, the post-cleaning flow rate was measured and found to be at only 1,500 GPM. This rate was far below the OEM specs. One month later, management decided to try mechanical tube cleaning and low-pressure water to mitigate the problem. To thoroughly clean all the tubes and restore the heat exchanger to maximum performance, the cleaning crew mounted a HydroDrill to the front of the tubesheet. The drill bits selected were custom engineered to be 0.005" below the minimum tube I.D.s. As the crew guided the rotating drill bit down the length of each tube, water was pumped through the hollow rigid shaft to orifices in the drill bit, pushing out deposits as they were loosened. The cleaning duration was 48 hours and all the tubes were opened fully. When the exchanger was returned to service, the flow rate had increased to >3,100 GPM, more than double the flow after acidizing. HydroDrilling has not only proven to be the only cleaning tactic able to open every tube, but was also the quickest of all the cleaning techniques employed.

SECTION 4 – CONCLUSION

Each year, fouled heat exchangers cost the processing industries billions in lost production, capacity and revenue. While traditional high-pressure water methods can be used to mitigate tube fouling, the large crew sizes, huge footprints, high volume of waste water and lengthy time required by hydroblasting operations to thoroughly clean a heat exchanger are causing many plant managers to rethink how they are cleaning their tubes. To improve the performance of fouled heat exchangers, operators must first understand the nature of the fouling, and then commit to better stewardship of their heat exchangers. Fortunately, there has never been a better time for heat exchanger maintenance. Innovation of new and improvements of existing cleaning technologies are resulting in better outcomes with less waste, which is good for business and the environment. Low-pressure water mechanical cleaning methods featuring minimal crew sizes, small footprints, low volumes of waste water and short cleaning times offer a next generation tool kit of highly effective and responsive cleaning for any plant that operates heat exchangers.

ACKNOWLEGEMENTS

The authors acknowledge Conco Services Corporation for their willingness to share their technology and experience with industry.

REFERENCES

Saxon, George E., Jr. and Putman, Richard E., "The Practical Application and Innovation of Cleaning Technology for Heat Exchangers," Edited by Paul Watkinson, University of British Columbia, Canada; Hans Müller-Steinhagen, German Aerospace Centre (DLR) and University of Stuttgart; and M. Reza Malayeri, German Aerospace Centre (DLR), Heat Exchanger Fouling and Cleaning: Fundamentals and Applications, ECI Symposium Series, Santa Fe, New Mexico, USA, May 18-22, 2003, Volume RP1.

Wilburn, J., "Low Pressure Mechanical Tube Cleaning… The Safe, Fast and Efficient Alternative to Hydroblasting," a BIC Alliance Webcast, Baton Rouge, Louisiana, USA, April 14, 2015.

Meyer, T., "Emerging Trends in Heat Exchanger Cleaning Improves Safety and Reduces Wastewater," Environmental, Health and Safety Seminar, Galveston, Texas, USA, June 1-4, 2015.

Saxon, George E., Jr, "Innovation and Efficiency: Low-Pressure Water Mechanical Cleaning of the Heat Exchanger Cases, Cases in Heat Exchanger Cleaning," Heat Exchanger Fouling and Cleaning XI, Enfield, Ireland, June 7-12, 2015.

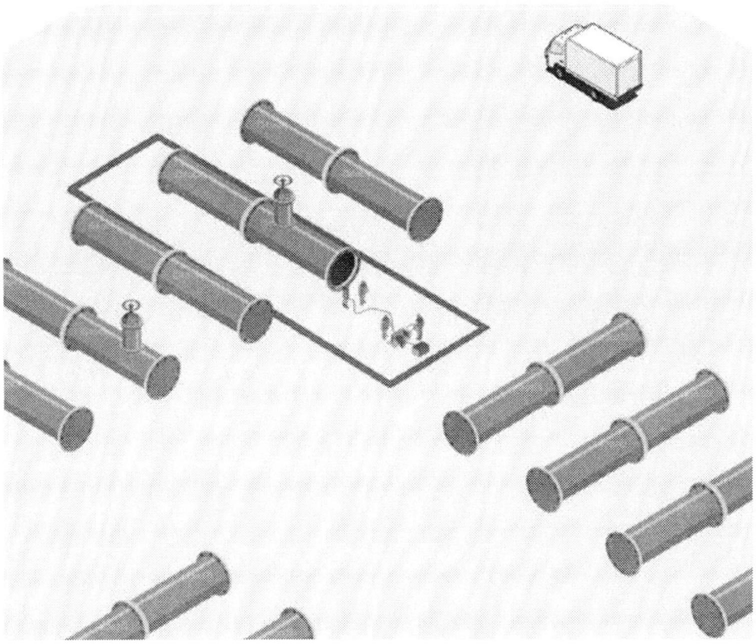

Figure 1. Low-Pressure Water Footprint

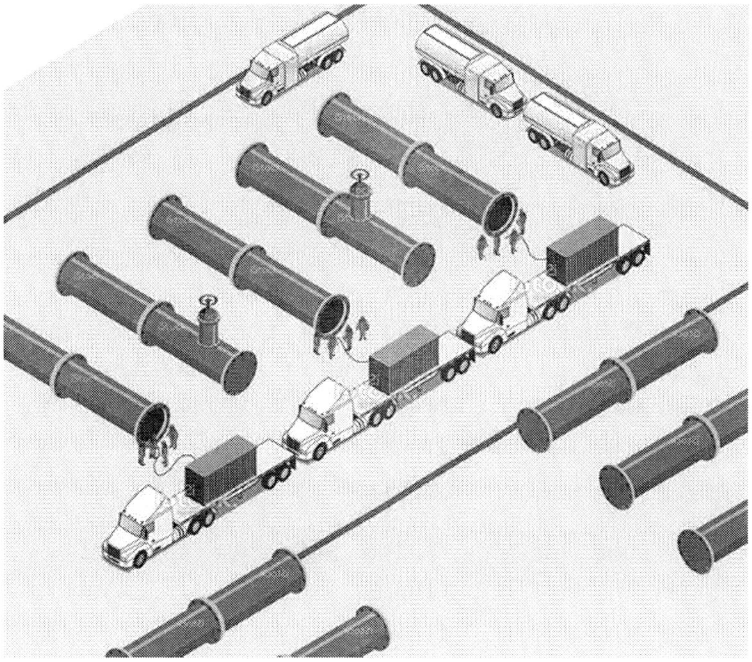

Figure 2. High-Pressure Water Footprint

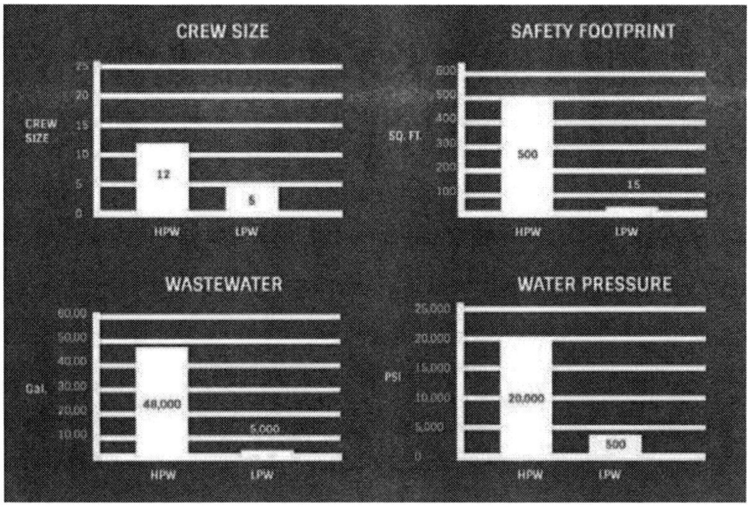

Figure 3. HPW vs. LPW Charts

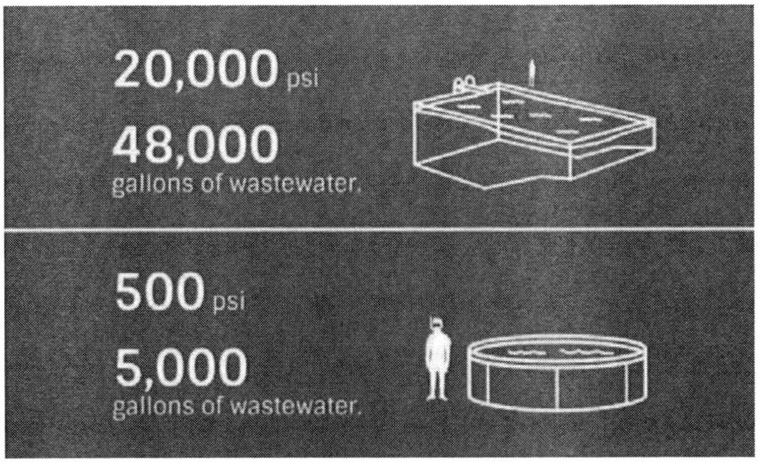

Figure 4. HPW vs. LPW Wastewater

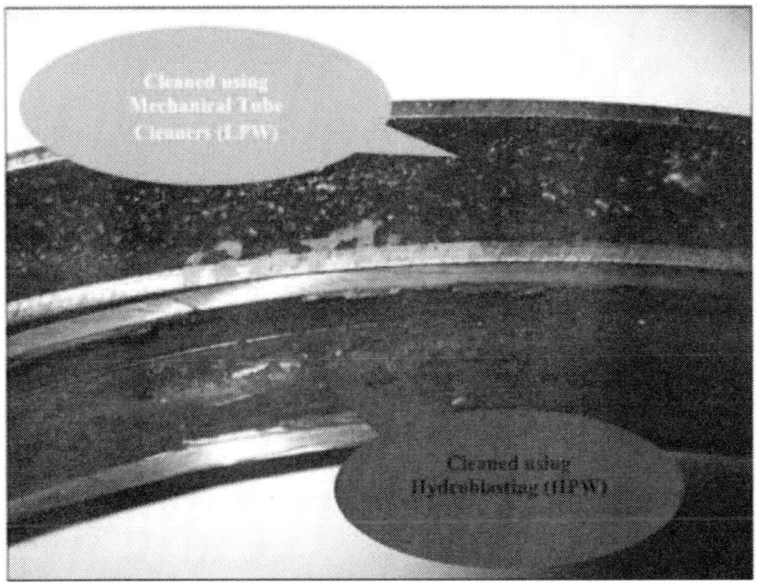

Figure 5. HPW vs. LPW Cleaning Results

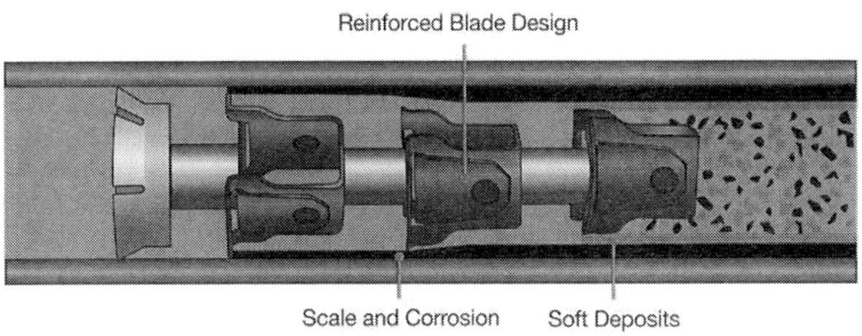

Figure 6. Tube Shooting

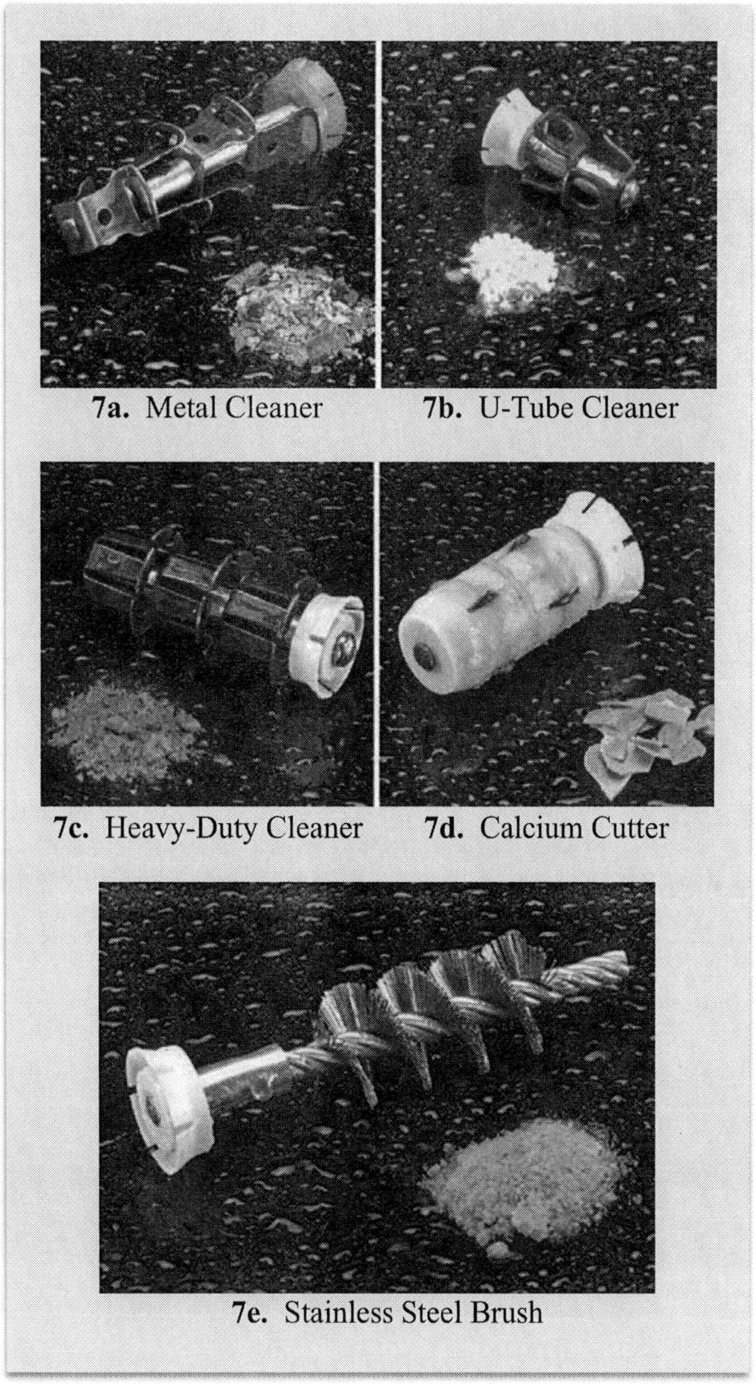

7a. Metal Cleaner **7b.** U-Tube Cleaner

7c. Heavy-Duty Cleaner **7d.** Calcium Cutter

7e. Stainless Steel Brush

Figure 7. Mechanical Tube Cleaners

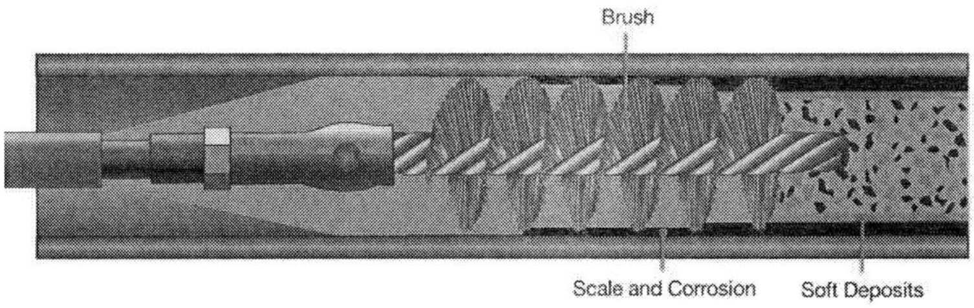

Figure 8. Tube Brushing

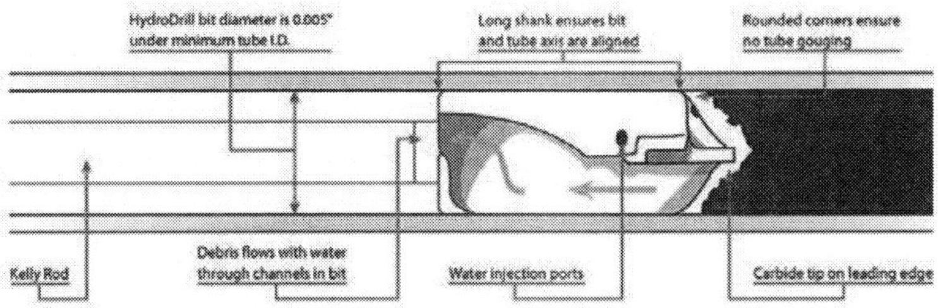

Figure 9. Tube Drilling

Figure 10. Tube Bridge

All figures courtesy of Conco Services Corporation.